Biomimetics in Photonics

SERIES IN OPTICS AND OPTOELECTRONICS

Series Editors: **E. Roy Pike**, Kings College, London, UK
Robert G. W. Brown, University of California, Irvine, USA

Recent titles in the series

Handbook of 3D Machine Vision: Optical Metrology and Imaging
Song Zhang

Handbook of Optical Dimensional Metrology
Kevin Harding

Laser-Based Measurements For Time And Frequency Domain Applications: A Handbook
Pasquale Maddaloni, Marco Bellini, and Paolo De Natale

Handbook of Silicon Photonics
Laurent Vivien and Lorenzo Pavesi

Biomimetics in Photonics
Olaf Karthaus

Optical Properties of Photonic Structures: Interplay of Order and Disorder
Mikhail F Limonov and Richard De La Rue (Eds.)

Nitride Phosphors and Solid-State Lighting
Rong-Jun Xie, Yuan Qiang Li, Naoto Hirosaki, and Hajime Yamamoto

Molded Optics: Design and Manufacture
Michael Schaub, Jim Schwiegerling, Eric Fest, R Hamilton Shepard, and Alan Symmons

An Introduction to Quantum Optics: Photon and Biphoton Physics
Yanhua Shih

Principles of Adaptive Optics, Third Edition
Robert Tyson

Optical Tweezers: Methods and Applications
Miles J Padgett, Justin Molloy, and David McGloin (Eds.)

Thin-Film Optical Filters, Fourth Edition
H Angus Macleod

Laser Induced Damage of Optical Materials
R M Wood

Principles of Nanophotonics
Motoichi Ohtsu, Kiyoshi Kobayashi, Tadashi Kawazoe, Tadashi Yatsui, and Makoto Naruse

Biomimetics in Photonics

Edited by

Olaf Karthaus

CRC Press is an imprint of the
Taylor & Francis Group, an **informa** business
A TAYLOR & FRANCIS BOOK

CRC Press
Taylor & Francis Group
6000 Broken Sound Parkway NW, Suite 300
Boca Raton, FL 33487-2742

First issued in paperback 2020

Version Date: 20120510

ISBN 13: 978-0-367-57665-3 (pbk)
ISBN 13: 978-1-4398-7746-3 (hbk)

Library of Congress Cataloging-in-Publication Data

Biomimetics in photonics / [edited by] Olaf Karthaus.
p. cm. -- (Series in optics and optoelectronics ; 13)
Includes bibliographical references and index.
ISBN 978-1-4398-7746-3 (hardback)
1. Photonics. 2. Biomimicry. I. Karthaus, Olaf.

TA1520.B544 2012
571.4--dc23 2012016466

Visit the Taylor & Francis Web site at
http://www.taylorandfrancis.com

and the CRC Press Web site at
http://www.crcpress.com

Contents

Preface vii
Contributors xiii

1. **Photonic Structures in Plants** 1
Silvia Vignolini, Beverley Glover, and Ullrich Steiner

2. **Biomineralization and Photonics** 19
Thomas Fuhrmann-Lieker

3. **Biomimetics of Optical Nanostructures** 55
Andrew R. Parker, Torben Lenau, and Akira Saito

4. **Photomechanic IR Receptors in Pyrophilous Beetles and Bugs** 117
Herbert Bousack, Helmut Budzier, Gerald Gerlach, and Helmut Schmitz

5. **Toward Industrial Production of Biomimetic Photonic Structures** 141
Hiroshi Fudouzi, Tsutomu Sawada, and Yoshihiro Uozu

6. **A Night Vision Algorithm Inspired by the Visual System of a Nocturnal Bee** 165
Eric Warrant, Magnus Oskarsson, and Henrik Malm

7. **Modeling and Simulation of Structural Colors** 191
Shuichi Kinoshita, Dong Zhu, and Akira Saito

Index 243

Preface

Biomimetic photonics is a fascinating research area that is rapidly developing, as is evident from the increase in the number of scientific publications. It is a unique research area in the sense that scientists from different backgrounds contribute to it.

Biologists are finding and describing a whole "zoo" of unique and astonishingly complex nano- and microstructures in fauna and flora on moth eyes, in insect cuticles, on butterfly wings, in diatoms, and in shells, just to name a few.

Material scientists, on the other hand, are developing novel multifunctional and hierarchical structures that have a wide variety of post–nano era photonics applications.

Mathematicians and computer experts are using computer models and simulations for a better understanding of the underlying principles of biomimetic structures.

I know from personal experience and from speaking with colleagues that there are gaps between these three communities. Concepts, structures, and phenomena that are well known in one community are quite unknown in others.

This book is intended for researchers and educators in the field of nanomaterials, surfaces, photonics, and biosciences. It can be used as a textbook for advanced courses in undergraduate school or as general teaching material for graduate students. Postdoctoral researchers will find a stimulating description of state-of-the-art research in the field of biomimetic photonics.

In this introduction, I sketch the background of both biomimetics and photonics, followed by a short description of each chapter to "wet the reader's appetite."

Biomimetics

Biology is the "closest to home" science for us. We humans have used biomaterials since ancient time for clothing, as tools, and for housing.

The 16th and 17th centuries saw the development of scientific methodology that led to new ideas and knowledge about nature, and one cornerstone for biology was the formulation of the Linnean classification system. Until today over 2.5 million distinct species have been identified.

The industrial revolution in the 19th century was made possible, not only due to a deeper understanding of the physical underpinnings of inanimate nature (thermodynamics), but also by applying the design principles of nature to machines (for example, bird-wing-shaped aircraft wings) and architectural structures (lightweight construction materials, bridges).

The discovery in the early 20th century that many biological materials are polymers opened the way for a multitude of synthetic polymers. Among the early synthetic polymers, Bakelite™ can be regarded as a model for the phenolic urushiol polymer, and Nylon™ is a polyamide analogue for polypeptides.

Thus, on both the molecular and the macroscopic scale, biologically inspired materials and structures are put into use.

In the late 1950s Otto Schmitt, an American biophysicist and electrical engineer, coined the word *biomimetics,* which nowadays is defined as the formal study of biological processes and systems as a model for creating synthetic materials and structures similar to those produced in nature. The word *biomimetics* is almost synonymous with *biomimicry* and *bionics,* which are subsets under the larger term *bioinspiration.*

According to the ISI Web of Knowledge, the number of annually published papers relating to biomimetics has increased linearly from around 100 in 1990 to 300 in 2000. The decade that followed then showed a disproportionate increase, and in 2009 alone more than 1,200 papers have been published. There is no sign of saturation.

This recent increase in popularity of biomimetics has also been fueled by the drive to develop environmentally friendly materials, production processes, and energy conversion and storage concepts to curb anthropogenic effects on the ecosystem.

With such a recent increase in the scientific output and technological advance, it is becoming necessary to categorize and formalize biomimetics. Hence, in May 2011 the Central Secretariat of the International Organization for Standardization (ISO) submitted the ISO/TS/P 222 Biomimetics technical specification (TS) proposal to the member bodies. The reason for this new ISO/TS is the that "bio-inspired materials and design are becoming of increasing interest in many fields of practical applications. In contrast to man-made materials, natural materials such as wood, bone and shells are composed of only a limited number of basic components. They gain their diversity in mechanical properties by hierarchical structuring which allows them to fulfill a variety of functions e.g., self-healing, mechanical stability, high toughness."

Even though the ISO proposal seems to be limited to *mechanical* properties, the key point that biomimetic structures use a limited number of materials in a hierarchical fashion to gain multiple functions is also valid for other applications besides structural reinforcement.

Photonics

The word *photonics* is derived from the Greek *photos*, meaning "light," and the ending *ics*, to indicate a research field. The word was coined around the same time as *biomimetics*, in the 1960s. It describes a research area that uses photons (light particles) instead of electrons to perform functions in telecommunication, information processing/storage, and sensing. Photonics is concerned with generation, modification, and detection of photons in a narrow band of electromagnetic radiation from ultraviolet to infrared wavelength (approximately 300 nm to 3 μm) by employing the optical phenomena of reflection, refraction, scattering, and absorption.

Reflection and *refraction* are caused by refractive index changes at the interface between two or more materials.

Light scattering is caused by material inhomogeneity. There are various types of light scattering depending on the size of the scattering particle. Rayleigh scattering is the elastic scattering of light by molecules and particles much smaller than the wavelength of the incident light. Mie scattering is scattering of light by spherical particles, and its intensity is sensitive to the particle size. Tyndall scattering is similar to Mie scattering without the restriction to spherical geometry and is particularly applicable to colloidal mixtures and suspensions. Brillouin scattering and Rayleigh scattering are inelastic and occur from the interaction of photons with acoustic and optical phonons, respectively.

Absorption is the uptake of the photon energy, which is then transformed to other types of energy, for example, heat, to fuel a chemical reaction or luminescence.

When the material has a spatial dimension close to the wavelength of light, *diffraction* and *interference* may be observed. One-dimensionally ordered, layered structures often show interference colors. Multilayered, periodic structures show enhanced reflectivity for selected wavelengths. Higher-dimensionally ordered structures may show other photonic effects, such as reduced reflection, photonic bandgaps, and low-loss waveguiding.

Overview of This Book

The first chapter, written by Silvia Vignolini, Beverley Glover, and Ullrich Steiner at the University of Cambridge, UK, gives an overview over photonic structures in plants, which dwell a little bit in the shadow of the more prominent photonic structures of insects. I hope that their description of photonic structures in flowers, leaves, and fruits will wake the interest of the reader to this topic.

Chapter 2, written by Thomas Fuhrmann-Lieker at the University of Kassel, Germany, deals with inorganic photonic structures produced in the aquatic environment by diatoms, sponges, and shells. The author describes mechanisms for biomineralization and gives various examples of how natural structures can be synthetically modified or even used as templates for artificial photonic materials.

Chapter 3 covers the most prominent class of biological photonic structures that are found in insects (beetles and butterflies). In the first section, Andrew R. Parker of the Natural History Museum in London gives an overview over photonic structures, including antireflecting surfaces (more on that in Chapter 5) and iridescent viruses. He gives several examples of how to artificially produce such structures. Section 2 of this chapter, by Torben Lenau at the Technical University of Denmark, describes light reflection and metallic effects in beetles and shows how the natural design principles can be used to produce structural colors and even metallic reflection from nonmetallic multilayers. Section 3 by Akira Saito at Osaka University, Japan, describes the beauty of the *Morpho* butterfly. He shows that the unique metallic blue appearance of the butterfly wings is due to an intricate interplay between order and random arrangement of nanoscale structures. He ends his section with an outlook for the mass production of artificial *Morpho* blue colors.

Chapter 4 highlights that biomimetic photonics is not limited to visible wavelength of electromagnetic radiation. Herbert Bousack, Helmut Budzier, Gerald Gerlach, and Helmut Schmitz at the Forschungszentrum Jülich, Germany, describe the incredibly sensitive infrared sensor of a fire-seeking beetle. Here too, structural principles in the biological original were used to design artificial sensors for infrared radiation.

Chapter 5 shows the possibilities and challenges to move from a laboratory environment to industrial-scale production of biomimetic photonic structures. Section 1 by Hiroshi Fudouzi and Tsutomu Sawada describes how the one-dimensional and three-dimensional photonic bandgap structures that produce structural colors in nature can be emulated artificially. The gist of this section is the rapid and reversible change of color in natural and artificial photonic structures, which gives rise to interesting and important applications. Section 2 by Yoshihiro Uozu at Mitsubishi Rayon Co. Ltd., Japan, is a short description of how to mass-produce antireflective surfaces that structurally are based on the moth eye.

Chapter 6 by Eric Warrant, Magnus Oskarsson, and Henrik Malm at the University of Lund, Sweden, describes the visual processing in the compound eye of a nocturnal bee to develop a bioinspired night vision algorithm that uses spatial and temporal integration of the light signal implemented on a graphics processing unit for enhanced color night vision.

The book finishes with two sections in Chapter 7 that deal with the modeling of structural colors found in the *Morpho* butterfly. One section is by Shuichi Kinoshita and Dong Zhu at Osaka University, in which they introduce the finite-difference time-domain (FDTD) method, which can be used to mathematically analyze the *Morpho* color. Akira Saito at Osaka University in the second section also uses the FDTD method and shows that different parameters can be fine-tuned to emulate the *Morpho* blue color. He highlights that both order and randomness of the nanostructures are crucial for the metallic appearance of the blue color over a wide viewing angle.

Summary

With this book I want to introduce the fascinating field of biomimetic photonics to the interested reader. During the discussions with the authors and by reading and editing their valuable contributions I pondered a few notions:

1. Nature does so many different things with so few building blocks, using so little energy. Perfect order often does not give the best results. Randomness, fuzziness, and even chaos are integral parts of how nature does things most effectively and efficiently.
2. We are the only conscious species on earth that is able to use technology to a point at which our carelessness and thoughtlessness lead to a significant global impact on the ecosystem. Still, we can learn from the design principles that are used in plants and animals to produce structures and devices that might help slow down and even reverse anthropogenic changes in the environment.
3. I am convinced that animate nature contains many yet undiscovered clues that will help scientists and engineers to solve some of the most prominent problems we face.

Finally, I am grateful to all the contributing authors who took time from their busy schedules to write their chapters and share their knowledge, expertise, and vision with the scientific community. I also thank John Navas, my editor at Taylor & Francis, and his staff, Laurie Schlags and Rachel Holt, for initiating this project and for providing enthusiastic support throughout the various stages of publication.

Olaf Karthaus

Contributors

Herbert Bousack
Biomimetic Sensors Group
Peter Grünberg Institut, PGI-8
Forschungszentrum Jülich
Jülich, Germany

Helmut Budzier
Institute for Solid State Electronics
Technical University
Dresden, Germany

Hiroshi Fudouzi
Applied Photonic Materials Group
Advanced Key Technologies Division/Photonic Materials Unit
National Institute for Materials Science
Tsukuba, Japan

Thomas Fuhrmann-Lieker
Institute of Chemistry and Center for Interdisciplinary Nanostructure Science and Technology (CINSaT)
University of Kassel
Kassel, Germany

Gerald Gerlach
Institute for Solid State Electronics
Technical University
Dresden, Germany

Beverley Glover
Department of Plant Sciences
University of Cambridge
Cambridge, United Kingdom

Olaf Karthaus
Department of Bio- and Material Photonics
Chitose Institute of Science and Technology
Chitose, Japan

Shuichi Kinoshita
Graduate School of Frontier Biosciences
Osaka University, Japan

Torben Lenau
Department of Mechanical Engineering
Technical University of Denmark
Kongens Lyngby, Denmark

Henrik Malm
Centre for Mathematical Sciences and
Department of Cell and Organism Biology
Lund University
Lund, Sweden

Magnus Oskarsson
Centre for Mathematical Sciences
Lund University
Lund, Sweden

Andrew R. Parker
Department of Zoology
Natural History Museum
London, United Kingdom
and
Green Templeton College
University of Oxford
Oxford, United Kingdom

Akira Saito
Department of Precision Science and Technology
Osaka University
Osaka, Japan

Tsutomu Sawada
Applied Photonic Materials Group
Advanced Key Technologies Division/Photonic Materials Unit
National Institute for Materials Science
Tsukuba, Japan

Helmut Schmitz
Institute for Zoology
University of Bonn
Bonn, Germany

Ullrich Steiner
Cavendish Laboratory
Department of Physics
University of Cambridge
Cambridge, United Kingdom

Yoshihiro Uozu
Mitsubishi Rayon Co., Ltd.
Yokohama Corporate Research Laboratories
Yokohama, Japan

Silvia Vignolini
Cavendish Laboratory
Department of Physics
University of Cambridge
Cambridge, United Kingdom

Eric Warrant
Department of Biology
Lund University
Lund, Sweden

Dong Zhu
Department of Physics
Graduate School of Frontier Biosciences
Osaka University
Osaka, Japan

1

Photonic Structures in Plants

Silvia Vignolini, Beverley Glover, and Ullrich Steiner

CONTENTS

1.1 Photonic Structures in Flowers 2
1.1.1 Anatomy and Optical Response 2
1.1.2 Directional Scattering and Glossiness: *Ranunculus repens* (Buttercup) 5
1.1.3 Iridescence in Flowers: *Hibiscus trionum* (Venice Mallow) and the Queen of the Night Tulip 8
1.2 Photonic Structures in Leaves 10
1.2.1 UV Protection Mechanism 10
1.2.2 Blue Iridescence in Understory Plants: *Selaginella willdenowii* and *Danaea nodosa* 11
1.3 Photonic Structures in Fruits 13
1.4 Summary 14
References 15

The appearance of structural color in nature dates back to the Cambrian epoch [Kobluk and Mapes 1989]. While widespread in the animal kingdom [Kinoshita 2008, Lee 2007], structural color in plants was until recently thought to occur only very rarely. With a vast literature on the use of photonics structures in butterflies, beetles, weevils, birds, and many other animal species, reports of color-generating structural elements in plants were limited to the iridescence in *Selaginella,* fern leaves, and several fruits, Rayleigh scattering in blue spruce and chalk dudleya, and enhanced ultraviolet (UV) reflection in Edelweiss [Lee 2007, Vigneron et al. 2005].

In this chapter, recent research regarding structural colors in plants is summarized, including gloss and iridescence in petals, and multilayer interference in leaves and fruits. The description of these optical phenomena will be linked to the surface structures and the plant anatomy. The chapter is divided into three sections.

The first section focuses on flower petals, discussing the shape of petal epidermal cells and their optical response in different plant species. The mechanism behind the exceptional glossiness of *Ranunculus repens* (Buttercup) is described in detail. This is followed by a description of the iridescence produced by diffraction gratings on the petals of *Hibiscus trionum* (Venice Mallow), *Tulipa kolpakowskiana*, and the Queen of the Night tulip cultivar.

The second section presents several photonic structures in leaves, reviewing the UV screening mechanism of the bracts of *Leontopodium nivale* and the structural colors produced from multilayers of the tropical understorey plants, *Selaginella willdenowii* and *Danaea nodosa*.

The final section provides an overview over the occurrence of structural color and iridescence in fruits, focusing on *Elaeocarpus angustifolius* (Blue Marble Tree) and *Margaritaria nobilis*.

1.1 Photonic Structures in Flowers

Before reviewing the main mechanisms flowers use to generate color, it is important to recall the biological role of flowers [Glover 2007]. Flowers are the reproductive structures in angiosperms [Soltis and Soltis 2004]. The reproduction of flowering plants involves pollination, that is, the transfer of the pollen from anther (male part that produces the sperm) to stigma (female part that produces the eggs). Some plants self-fertilize, but because the pollen from an anther fertilizes the eggs of the same flower, their genetic diversity does not increase. Cross-fertilization by two individuums, however, generates genetic diversity and implies that the pollen produced in one flower has to be transferred to the stigma of a different plant. For some plants abiotic carriers (air and in some rare cases water) take that role of a transfer agent, but often the transfer of the pollen from flower to flower is achieved by animals (insects and small vertebrates, such as birds and bats). To increase the chances of pollination, flowers have to be attractive to such pollinators and they have to be recognizable with respect to the background. They therefore develop brilliant color, or additional cues, such as iridescence and glossiness. Although many different factors influence the relationships between pollinators and flowers, such as odor, shape, and size, the visual appearance of flowers is one of the most important factors in this communication. The study of color of flowers and how the anatomy of the petal influences their optical behavior also allows an insight into the color perception of the pollinators [Kevan et al. 1996, Kevan and Backhaus 1998, Gorton and Vogelmann 1996, Dyer et al. 2007, Whitney et al. 2009a, Stavenga 2002, Briscoe and Chittka 2001].

1.1.1 Anatomy and Optical Response

In the most angiosperms the flower coloration is provided by three classes of pigments: anthocyanins, flavonoids, and carotenoids [Tanaka et al. 2008, Brockington et al. 2011]. By combining these pigments or varying their concentration, a wide color palette can be obtained. However, flower optical response depends not only on the pigments contained in its petals, but also on the petal anatomy [Kay et al. 1981]. An interesting example is shown in Figure 1.1.

Wild-type *Antirrhinum maius* has petals with conical epidermal cells. In a recent genetic mutant the petal epidermal cells are flat [Gorton and Vogelmann 1996, Noda et al. 1994, Glover and Martin 1998]. This change in the cell shape is reflected in the appearance of the flower itself, as shown in the photographs and the scanning electron microscopy (SEM) images in Figure 1.1. Despite the same amount of pigment per cell, the mutant form of the flower is less colored. This can be understood in terms of simple geometrical optics. The dimensions of the cells are on the order of a few tens of microns, and the conical shape of the cell gives rise to a microlens that focuses light into the cell, increasing its interaction with the pigment. Moreover, light that is reflected from the air-epidermis interface is redirected toward a neighboring cell, while for the flat surface the reflected light is lost. Figure 1.1 shows a simplified sketch of the light path for conical (c) and flat (f) cells. A more detailed and quantitative calculation of the propagation of light inside the petal can be obtained by ray tracing optics once the shape of the cells is properly determined from cross-sectional microscopy, assuming a constant refractive index inside the cell [Gorton and Vogelmann 1996, Kay et al. 1981, Bone et al. 1985, Vogelmann 1993]. This description works well when the inhomogeneities inside the cells and reflection from the cuticle are negligible.

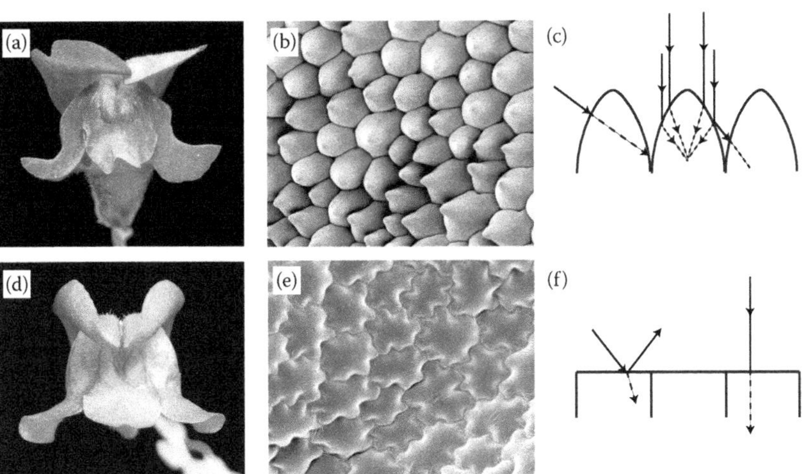

FIGURE 1.1 (See color figure at http://www.crcpress.com/product/isbn/9781439877463)
Influence of epidermal cell shape on petal coloration. (a) A wild-type *Antirrhinum majus* and (b) SEM image of the petal's adaxial epidermal layer of cells. (c) Scheme of the ray path through conically shaped cells. (d) The *mixta* mutant. (e) SEM of the petal's adaxial epidermal layer of cells and (f) scheme of the light rays impinging on flat cells. (Courtesy of Alison Reed.)

Even though the majority of cells in the petal epidermis are conical, the slope at the base, the height of the cone, their overall dimension, and the geometry of the base of the cell can vary significantly [Whitney et al. 2011, Bradshaw et al. 2010]. In Figure 1.2 pictures of several flowers and the respective SEM images of their petal's adaxial epidermis are compared. In particular, it is interesting to note that some flowers have isodiametric cells, e.g., in Figure 1.2 (e) and (f), similar to the ones shown in Figure 1.1, and elongated cells, as in Figure 1.2 (g) and (h). The overall surface morphology can also vary dramatically, ranging from completely flat cells, as in the buttercup (a) or the yellow poppy (c), to conical cells with sharp tips, as in the Common Bugloss (b) or the *Cistus cyprius* (d). The cell dimensions also vary

FIGURE 1.2 (See color figure at http://www.crcpress.com/product/isbn/9781439877463) Variation in petal epidermal cell shape. *Ranunculus repens* (Buttercup) (a), *Anchusa oficinalis* (Common Bugloss) (b), *Dicranostigma leptopodum* (Yellow Poppy) (c), *Cistus cyprius* (d), and corresponding SEM images (150 μm × 150 μm) of the adaxial epidermis (e–h). Photographs of *Aubrieta deltoidea* (i), *Geranium endressii* (j), *Gazania tenuifolia* (k), *Glaucium avum* (Yellow Horned Poppy) (l), and corrsponding SEM images (50 μm × 50 μm) of the adaxial epidermis (m–p). (SEM images are courtesy of Murphy M. Thomas.)

substantially; for instance, the dimension of the images in Figure 1.2 (e–h) is three times larger that the ones in Figure 1.2 (m–p). Finally, the surface structure of the cells is variable, going from smooth in Figure 1.2 (e–h), (j), and (l) to striated in (m) and (o). These striations can be arranged in many ways depending also on the cell geometry.

1.1.2 Directional Scattering and Glossiness: *Ranunculus repens* (Buttercup)

An intriguing example of how the anatomy of the petal can produce striking optical effects, such as directional scattering and extreme gloss, is the *Ranunculus repens* (Buttercup) flower. Even though these effects are not strictly classified as structural color (the yellow color arises from carotenoid pigments), it is possible to recognize a micrometer-scale layered structure in the epidermis that is responsible for the unique optical response of the flower. A way to test the striking optical behavior of the buttercup flower consists of playing the childhood game of holding a buttercup under the chin, as shown in Figure 1.3 (a). As the sun illuminates the buttercup, a strong yellow reflection from the petals illuminates the chin of the person holding it. The resulting bright yellow spot means, in the game, that the person likes butter. This childhood game observation raises interesting questions concerning the origin of gloss, and the optically outstanding nature of *Ranunculus* flowers has been long noted in the literature [Brett and Sommerard 1986, Galsterer et al. 1999, Hörandl et al. 2005, Parkin 1928, 1931, Vignolini et al. 2012].

In order to understand the origin of such a peculiar optical effect it is important to link the optical response to the anatomy of the buttercup petal [Vignolini et al. 2012]. Figure 1.3 (b) shows the SEM image of an entire buttercup petal. All the cells of the adaxial epidermis are extremely flat. A zoom into a smaller area is shown in Figure 1.2 (e). An optical microscopy image of a transverse section in Figure 1.3 (c) visualizes the layered structure of the petal. Focusing on the glossy adaxial epidermis, the transmission electron microscopy (TEM) cross-sectional image in Figure 1.3 (d) reveals a 5 μm thick pigment-bearing epidermal layer covered with 500 nm of wax. At the bottom of this layer, separated by an air gap, a 7 μm thick starch layer is recognizable.

The waxy layer at the top of the epidermis is responsible for the glossiness of the petal. It is important to recall that such glossiness is just specularly reflected light from the air-epidermis interface and consequently has no wavelength variation. Figure 1.4 (a) and (b) show the glossy effect both in the visible and in the ultraviolet range. The fact that glossy petals change the appearance of the flower as a function of viewing angle is crucial for pollination because it can help the pollinators with their long-distance orientation, and furthermore, the glossiness can mimic the presence of nectar drops [Galsterer et al. 1999].

It is also important to recall that the reflection of the *Ranunculus* petal is the same across the visible range (also outside the pigment absorption

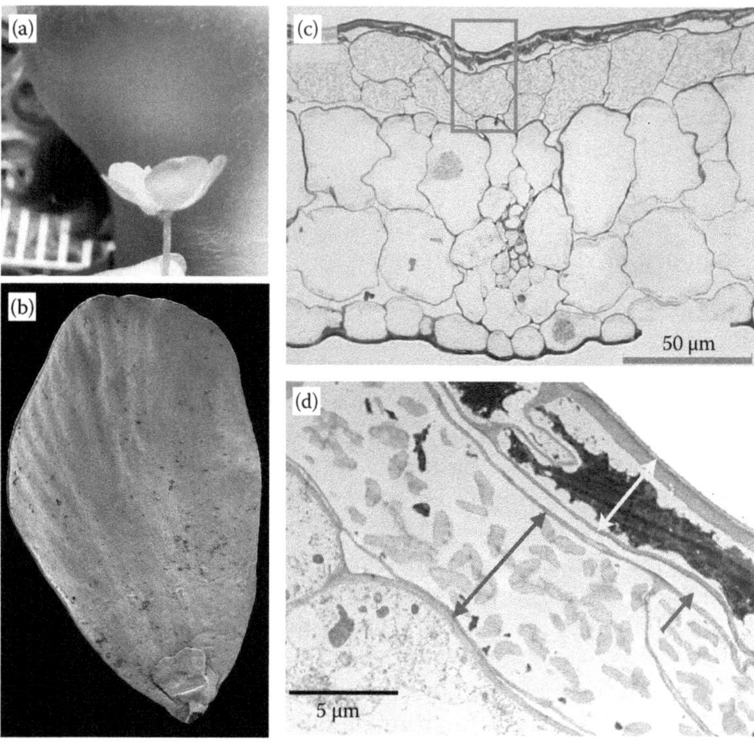

FIGURE 1.3 (See color figure at http://www.crcpress.com/product/isbn/9781439877463)
Buttercup chin illumination effect and petal anatomy. (a) *Ranunculus repens* (Buttercup) under solar illumination. If the flower is held under the chin a yellow light illuminates the skin. (b) SEM image of an entire *Ranunculus* petal. (c) Optical microscopy image of a transverse section of an entire *Ranunculus* petal; the red square indicates the region where the TEM image in (d) has been collected. The yellow arrow in the image spans the epidermal layer of cells, the blue arrow shows the starch layer, while the red arrow indicates the air gap between the two layers. (From Vignolini, S., et al., *Journal of the Royal Society, Interface*, 9:1295–301, 2012.)

wavelength range) and in the UV region (Figure 1.4 (c)). Many insect and bird eyes are sensitive to part of the UV spectrum. Consequently, the reflection from flowers in the UV range plays an important role in the synergetic interactions of pollinators and plants [Kevan et al. 1996, Gronquist et al. 2001]. By illuminating the petal at a fixed angle (Figure 1.4 (d)) and collecting the reflected light in different directions in the plane of the illumination, it is possible to characterize the scattering properties of the flower. The reflectance (red curve in Figure 1.4 (d)) of the entire buttercup petal shows a strong peak at the specular reflection direction superimposed on a weaker diffusive signal. By analyzing the different parts of the petal it is possible to directly link this optical signature to the anatomy of the petal, as shown in the cross section in Figure 1.4 (d). The starch layer (blue curve) acts in a similar way as a standard diffuser with an intensity that is

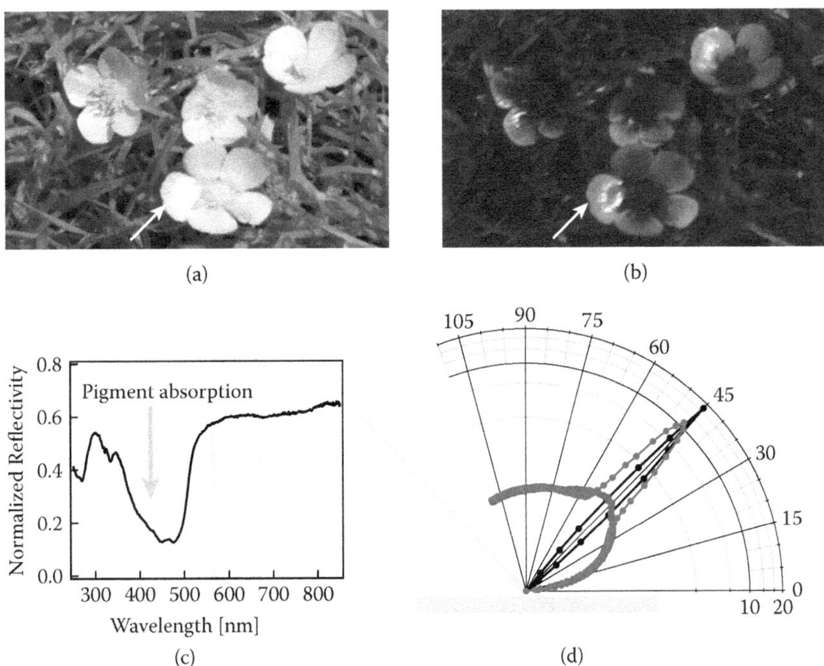

FIGURE 1.4 (See color figure at http://www.crcpress.com/product/isbn/9781439877463)
UV-VIS optical characterization of buttercup reflection. Pictures of a group of *Ranunculus* flowers taken (a) by a standard camera and (b) by a UV-sensitive camera with a bandpass filter for UV light. The arrow indicates the region of the petal that looks glossy from the direction in which the image was taken. The gloss is clearly recognizable both in the visible and in the UV image. (c) Optical response of the buttercup petals. The spectrum is obtained in specular reflection (see Vignolini et al. [2012] for details). The green arrow indicates the region where the pigment is absorbing. Note that intensity of the UV signal is comparable to the signal at visible wavelengths. (d) The polar graph shows the intensity of the scattered light from the buttercup petal integrated in the spectral region between 580 and 680 nm on a logarithmic scale. The red curve is the scattered intensity as a function of the collection angle for the entire flower, the black curve is the epidermal layer on a glass substrate, and the blue curve is the scattering from the starch layer once the epidermis has been peeled off. All spectroscopy measurements are normalized to a white diffuser standard. (From Vignolini, S., et al., *Journal of the Royal Society, Interface* 9:1295–1301, 2012.)

nearly constant and not a function of the detection angle, while the strong peak at the specular reflection direction clearly stems from the epidermal layer. In this flower the presence of an air gap between the epidermal and the starch layer provides a high directional reflectivity since it acts as a second planar interface from which light can be specularly back-reflected [Vignolini et al. 2012]. This layer sequence is therefore responsible for the peculiar optical response of these flowers. The transparent, pigment-bearing epidermal layer, the white starch layer, and the air gap between these two give rise to a highly directional yellow reflection that is responsible for the intense gloss and the chin illumination.

1.1.3 Iridescence in Flowers: *Hibiscus trionum* (Venice Mallow) and the Queen of the Night Tulip

Very recently a further interesting optical effect was observed in flowers that can act as a cue for pollinators [Whitney et al. 2009a, 2009b]. A range of flowers such as *Tulipa kaufmanniana, Tulipa kolpakowskiana,* and *Hibiscus trionum* produce iridescence via diffraction gratings [Whitney et al. 2009a]. In these flower species, the presence of surface striations in the epidermal layer gives rise to an angular color variation similar to an illuminated compact disk [Kettler 1991]. Figure 1.5 (a) shows a photograph of the *Hibiscus trionum* flower. Iridescence is observed at the base of the petal, where the adaxial epidermal cells have regularly spaced surface striations.

Similar structures are found in the petals of the *Tulipa kolpakowskiana* (Figure 1.6). Also in this case, the SEM image and the cross section in Figure 1.6 (d) show regular striations on the surface of the cells with a lattice constant of about 1.2 μm, which are responsible for the diffraction pattern in Figure 1.6 (b). Figure 1.6 (f) shows the spectroscopic characterization of *Tulipa kolpakowskiana.* The sample is illuminated at a fixed angle of 30° and the reflected spectrum is collected for different directions in the plane of illumination. The reflectivity peak red shifts with increasing detection angle, as expected for diffraction of light from a grating-like structure.

FIGURE 1.5
Hibiscus trionum (Venice Mallow). (a) *Hibiscus trionum* (Venice Mallow) flower. (b) The petal base. (c) SEM image of the petal; the upper part of the picture corresponds to the white part of the petal where the cells are smooth, while the lower half spans the red region, where the cells are regularly striated. (From Whitney, H.M., et al., *Communicative and Integrative Biology*, 2:230–232, 2009b.)

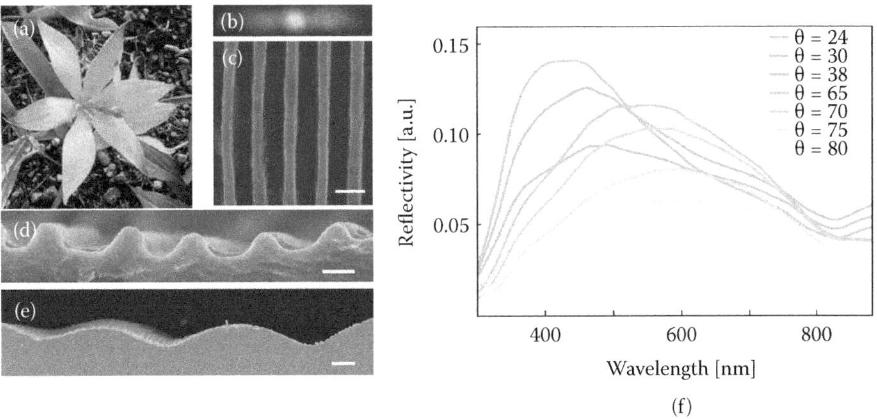

FIGURE 1.6
Tulipa kolpakowskiana. (a) *Tulipa kolpakowskiana* flower. (b) Diffraction pattern of the grating in transmission. SEM images of the adaxial epidermis in top view (c) and in side view at different magnifications. (d, e) Angularly resolved spectra of a transparent epoxy cast of the flower surface. The measurements are obtained by illuminating the sample with a white light at a fixed angle of 30° with respect to the surface normal and collecting the reflected light at different angles. The Θ = 30° curve corresponds to specular reflection. (From Kolle, M., *Photonic Structures Inspired by Nature*, Springer, Berlin, 2011. Reproduced with permission.)

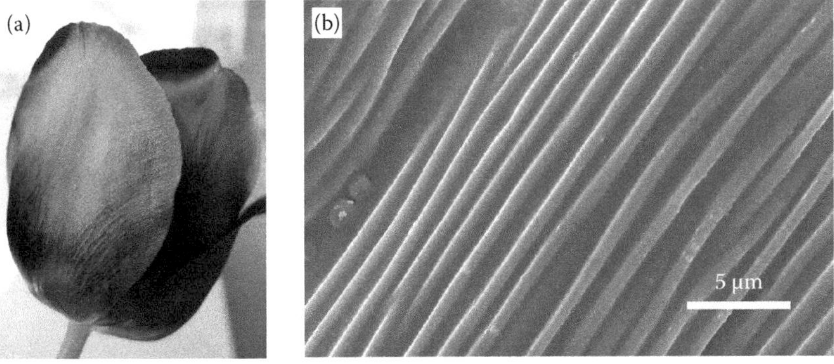

FIGURE 1.7
Queen of the Night Tulip. (a) Queen of the Night Tulip. (b) SEM image of the abaxial epidermis in top view.

This iridescence effect is not only visible at the base of flower petals but is also recognizable in other petal parts, depending on the geometry of the flower itself. In the case of the Queen of the Night Tulip, the part of the petal that shows strongest iridescence is not the adaxial epidermis but the abaxial (lower) region, mainly recognizable at the petal tip, as shown in Figure 1.7.

Similar to the animal kingdom [Shawkey et al. 2009], iridescence in flowers superimposed on a pigment-based coloration may have a range of advantages.

For example, bumblebees are able to see iridescence, enhancing target detectability [Whitney et al. 2009b]. Similar to floral patterns, the intrinsic directionality of grating interference may help pollinators to orientate themselves on flowers, increasing their foraging efficiency [Whitney et al. 2009a, 2009b]. Moreover, since some of the attractive nature of the floral patterns relies on the color contrast between the two pigments, an additional overlying color effect that highlights or enhances the difference in one area could increase the attractiveness of the entire flower, producing a uniquely enhanced color.

1.2 Photonic Structures in Leaves

The main function of leaves is to interact with light as the energy source for photosynthesis [Lee 2007]. In some cases, however, leaves use ambient light as an environmental signal. Depending on the condition of illumination leaves can develop sophisticated mechanisms to redirect and optimize the illumination of their photosynthetic organs [Bone et al. 1985]. Both experimental approaches [Vogelmann and Björn 1984] and theoretical models [Jacquemoud and Baret 1990] have been developed in order to quantitatively characterize the amount of light penetrating throughout the epidermal layer to different tissues [Vogelmann 1993]. This section reviews several examples of photonic structures in leaves and discusses their functions.

1.2.1 UV Protection Mechanism

The *Leontopodium nivale* (Edelweiss) flower, Figure 1.8 (a), grows in mountains at an altitude up to 3,400 m, where the high flux of UV radiation has

FIGURE 1.8
Leontopodium nivale (Edelweiss). (a) The flowers; the white parts are not petals but the bracts in which the UV screening structure is found. (b) SEM image of the filaments covering the leaflets surrounding the Edelweiss bracts (inset: high-magnification image of the filament surface). (From Vigneron, P. (2005). *Physical Review* 71:011906. Reproduced with permission. Copyright © 2005 by the American Physical Society.)

to be efficiently screened before it reaches and damages the cellular tissues. To this end the Edelweiss developed small leaves surrounding the flower (bracts) that are covered with a wooly layer composed of nanostructured fibers (Figure 1.8 (b)). The presence of the structured fibers has been demonstrated to be particularly efficient in screening harmful UV radiation [Vigneron et al. 2005], preventing damage to the living tissue. Other species have developed different methods to screen UV radiation. Some coniferous species such as *Pseudotsuga menziesii* (Blue Spruce) [Clark and Lister 1975] and other plants such as *Dudleya brittonii* [Mulroy 1979] use a thin wax layer, which is particularly efficient in UV screening.

1.2.2 Blue Iridescence in Understory Plants: *Selaginella willdenowii* and *Danaea nodosa*

The leaves of several species of tropical rainforest understory species, such as *Selaginella willdenowii* [Thomas et al. 2010, Lee and Lowry 1975, Lee 1997], produce brilliant blue color caused by surface multilayer structures. TEM images of the sectioned blue leaves reveal a layered structure of the outer edge of the cell wall of the upper epidermis (Figure 1.9). These lamellae are

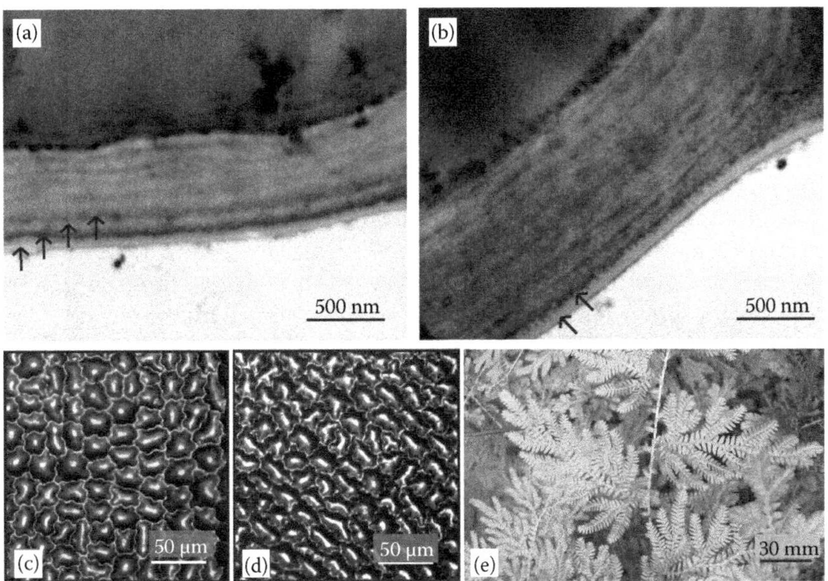

FIGURE 1.9 (See color figure at http://www.crcpress.com/product/isbn/9781439877463)
Selaginella willdenowii. (a, b) TEM images of the outer cell wall and the cuticle from the upper epidermis of a juvenile blue leaf and an older green leaf, respectively. The arrows point to the layers that constitute the photonic structure. (c, d) Optical micrographs of the leaf surface of a juvenile blue leaf and a mature green leaf, respectively. (e) Juvenile *Selaginella willdenowii* leaves. (With permission of the Royal Society.)

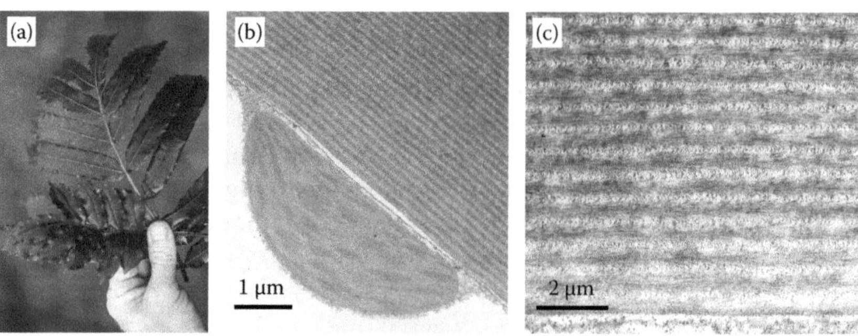

FIGURE 1.10 (See color figure at http://www.crcpress.com/product/isbn/9781439877463) *Danaea nodosa.* (a) Juvenile (blue) and adult leaves of the plant. (b) TEM cross-sectional image of the outer cell wall and cytoplasm of outer epidermis and (c) a zoomed image of the cell wall. (From Lee, D.W., *Nature's Palette,* Chicago University Press, Chicago, 2007. With permission.)

not found in older leaves, which appear green. Iridescence was reported to depend also on the growing conditions of the plant [Lee and Lowry 1975]. The microscopy images in Figure 1.9 (c, d) show that the epidermal cells are dome shaped for both juveniles and older leaves. Similar blue coloration has also been observed in various other understory plants, such as in *Diplazium tomentosum, Lindsaea lucida, Phyllagathis rotundifolia, Begonia pavonina,* and *Danaea nodosa* [Graham et al. 1993, Gould and Lee 1996]. The latter has a rather complicated structure that produces the described multilayer effect. Figure 1.10 (a) shows a photograph of a juvenile and an adult leaf of *Danaea nodosa.* Similar to *Selaginella willdenowii* the color of the leaves changes with aging. Figure 1.10 (b) and (c) shows a TEM transverse section of the outer cell wall of a juvenile leaf. Here, the multilayer structure consists of differently aligned layers of cellulose microfibrils that make up the plant cell wall, similar to the chiral nematic liquid crystal phase [Neville and Caveney 1969, Neville and Levy 1985]. The contrast in the TEM images corresponds to differing densities of the microfibrils in each layer, performing a rotation of 180° over a distance of about 160 nm. From an evolutionary point of view, it is not clear what the function of the blue iridescence in such plants is. Similar blue iridescence has been observed in algae [Pederson et al. 1980, Gerwick and Lang 1977] that grow in low-illumination conditions. In Thomas et al. [2010] it was demonstrated that, in the case of *Selaginella willdenowii,* the capture of photosynthetically important wavelengths is reduced, and consequently this structure does not provide an adaptive advantage in terms of photosynthesis efficiency. Leaf iridescence may result, for example, in a photoprotective mechanism of shade-adapted plants against sun flecks breaking through the canopy, or it may discourage herbivores from eating the young shoots of the plant [Gerwick and Lang 1977].

1.3 Photonic Structures in Fruits

Animal-dispersed plants invest in the attraction of dispersers in two ways: giving them nutritious rewards or using colorful displays to attract them. In this relationship both species gain something: the animals benefit from eating the tissues surrounding the seeds, while the plants benefit from the dispersal of their seeds [Jordano 2000]. However, some plants deceive seed dispersers by producing fruits that visually mimic fresh fruits without fleshy pulp or any nutritional reward [Galetti 2002]. *Margaritaria nobilis* in Figure 1.11 is one of these examples. The fruits of this plant show a strong iridescence in the blue-green part of the spectrum [Cazetta et al. 2008] that gives the fruit a strong metallic appearance.

Another example of a blue iridescent fruit is *Elaeocarpus angustifolius*. Figure 1.12 (a–c) shows photographs of the plant's flowers and fruits. Here, the coloration of the fruit is provided by the presence of iridosomes, which consist of cellulose layers that are, in mature fruits, secreted by the epidermal cell [Lee 1991]. Figure 1.12 (d, e) shows optical and TEM images of the transverse section of the fruit epidermis, clearly showing the location of the iridosomes inside the cell wall but outside the cell membrane. Similar iridosomes produce the blue coloration in *Delarbrea michieana* fruits [Lee et al. 2000].

Such iridescent blue coloration in fruits can be advantageous: the plant saves energy because it does not have to produce nutritious tissue surrounding its seeds and it attracts seed dispersers only using its striking visual appearance. To achieve this, structural coloration might be advantageous over pigment color because the reflected light intensity can be larger by up to one order of magnitude.

FIGURE 1.11 (See color figure at http://www.crcpress.com/product/isbn/9781439877463) *Margaritaria nobilis*. The fruit falls to the ground and starts opening the green dehiscent capsule, exposing the blue metallic epidermis. (From Cazetta, E., *Revista Brasileira de Botânica*, 31:303–308, 2008. With permission.)

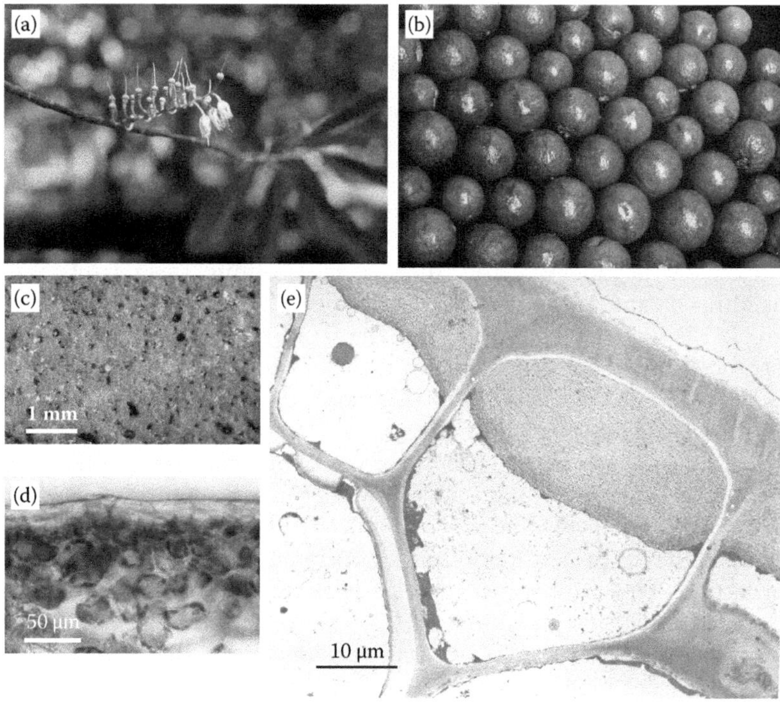

FIGURE 1.12 (See color figure at http://www.crcpress.com/product/isbn/9781439877463) *Elaeocarpus angustifolius.* (a) A branch of flowers with tiny juvenile fruits. (b) Iridescent blue fruits. The diameter of each fruit is about 2.2 cm. (c) Blue fruit surface. (d) Microscopy image of the cross section of a fruit. (e) TEM cross section image of the epidermal cell wall showing the multilayer structure responsible for the blue coloration. (From Lee, D.W., *Nature's Palette,* Chicago University Press, Chicago, 2007. With permission.)

1.4 Summary

In this chapter we reviewed several important examples of photonic structures in plants. In addition to the earlier reported presence of structural color in leaves and fruits, a large range of species have recently been discovered showing photonic structures in flowers. These consist of diffraction gratings or thin films, which provide optical effects that, together with the pigment-based coloration, provide a strong visual effect to improve the chances of pollination. The research on photonic color in plants is ongoing, and it now seems that coloration based on structural motives in plants is much more widespread than previously assumed.

References

Bradshaw, E., Rudall, P.J., Devey, D.S., Thomas, M.M., Glover, B.J., and Bateman, R.M. (2010). Comparative labellum micromorphology of the sexually deceptive temperate orchid genus *Ophrys*: diverse epidermal cell types and multiple origins of structural colour. *Botanical Journal of the Linnean Society*, 162:504–540.

Brett, D.W., and Sommerard, A.P. (1986). Ultrastructural development of plastids in the epidermis and starch layer of glossy *Ranunculus* petals. *Annals of Botany*, 58:903–910.

Briscoe, A.D., and Chittka, L. (2001). The evolution of color vision in insects. *Annual Review of Entomology*, 46:471–510.

Bone, R.A., Lee, D.W., and Norman, J.M. (1985). Epidermal cells functioning as lenses in leaves of tropical rain-forest shade plants. *Applied Optics*, 24:1408.

Brockington, S.F., Walker, R.H., Glover, B.J., Soltis, P.S., and Soltis, D.E. (2011). Complex pigment evolution in the *Caryophyllales*. *New Phytologist*, 190:854–864.

Cazetta, E., Zumstein, L.S., Melo-Júnior, T.A., and Galetti, M. (2008). Frugivory on *Margaritaria nobilis* L.f. (Euphorbiaceae): poor investment and mimetism. *Revista Brasileira de Botânica*, 31:303–308.

Clark, J.B., and Lister, G.R. (1975). Photosynthetic action spectra of trees. II. The relationship of cuticle structure to the visible and ultraviolet spectral properties of needles from four coniferous species. *Plant Physiology*, 55:407–413.

Dyer, A.G., Whitney, H.M., Arnold, S.E., Glover, B.J., and Chittka, L. (2007). Mutations perturbing petal cell shape and anthocyanin synthesis influence bumblebee perception of *Antirrhinum majus* flower colour. *Arthropod-Plant Interactions*, 1:45–55.

Galetti, M. (2002). Seed dispersal of mimetic seeds: parasitism, mutualism, aposematism or exaptation? In *Seed dispersal and frugivory: ecology, evolution and conservation*, 177–192. CABI Publishing, Oxon.

Galsterer, S., Musso, M., Asenbaum, A., and Fürnkranz, D. (1999). Reflectance measurements of glossy petals of *Ranunculus lingua* (Ranunculaceae) and of non-glossy petals of *Heliopsis helianthoides* (Asteraceae). *Plant Biology*, 1:670–678.

Gerwick, W.H., and Lang, N.J. (1977). Structural, chemical and ecological studies on iridescence in *Iridaea* (Rhodophyta). *Journal of Phycology*, 13:121–127.

Glover, B.J., and Martin, C. (1998). The role of petal cell shape and pigmentation in pollination success in *Antirrhinum majus*. *Heredity*, 80:778–784.

Glover, B. (2007). *Understanding flowers and flowering: an integrated approach*. Oxford University Press, Oxford.

Gorton, H.L., and Vogelmann, T.C. (1996). Effects of epidermal cell shape and pigmentation on optical properties of antirrhinum petals at visible and ultraviolet wavelengths. *Plant Physiology*, 112:879–888.

Gould, K.S., and Lee, D.W. (1996). Physical and ultrastructural basis of blue leaf iridescence in four Malaysian understory plants. *American Journal of Botany*, 83:45–50.

Graham, R.M., Lee, D.W., and Norstog, K. (1993). Physical and ultrastructural basis of blue leaf iridescence in two neotropical ferns. *American Journal of Botany*, 80:198–203.

Gronquist, M., Bezzerides, A., Attygalle, A., Meinwald, J., Eisner, M., and Eisner, T. (2001). Attractive and defensive functions of the ultraviolet pigments of a flower (*Hypericum calycinum*). *Proceedings of the National Academy of Sciences of the United States of America*, 98:13745–13750.

Hörandl, E., Paun, O., Johansson, J.T., Lehnebach, C., Armstrong, T., Chen, L., and Lockhart, P. (2005). Phylogenetic relationships and evolutionary traits in *Ranunculus* s.l. (Ranunculaceae) inferred from ITS sequence analysis. *Molecular Phylogenetics and Evolution*, 36:305–327.

Jacquemoud, S., and Baret, F. (1990). PROSPECT: a model of leaf optical properties spectra. *Remote Sensing of Environment*, 34:75–91.

Jordano, P. (2000). Seed disperser effectiveness: the quantity component and patterns of seed rain for *Prunus mahaleb*. *Ecological Monographs*, 70:591–615.

Kay, Q.O.N., Daoud, H.S., and Stirton, C.H. (1981). Pigment distribution, light reflection and cell structure in petals. *Botanical Journal of the Linnean Society*, 83:57–84.

Kettler, J.E. (1991). The compact disk as a diffraction grating. *American Journal of Physics*, 59:367.

Kevan, P., Giurfa, M., and Chittka, L. (1996). Why are there so many and so few white flowers? *Trends in Plant Science*, 1:252.

Kevan, P.G., and Backhaus, W.G.K. (1998). Color vision: ecology and evolution in making the best of the photic environment. In Backhaus, W., Kliegl, R., and Werner, J.S. (eds.), *Color vision: perspectives from different disciplines*, 163–83. Walter de Gruyter & Co., New York.

Kinoshita, S. (2008). *Structural colors in the realm of nature*. World Scientific Publishing Co., Singapore.

Kobluk, D.R., and Mapes, R.H. (1989). The fossil record, function, and possible origins of shell color patterns in paleozoic marine invertebrates. *PALAIOS*, 4:63–85.

Kolle, M. (2011). *Photonic structures inspired by nature*. Springer, Berlin.

Lee, D.W. (1991). Ultrastructural basis and function of iridescent blue colour of fruits in *Elaeocarpus*. *Nature*, 394:260–262.

Lee, D.W. (1997). Iridescent blue plants. *American Scientist*, 85:56–63.

Lee, D.W. (2007). *Nature's palette*. Chicago University Press, Chicago.

Lee, D.W., and Lowry, J.B. (1975). Physical basis and ecological significance of iridescence in blue plants. *Nature*, 254:50–51.

Lee, D.W., Taylor, G.T., and Irvine, A.K. (2000). Structural fruit coloration in *Delarbrea michieana* (Araliaceae). *International Journal of Plant Sciences*, 161:29–300.

Mulroy, T.W. (1979). Spectral properties of heavily glaucous and non-glaucous leaves of a succulent rosette-plan. *Oecologia*, 357:349–357.

Neville, A.C., and Caveney, S. (1969). Scarabaeid beetle exocuticle as an optical analogue of cholesteric liquid crystals. *Biological Reviews of the Cambridge Philosophical Society*, 44:531–562.

Neville, A.C., and Levy, S. (1985). The helicoidal concept in plant cell wall ultrastructure and morphogenesis. In Brett, C.T., and Hillman, J.R. (eds.), *Biochemistry of plant cell walls*, 99–123. Press Syndicate of the University of Cambridge, Cambridge.

Noda, K., Glover, B.J., Linstead, P., and Martin, C. (1994). Flower colour intensity depends on specialized cell shape controlled by a Myb-related transcription factor. *Nature*, 369:661–664.

Parkin, J. (1928). The glossy petal of *Ranunculus*. *Annals of Botany*, XLII:739–755.

Parkin, J. (1931). The structure of the starch layer in the glossy petal of *Ranunculus*. *Annals of Botany*, XLV:201–205.
Pederson, M., Roomans, G.M., and Hofsten, A. (1980). Blue iridescence and bromine in the cuticle of the red alga *Chondrus crispus Stackh*. *Botanica Marina*, 23:193–196.
Shawkey, M.D., Morehouse, N.I., and Vukusic, P. (2009). A protean palette: colour materials and mixing in birds and butterflies. *Journal of the Royal Society, Interface*, 6 (Suppl 2):S221–231.
Soltis, P.S., and Soltis, D.E. (2004). The origin and diversification of angiosperms. *American Journal of Botany*, 91(10):1614.
Stavenga, D.G. (2002). Colour in the eyes of insects. *Journal of Comparative Physiology A*, 188:337–348.
Tanaka, Y., Sasaki, N., and Ohmiya, A. (2008). Biosynthesis of plant pigments: anthocyanins, betalains and carotenoids. *The Plant Journal*, 54:733–749.
Thomas, K.R., Kolle, M., Whitney H.M., Glover, B.J., and Steiner, U. (2010). Function of blue iridescence in tropical understorey plants. *Journal of the Royal Society, Interface*, 7:1699–1707.
Vigneron, J.P., Rassart, M., Vértesy, Z., Kertész, K., Sarrazin, M., Biró, L.P., Ertz, D., and Lousse V. (2005). Optical structure and function of the white filamentary hair covering the Edelweiss bracts. *Physical Review E*, 71:011906.
Vignolini, S., Thomas, M.M., Kolle, M., Wenzel, T., Rowland, A., Rudall, P., Baumberg, J.J., Glover, B.J., and Steiner, U. (2012). Directional scattering from the glossy flower of *Ranunculus*: how the buttercup lights up your chin. *Journal of the Royal Society, Interface*, 9:1295–1301.
Vogelmann, T.C. (1993). Plant tissue optics. *Annual Review of Plant Biology*, 44:231–251.
Vogelmann, T.C., and Björn, L.O. (1984). Measurement of light gradients and spectral regime in plant tissue with a fiber optic probe. *Physiologia Plantarum*, 60:361–368.
Whitney, H.M., Bennett, K.M.V., Dorling, M., Sandbach, L., Prince, D., Chittka, L., and Glover, B.J. (2011). Why do so many petals have conical epidermal cells? *Annals of Botany*, 108:609–611.
Whitney, H.M., Kolle, M., Alvarez-Fernandez, R., Steiner, U., and Glover, B.J. (2009b). Contributions of iridescence to floral patterning. *Communicative and Integrative Biology*, 2:230–232.
Whitney, H.M., Kolle, M., Andrew, P., Chittka, L., Steiner, U., and Glover, B.J. (2009a). Floral iridescence, produced by diffractive optics, acts as a cue for animal pollinators. *Science*, 323:130–133.

2

Biomineralization and Photonics

Thomas Fuhrmann-Lieker

CONTENTS

2.1 Optical Properties of Biominerals and Artificial Materials 20
 2.1.1 Calcium Carbonate 21
 2.1.2 Silica 22
 2.1.3 Artificial Optical Biominerals 23
2.2 General Principles of Biomineralization 23
2.3 Examples from Nature 26
 2.3.1 Mother-of-Pearl 26
 2.3.2 Calcite-Based Imaging 28
 2.3.3 Tabashir and Other Plant Silica 29
 2.3.4 Diatoms 30
 2.3.5 Sponge Fibers 36
2.4 Functionalizing Biological Structures 38
 2.4.1 Replacing Silica 39
 2.4.2 In Vivo Fluorochromation 39
 2.4.3 Surface Functionalization 41
 2.4.4 Biomineral Replicas 42
2.5 Biomimetic Approaches Toward Optical Materials 43
 2.5.1 Calcium Carbonate 44
 2.5.2 Silica 44
 2.5.3 Titanium Dioxide 45
 2.5.4 Zinc Oxide 45
 2.5.5 Metal Sulfides and Metals 45
2.6 Summary 46
Acknowledgments 46
References 46

Biominerals are formed in a large variety by organisms, giving them primarily mechanical protection and strength. Little is known up to now about their photonic properties and functions. Biomineralization is interesting for photonic scientists and engineers for several reasons. First, contrast in the refractive index is an intrinsic property of a biomineralized structure, leading to color effects due to selective reflection (Figure 2.1) and optical

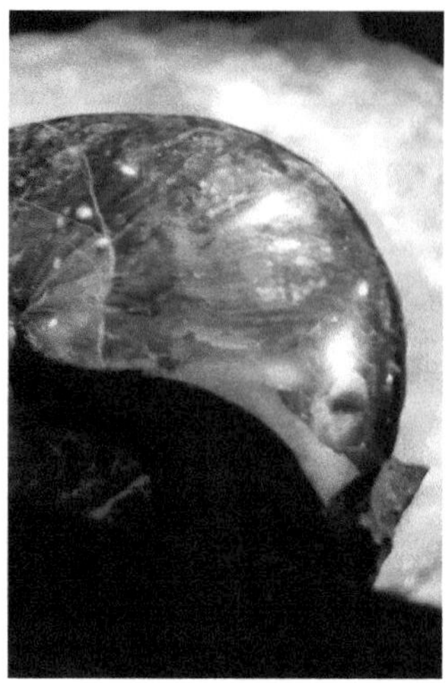

FIGURE 2.1 (See color insert.)
A fossil ammonite (Museum Zeche Zollverein, Essen, Germany). Such gemstones with color effects are called ammolites.

waveguiding that is desired for a broad range of applications. Second, the principle of biomineralization involves self-organization on several hierarchical length scales, including submicrometer patterning, which is the length scale of nanophotonics. As bottom-up strategies for the construction of nanostructures, these self-organization principles may be transferred in a biomimetic approach to other photonic materials and other nanooptical structures if the key processes taking place in biomineralization are well understood.

2.1 Optical Properties of Biominerals and Artificial Materials

The range of different biominerals formed by organisms is very broad. Fascinating examples are the formation of magnetic iron oxide chains by bacteria, the capability of precipitating cadmium sulfide quantum dots by yeasts, and skeletons made of strontium sulfate by a group of radiolarians called the acantharians [Lowenstam and Weiner 1989, Mann 2006]. The major part of biominerals, however, consists of three material systems: calcium hydroxyl phosphate, calcium carbonate, and silica. Since photonic applications of calcium phosphates are rare, we will focus mainly on the latter two materials. Considering the inventiveness of nature, one may ask why other minerals that are interesting for photonics seem not to exist in living nature at all, for example, the high refractive index pigment titania (i.e., titanium dioxide).

Here, the toolbox developed by organisms for controlled mineralization may be used to synthesize these materials nevertheless, either in vivo or in vitro. Therefore, photonic target materials that are feasible by biomimetic strategies are included in this overview as well. The occurring polymorphs and their optical properties are discussed.

2.1.1 Calcium Carbonate

Calcium carbonate, $CaCO_3$, occurs in various polymorphous crystal structures, of which the most stable is calcite, but also aragonite and the less stable vaterite can be found in organisms. Calcite forms trigonal crystals of space group R3c with unit cell dimensions of a = 4.99 Å and c = 17.06 Å and a density of 2.71 g cm^{-3} [Graf 1961]. The most significant optical property of calcite is its uniaxial negative birefringence with an ordinary refractive index of 1.658 and an extraordinary refractive index of 1.486 (at 589.3 nm). The optical axis that coincides with the crystallographic *c* axis constitutes an axis of threefold symmetry and is oriented perpendicular to the planes of the carbonate ions. Thus the polarizability for light waves with their electric field perpendicular to that axis, the ordinary waves, is highest [Hecht 1987].

In contrast, aragonite has an orthorhombic structure (space group Pmcn) with unit cell dimensions of a = 4.96 Å, b = 7.97 Å, and c = 5.74 Å and a density of 2.93 g cm^{-3} [Dal Negro and Ungaretti 1971]. It has to be noted, however, that recent studies revealed a lower symmetry [Bevan et al. 2002]. Also in aragonite, assuming the Pmcn space group, all carbonate ions are located in planes perpendicular to the crystallographic *c* axis, which explains the occurrence of one low and two high refractive indices in the biaxially birefringent material. For comparison, the crystal structures of both calcite and aragonite are displayed in Figure 2.2.

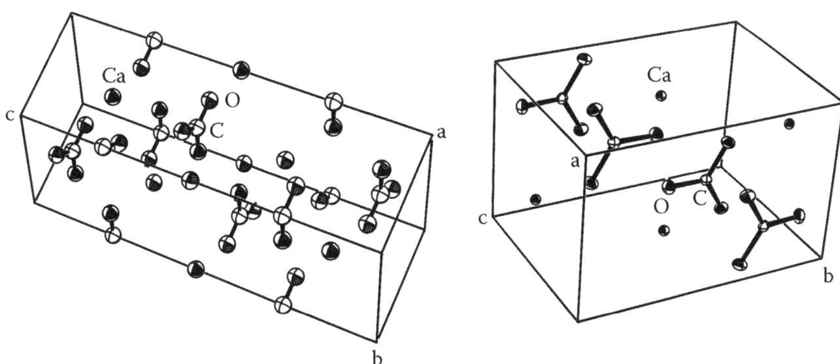

FIGURE 2.2
Crystal structures of calcite (left) and aragonite (right). Thermal ellipsoids with 50% probability are displayed. (Drawn with ORTEP-3 [Farrugia, L.J., *J. Appl. Cryst.* 30:565, 1997] using structure data from Graf [*Am. Mineral.* 46:1283–316, 1961] and Dal Negro and Ungaretti [*Am. Mineral.* 56:768–772, 1971], respectively.)

TABLE 2.1

Dispersion Curve Parameters for Calcite and Aragonite

	c_1	c_2	c_3	c_4	$\lambda_4/\mu m^2$
Calcite, n_ω	0.43257	0.82932	0.43376	0.61855	6.7
Calcite, n_ε	0.45899	0.69835	0.02680	0.30018	11.3
Aragonite, n_α	0.51143	0.77665	0.02250	0.33671	11.55
Aragonite, n_β	0.51091	0.88367	0.37976	0.54277	6.6
Aragonite, n_γ	0.49082	0.89499	0.40139	0.53266	6.6

Source: Ramachandran, G.N., *Proc. Indian Acad. Sci.* 26:77–92, 1947.

Note: The dispersion formula is given in the text. n_ω and n_ε denote ordinary and extraordinary refractive indices, respectively.

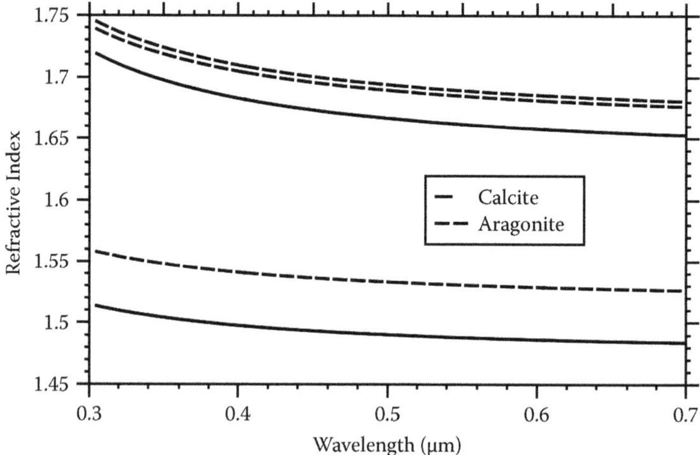

FIGURE 2.3
Principal refractive indices of calcite and aragonite in the visible spectral range. (Data from Ramachandran, G.N., *Proc. Indian Acad. Sci.* 26:77–92, 1947.)

The dependence of the refractive indices on the wavelength λ can be described with a dispersion formula of the type

$$n^2 - 1 = \frac{c_1\lambda^2}{\lambda^2(0.05\mu m)^2} + \frac{c_2\lambda^2}{\lambda^2(0.1\mu m)^2} + \frac{c_3\lambda^2}{\lambda^2(0.1535\mu m)^2} + \frac{c_4\lambda^2}{\lambda^2\lambda_4^2}$$

with parameters given in Table 2.1 [Ramachandran 1947]. The corresponding dispersion in the visible spectral range is plotted in Figure 2.3.

2.1.2 Silica

Silicon dioxide as a biomineral differs fundamentally from the carbonates since it occurs not in a crystalline structure but as amorphous hydrated silica. As a mineral, it may best be referred to as opal, $SiO_2 \cdot nH_2O$. The mean refractive index of opal silica with a density around 1.9 g cm^{-3} is 1.44 at 589.3 nm, exhibiting a normal dispersion curve with lower refractive index for larger wavelengths, but the exact value highly depends on the density and the water content [Kokta 1930]. Here, a higher hydration level leads to a lower refractive index. For biosilica derived from diatoms, we measured a refractive index of 1.43 at 632.8 nm. Other silica forms that may be compared with biosilica are synthetic amorphous quartz (fused silica) and the crystalline silica polymorphs tridymite, cristobalite, and natural quartz with increasing density and refractive index, in that order.

2.1.3 Artificial Optical Biominerals

The first important photonic material to be considered that has not been found in natural biominerals is titanium dioxide or titania, TiO_2. Titania exhibits an exceptionally high refractive index and is used as a white pigment. As a semiconductor it is used in hybrid solar cells [O'Regan and Graetzel 1991], and it shows photocatalytic activity [Fujishima and Honda 1972]. Comparing it with natural biominerals, it resembles silica in its sol-gel chemistry and calcium carbonate in the occurrence of different polymorphs. The most stable form in the bulk is rutile, followed by anatase and the least stable, brookite. Both rutile and anatase have tetragonal unit cells [Baur 1956, Horn et al. 1972], but they differ in the sign of birefringence: rutile, which is uniaxial positive, and anatase, which is uniaxial negative. Both are semiconductors with high bandgaps of 3.25 and 3.75 eV, respectively, measured in thin films [Park and Kim 2005].

Another important high-bandgap semiconductor that is accessible by biomimetics is zinc oxide, ZnO. Zinc oxide is an emitter at 390 nm with a direct bandgap of 3.37 eV [Zu et al. 1997]. It is an n-type electron conductor and interesting for luminescent sensors and transparent electrodes. The mineral zincite, which can also be synthesized artificially, occurs in two polymorphs, the hexagonal wurtzite structure and the cubic zincblende structure.

Other II/VI semiconductors, with a lower bandgap, are interesting due to their emission properties in the visible range that can be tuned by the size of their nanostructures. One prominent example that is important in the context of biomineralization is cadmium sulfide, CdS. The bandgap of cadmium sulfide is around 2.5 eV, depending on polymorph and size [Roessler 1999].

The most important properties of the bulk materials are compiled in Table 2.2 for easy reference.

2.2 General Principles of Biomineralization

Before discussing how these photonic materials are actually formed in or with the help of organisms, some general remarks on the principles of biomineralization will be given. The formation of biominerals is a fascinating example

TABLE 2.2

Overview of Polymorphs of Interesting Minerals for Biophotonics

	Mineral	$\Delta_f G^0$ (kJ/mol)	n_α n_ω	n_β n_ε	n_γ
$CaCO_3$	Vaterite	–1,030.6[a]	1.550	1.650	
	Aragonite	–1,032.8[a]	1.530	1.681	1,685
	Calcite	–1,033.9	1.658	1.486	
SiO_2	Opal		1.44		
	Fused silica	–727.9	1.46		
	Tridymite	–735.9	1.477	1.477	1.481
	α-Cristobalite	–736.4	1.454		
	Quartz	–737.6	1.544	1.553	
TiO_2	Brookite	–811.4[a]	2.583	2.584	2.700
	Anatase	–813.3[a]	2.561	2.488	
	Rutile	–814.0	2.616	2.903	
ZnO	Zincite	–292.0	2.013	2.029	
CdS	Greenockite	–128.8	2.506	2.529	

Source: Schaefer, K., and Lax, E. (eds.), *Landolt-Boernstein—Zahlenwerte und Funktionen*, Vol. II-4, 6th ed., Springer, Berlin, 1961; Hellwege, K.-H., and Hellwege, A.M. (eds.), *Landolt-Boernstein—Zahlenwerte und Funktionen*, Vol. II-8, 6th ed., Springer, Berlin, 1962; Turnbull, A.G., *Geochim. Cosmochim. Acta* 37:1593–1601, 1973; and Navrotsky, A., *Geochem. Trans.* 4:34–37, 2003.

Note: Values of standard Gibbs free energy of formation at 298.15 K ($\Delta_f G^0$) and principal refractive indices of polymorphs are compiled from literature data [Schaefer and Lax 1961, Hellwege and Hellwege 1962].

[a] Calculated from transformation enthalpies to the most stable polymorph [Turnbull 1973, Navrotsky 2003]. All refractive indices are given for 589 nm (Na_D line).

of how physicochemical self-organization is applied for the controlled synthesis of nanostructures with a distinct morphology.

The first step in the morphogenesis of specific structures is the definition of biomineralization sites by geometrical confinement. Generally, these sites can be extracellular or intracellular, each requiring different transport and directing mechanisms for the mineral precursors to these sites [Mann 2006]. Intracellular biomineralization is controlled in most cases by vesicles that define the shape in interaction with the cytoskeleton, whereas extracellular biomineralization is directed by frameworks of organic biopolymers that are secreted in the extracellular space forming a precipitation matrix for the biominerals. In extracellular biomineralization, vesicles are also involved, but mainly for transport of the precursors by exocytosis. Once the precipitation site is defined, a supersaturation of the mineral has to be achieved by active transport of ions, followed by controlled nucleation and growth of the aggregates. The other possible mechanism for such a phase separation, spinodal decomposition, is generally not observed but may play a role in some special cases in the precipitation of amorphous minerals like silica.

For the transformation of the precursors into the final biomineral some subtle physicochemical processes are involved. First, the direction of crystal growth depends on the presence of organic additives such as distinct biopolymers. Since the different crystallographic facets of the crystal nuclei may interact specifically with organic components in the reaction space, the shape of the crystals grown under these anisotropic conditions may be totally different from crystals grown from solutions free of additives. This feature is used in many biomimetic syntheses of mineral nanoparticles with defined shapes. However, for the final morphology in biological systems, much more complex interactions may take place, breaking the crystal symmetry and even resulting in single crystals with very complex shapes, like the anvil-shaped calcite plates of coccoliths or the spicules of sea urchin larvae with unusual threefold rotational symmetry [Mann 2006].

The second point is related to the crystal structure. Phase transformation between polymorphs occurs during the biomineralization process, and the sequence of the different stages can be well described by Ostwald's step rule [Ostwald 1897]. According to this principle, the metastable polymorphs with the highest free energy are formed first under kinetic control, being subsequently transformed into the next stable modification (Figure 2.4). The three-step transformation series from amorphous calcium carbonate via vaterite and aragonite to calcite has been experimentally confirmed in many cases by x-ray diffraction studies. In some other biominerals, e.g., calcium phosphates, the water or hydroxide content may also decrease from one step to the next. This is also true for hydrated silica, but in this case analysis is much more difficult because of the amorphous character of the biomineral limiting the application of x-ray structure analysis to a large extent. Interestingly, the transformation can be stopped at a certain stage, and that is the reason why

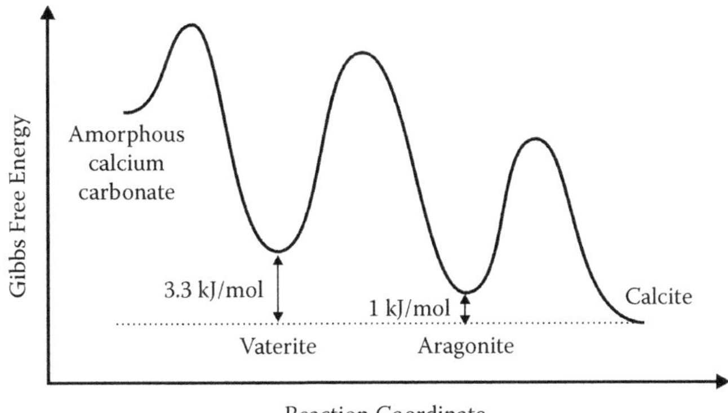

FIGURE 2.4
Relative stability of calcium carbonate polymorphs in bulk and their transformation sequence according to Ostwald's rule. Activation energies are not in scale, the values being several hundred kJ/mol. Direct transitions from, e.g., vaterite to calcite were observed experimentally [From Davies, P. et al. *J. Therm. Anal.* 13:473–487, 1978].

minerals that are not the most stable polymorph can be the predominant modification in biominerals. For example, aragonite and not calcite is found in many biominerals. An important role causing this behavior can be attributed to interface effects. From thermodynamic studies of titania and other polymorphs it is known that nanosized crystals exhibit a different order of stability because of the contribution of surface energy to the total free energy of the system [Navrotsky 2003]. Therefore, the bulk values given in Figure 2.4 can be substantially different in nanostructured biominerals.

In all cases, and this is perhaps the main point if one tries to deduce physical properties of biominerals from mineral data, it has to be taken into account that the directing biopolymers are part of the final biominerals, so strictly speaking, not neat minerals are formed but composite materials of inorganic and organic constituents with very complex nanostructures. This has many consequences not only for the mechanical properties, in which biogenic minerals are often superior to abiogenic minerals and crystals, but also for the photonic properties.

2.3 Examples from Nature

Since the focus of this book is photonics we select some examples of biominerals for which photonic properties can be derived.

2.3.1 Mother-of-Pearl

We start our discussion with nacre, or mother-of-pearl. The beautifully reflecting inner sides of the shells of some mollusks are among the best Bragg reflectors or dichroitic mirrors found in organisms. In contrast to the outer, prismatic layer of the shells, which is based on calcite, nacre consists of aragonite platelets that are stacked in periodic layers. This layered structure occurs in the three most important classes of the phylum Mollusca, namely, in Bivalvia, Gastropoda, and Cephalopoda. Important examples, some of them frequently used for the fabrication of jewelry (Figure 2.5), include the pearl oysters (Bivalvia, family Pteriidae, genus *Pinctada*), the abalones (Gastropoda, family Haliotidae, genus *Haliotis*), and the nautilus (Cephalopoda, family Nautilidae, genus *Nautilus*). But also in other classes, such as the monoplacophorans, aragonite stacks have been detected, but often with a different foliated structure called false nacre [Checa et al. 2009].

The structure of the composite material is given in Figure 2.6. The aragonite platelets are 0.4–0.5 μm thick, 10–20 μm wide, and ordered in layers. The crystallographic *c* axis is oriented perpendicular to the layers and in the direction of growth, although recent x-ray dichroism measurements show that there is some disorder in the orientation [Metzler et al. 2007].

FIGURE 2.5 (See color insert.)
Jewelry made from the nacreous parts of abalone shells.

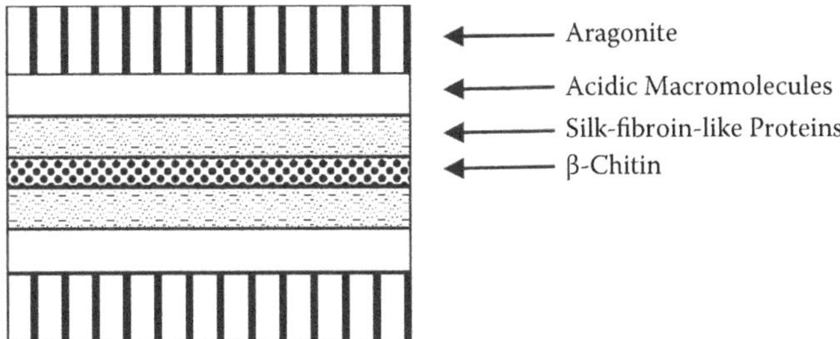

FIGURE 2.6
Composite structure of nacre.

Interestingly, the *a* and *b* axes are aligned as well in bivalves and in *Nautilus pompilius*, whereas in gastropods this orientation is missing [Wise 1970]. The organic matrix is located between these platelets causing a refractive index contrast to aragonite. The organic matrix consists of a β-chitin backbone with a multilayer cladding of hydrophobic proteins in an antiparallel β-sheet structure. The chitin and protein chains are oriented perpendicular to each other and are parallel to the *a* and *b* axes of aragonite, respectively [Weiner and Traub 1984]. Between the hydrophobic core and the aragonite crystallites there is a layer of acidic, aspartate-rich polypeptides as nucleation sites for the calcium carbonate precipitation. This composite arrangement consisting of a hydrophobic framework, polyelectrolytes, and the mineral is a common structural motif for biominerals that is found in many different systems [Addad and Weiner 1992]. The color effect of nacre results from constructive interference of light waves that are reflected at the interfaces with a relative phase shift of one to several wavelengths. Artificial dielectric mirrors are optimized for high reflectance by alternating layers of high and low refractive index with an optical thickness (physical thickness times refractive index) of one-quarter of a wavelength. In nacre, the organic layers with low refractive index are very thin in comparison with the aragonite platelets. With this approximation, the difference in optical pathlength between reflections from two adjacent platelets is given by $2d/n$, where d is the thickness of the plate, and n the relevant refractive index, in this case 1.68 for beam propagation in the direction of the c axis. For thicknesses between 400 and 500 nm, this difference is first order in the visible spectrum between blue (476 nm) and orange (595 nm), respectively. Variation of the reflectance maximum with thickness and the angle of incidence results in the iridescent appearance of nacre.

2.3.2 Calcite-Based Imaging

Whereas the color effect of nacre may be coincidence and does not serve a biological function, calcium carbonate-based lenses definitely have an optical purpose. The compound eye of trilobites, fossils with the oldest visual system, consists of calcite [Towe 1973]. Calcite eyes are usually divided in three types: the most common, *holochroal* eye, which consists of many small lenses; the *schizochroal* eye, with larger, separated lenses occurring in the suborder Phacopina; and the *abathochroal* eye, with small, separated lenses in the suborder Eodiscina [Clarkson et al. 2006]. We concentrate here briefly on the material aspects of these calcite lenses.

In holochroal eyes hexagonally arranged cylinders consisting of calcite lamellae radiate from the center. The lamellae in turn are formed from thin rods called *trabeculae*. In trabeculae crystals, the crystallographic c axis is oriented normal to the optical principal plane, which means that double refraction of light and thus bifocal imaging is avoided. This is an important point in using birefringent materials and shows how elegantly the control of biomineralization follows the optical function.

The same double-refraction-free orientation of crystals can also be seen in schizochroal eyes. For these eyes, internal structures such as intralensar "bowls" and "cores" are described that help minimize spherical aberrations [Clarkson and Levi-Setti 1975]. Recent studies show that the right refractive index contrast in these structures can be achieved by subtle mineral chemistry, modifying the mineral locally by replacing calcium with varying amounts of magnesium [Lee et al. 2007]. Interestingly, in a detailed examination of the eye of the trilobite *Dalmanitina socialis* it turned out that bifocal imaging was possible because of the lens shape, instead of birefringence [Gál et al. 2000]. It can only be speculated about the function of bifocal imaging in extinct species. In fact, bifocal lenses had been believed to be extinct until the recent discovery of nonmineralized bifocal lenses in the beetle *Thermonectus marmoratus* [Stowasser et al. 2010].

Calcite lenses can not only be found in proper eyes, but also were discovered in the photosensitive skin of some brittlestars (phylum Echinodermata, class Ophiuroidea) [Aizenberg et al. 2001]. The surface of the dorsal arm plates of the brittlestar *Ophiocoma wendtii* is covered with a calcite microlens array consisting of interconnected lenses with 40–50 μm diameter and 40 μm thickness. The calculated focal length of the lenses is 4–7 μm, the foci being located in a layer of photoreceptors that is responsible for photosensing. Detailed investigations showed that the exact shape of the lenses with a spherical and a nonspherical surface helps suppress spherical aberrations. As in the case of the trilobites, proper orientation of the crystallographic *c* axis suppresses the occurrence of birefringence. In brittlestars that are not photosensitive, such as *Ophiocoma pumila*, a similar lens array cannot be detected.

Very recently, a similar imaging system made of aragonite was reported for the chiton *Acanthopleura granulata* (phylum Mollusca, class Polyplacophora, family Chitonidae) [Speiser et al. 2011]. These mollusks have eight dorsal shell plates for protection, equipped with small *ocelli* of 65–80 μm diameter and 48 μm thickness. The angular resolution of this imaging system was determined to be 9–12°, which is approximately the same as in the case of the brittlestar *Ophiocoma wendtii*. The crystallographic *c* axes of aragonite in the chiton ocelli seem not to be oriented. Here, birefringence causes bifocal imaging. Interestingly, in water the focal point determined by the higher refractive indices falls on photoreceptors, whereas the second focus lies behind the eye. In air, the focal point determined by the lower refractive index falls on photoreceptors, whereas the other one lies in the lens (Figure 2.7). According to the authors this could be a hint for optimized vision in both environments.

2.3.3 Tabashir and Other Plant Silica

After the discussion of carbonate structures we now turn our attention to silica. Amorphous biosilica can be found in many organisms, including plants such as grasses, sponges, and in the beautiful cell architectures of radiolarians and diatoms [Simpson and Volcani 1981]. In many cases it serves a protective function. For example, its abrasive property has been applied technically in the case of horsetail (*Equisetum* sp.) that has been used by mankind for cleaning pots, especially those made of tin (therefore the German name *Zinnkraut*). In grasses (Poaceae), silica is especially frequent, and in some cases it exhibits interesting nanostructures. In the hollow stems of bamboo, the so-called tabashir is found, with a structure comparable to abiogenic opal [Jones et al. 1966]. It consists of nearly spherical particles of 10 nm diameter that give tabashir a milky, translucent appearance. Since the particle diameter is 10 times less than the diameter in the gemstone opal and the ordering is not as regular, iridescence is not observed. The refractive index of tabashir was measured to be 1.427 ± 0.02 for a composition of 86% SiO_2 with a density of 1.93 g/cm^3.

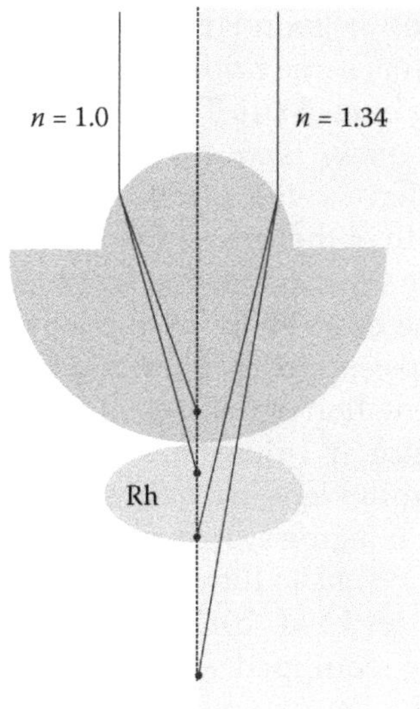

FIGURE 2.7
Schematic representation of the bifocal imaging of a chiton aragonite lens. Left: Focal points in air. Right: Focal points in seawater. The upper focal point corresponds to the ordinary beam. The area of photoreceptors is denoted as Rh (rhabdom region). (After Speiser, D.I., *Curr. Biol.* 21:665–670, 2011.)

2.3.4 Diatoms

The architecture of diatom cell walls has attracted much interest in the field of material science, as was envisioned in early communications [Parkinson and Gordon 1999]; thus we will devote some space in this chapter to this highly organized biosilica system. Diatoms (Bacillariophyta) are unicellular algae with a cell wall consisting of silica and various amounts of organic components. The cell wall (*frustule*) consists of two halves that are combined like a Petri dish, the upper half (*epitheca*) covering the lower half (*hypotheca*). Each *theca* in turn consists of a valve (*valva*, upper and lower part, respectively) and several girdle bands (*cingulum*) at the circumference of the cell [Simpson and Volcani 1981, Picket-Heaps et al. 1990]. During cell division, both halves separate, and a new hypotheca is formed in each half (Figure 2.8). By this process, the cells tend to become smaller in each generation, until sexual reproduction takes place, leading to new parent cells.

A regular array of pores is located within the cell wall. These pores (*areolae*) have an asymmetric cross section and are often covered by a perforated sieve plate on one side. Interestingly, for some genera this plate is at the outer surface, and for some genera at the inner surface. The thickness of the cell wall is around 1 μm, and the whole cell diameter is around 100 μm, but these values are subject to large variations from species to species. According to the overall shape of the cell, two groups of diatoms can be distinguished: the Centrales,

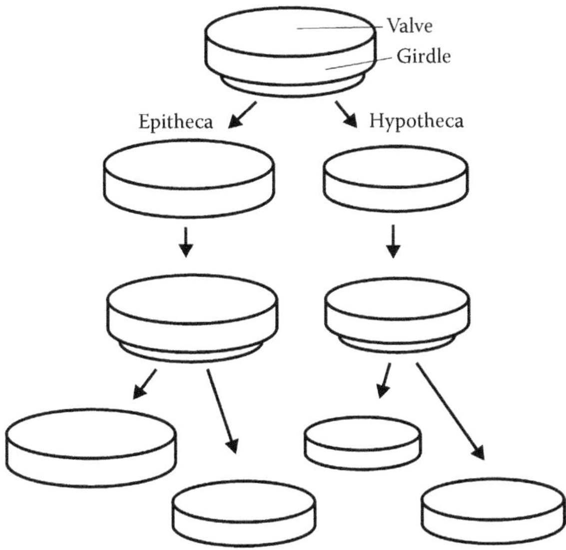

FIGURE 2.8
Schematic of the petri dish-like achitecture of diatom silica cell walls and the inheritance of the epitecae during cell division.

with a radially symmetric pattern, and the Pennales, with an elongated, bilateral shape. The taxonomy of the diatoms, however, is still under discussion, and may be clarified with the help of genetic analysis in the future.

In the last decade, many efforts have been made to characterize the organic components of the cell wall. Proteins called frustulines, pleuralins, silaffines, and silacidines have been found [Kröger et al. 1999, Kröger and Wetherbee 2000, Wenzl et al. 2008]. Whereas frustulines do not play an active role in silica precipitation, silaffines have been identified as an important matrix component for silica nucleation [Kröger 1999, Sumper and Brunner 2006]. They can be extracted by dissolving the purified frustules with fluoride in the form of anhydrous hydrogen fluoride (HF) or concentrated aqueous ammonium fluoride, and are characterized by a high degree of phosphorylation as well as by lysine residues to which long-chain polyamines are bound (Figure 2.9). Interestingly, such polyamines can also be found without attachment to a peptide as a component in diatom biosilica. Silacidines have phosphorylated serine residues as well, but contain mainly acidic amino acids such as aspartic and glutamic acid instead of lysine [Wenzl et al. 2008]. Recently, some further components of the silica biocomposite have been identified. Brunner and coworkers [2009] were able to show that chitin plays a role as an integral part of biosilica in *Thalassiosira pseudonana,* and Ingalls and coworkers [2010] identified UV-absorbing amino acids that may provide photoprotection of the cells.

The predominant contribution of silaffines and long-chain polyamines in silica precipitation led to a phase separation model that is able to predict

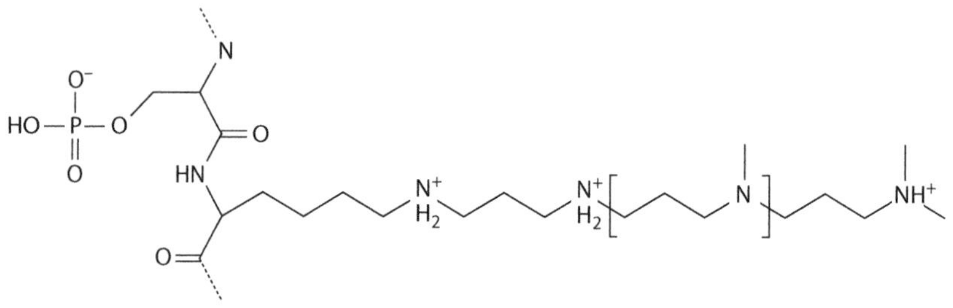

FIGURE 2.9
Typical structural motif from silaffins containing phosphorylated serine and polyamine-functionalized lysine in the chain sequence. (After Kröger, N., *Science* 286:1129–1132, 1999.)

the hierarchical pattern formation of silica networks in centric diatoms of the genus *Coscinodiscus* [Sumper 2002]. The central assumption of the model is the phase separation of an aqueous, silica-rich phase and ordered polyamine droplets that later form the pores of the silica network. By in vitro experiments it became evident that the morphology of a porous network can be achieved if the polyamines serve a second function as electrostatic stabilizers of precursor silica sol particles [Sumper 2004, Sumper and Brunner 2006]. The condensation of a precursor sol is also the base of a slightly different model [Vrieling et al. 2002]. Indeed it was demonstrated by atomic force microscopy that the fine structure of the cell wall consists of granular particles [Wetherbee et al. 2000, Noll et al. 2002]. These nanosized precursor silica particles seem to assemble into larger structures, for which pattern control at the 100 nm to micrometer level is needed. Here, the silica deposition vesicles (SDVs) that are responsible for silica uptake, condensation, and transport may play a major role as a structure-directing matrix. SDVs are intracellular vesicles that are characterized by a low pH value and a special membrane, the silicalemma [Schmid and Schulz 1979, Vrieling et al. 1999]. Recent models thus discuss a participation of membrane-bound proteins of the SDVs and the cytoskeleton in morphology control [Kröger 2007, Tesson and Hildebrand 2010].

Concerning the photonic properties of diatom frustules, the silica cell wall can be regarded as a glass frame in which the protoplasm is located. Certainly, such a frame structure acts as an optical waveguide because the refractive index of biosilica (1.43) is substantially higher than the seawater environment (1.34). The number of waveguide modes that may occur for a given wavelength depends on the thickness of the waveguide. Usually the thickness of the cell wall is in the range of 1 μm; thus only a few modes can propagate [Fuhrmann et al. 2004]. The number of modes for a given ratio of slab thickness and wavelength can be obtained from Figure 2.10. Each mode is characterized by an effective refractive index in the propagation direction, with an evanescent field into the outer medium that decays more slowly for small effective indices. The fine structure of the cell wall in the range

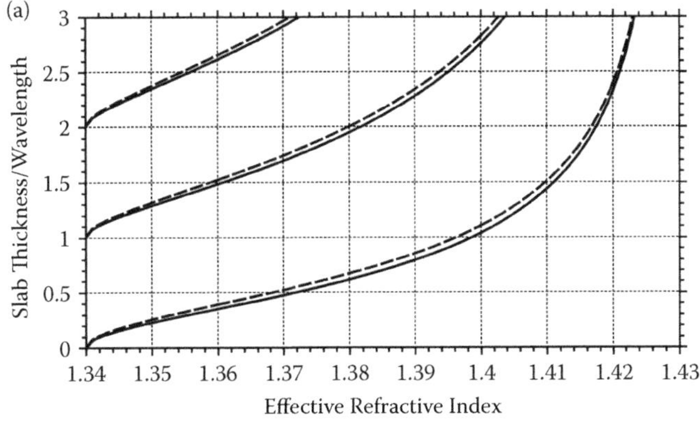

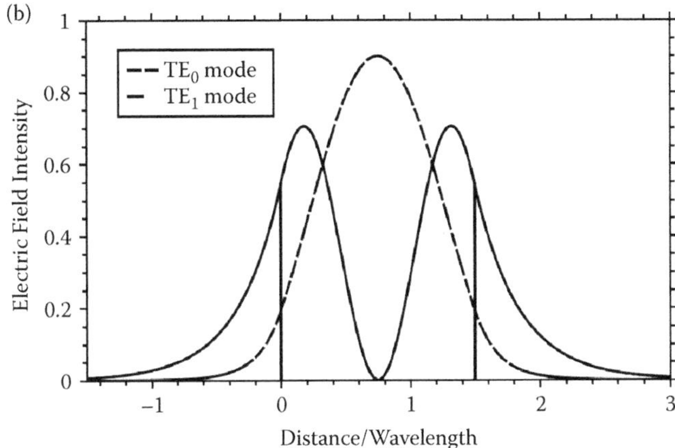

FIGURE 2.10
(a) Mode calculation for a biosilica slab with refractive index 1.43 in seawater. The effective index and the thickness onset for each mode can be extracted. Note that two distinct polarizations exist: transversal electric (TE) with the electrical field perpendicular to the plane spanned by the direction of propagation and the slab normal, and transversal magnetic (TM) with the electrical field in this plane. (b) Electric field profile for the two TE modes with effective indices 1.417 and 1.362 in a slab with a thickness of 1.5 λ (wavelength λ = 632.8 nm).

of optical wavelengths disturbs the waveguide, leading to complex scattering and coupling phenomena. The UV scattering properties are discussed in Chapter 3, Section 3.2, by Lenau.

Especially interesting are the optical modes in cell walls with regularly ordered pores that are found, as in some centric diatoms. In that case, an additional confinement of the waveguide modes is imposed by the optical band structure due to the periodic pattern. Therefore, the diatom cell wall can be regarded as a photonic crystal slab waveguide [Fuhrmann et al. 2004], with an optical mode structure identical to that of artificial slab photonic crystals [Johnson et al. 1999]. As an illustrative example, the genus *Coscinodiscus*

FIGURE 2.11
Scanning electron microscopy image of the pore patterns in the cell wall of the diatom *Coscinodiscus granii*. The transition from the hexagonal pattern in the valve to the square pattern in the girdle can be seen.

is discussed. In *Coscinodiscus granii* and *Coscinodiscus wailesii*, very distinct regular pore patterns in the cell wall can be observed (Figure 2.11). The valve consists of a hexagonal array of pores with mean lattice constants of 0.9 µm for *Coscinodiscus granii* and 1.8 µm for *Coscinodiscus wailesii*. Interesting is the formation of defect rows that end in silica appendages protruding from the slab, the rimoportulae. A hexagonal pattern is encountered frequently in nature and may be easily explained by a dense packing of colloidal drops. The girdle, in contrast, exhibits a square array of pores that is remarkable and indicates the influence of the curved geometry of the girdle or repulsive electrostatic forces during pattern formation. The lattice constant is significantly smaller than in the valve, with mean values of 270 nm for *Coscinodiscus granii* and 310 nm for *Coscinodiscus wailesii*, respectively [Kucki et al. 2006]. Calculations were performed in order to obtain the optical band structure for the photonic crystal slab waveguide. A typical band diagram is given in Figure 2.12, calculated for the girdle idealized as a perforated slab with a refractive index of 1.43 in seawater. The frequencies of allowed modes for given wave vectors in the reciprocal space are displayed. Because of the moderate refractive index contrast, the confined modes are found very near

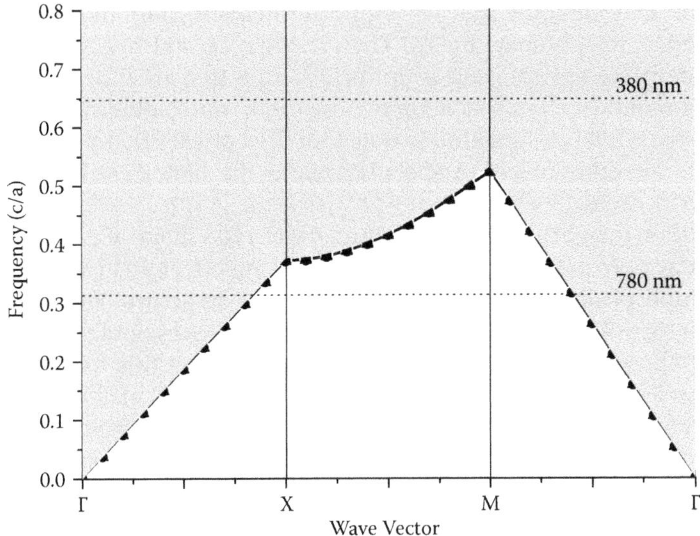

FIGURE 2.12
Optical band diagram for a diatom girdle in seawater. The frequency is given in units of vacuum light velocity divided by lattice constant. A lattice constant of a = 250 nm, a slab thickness of 0.8 a, and a pore radius of 0.36 a were assumed. All modes are located just below the light line. Again, different polarizations can be distinguished (triangles and circles).

to the "light line," separating the region of confined modes from the region of modes radiating into the medium (grey shade). At the distinguished edge points in the reciprocal space, for square patterns denoted as X and M, resonances can be found. Such modes correspond to stabilized standing waves, as applied in photonic crystal lasers [Notomi et al. 2001]. For a girdle with a lattice constant of 250 nm, they are located in the red and blue spectral region (X point and M point, respectively). If the surrounding medium is not water but air, the light line shifts upwards, the modes separate, and the resonances are found in the green and violet spectral regions, respectively [Kucki et al. 2006]. Thus, for the girdle, the periodicity of the pattern is just in the range suitable for manipulating visible light. For the valve, optical resonances are only possible in a higher order for visible light since the lattice constant is in the micrometer range [Fuhrmann et al. 2004]. Those modes may propagate in the waveguide but are strongly coupled to waves radiating from the slab.

The silica cell wall is certainly a photonic crystal slab per definition, but it is unclear whether there is any biological function resulting from that property. However, the whole structure can be regarded as a photonic box with a well-defined set of optical modes. Since the waveguide modes decay far into the adjacent medium (Figure 2.10 (b)), they can in principle couple to the chloroplasts that are located quite close to the cell wall [Sumper and Brunner 2006].

Thus, one may imagine that waveguiding helps to trap and distribute the light needed for photosynthesis. The photonic crystal fine structure could help by defining optical wave coupling schemes that are independent on the orientation of the cell in space. Up to now there is no biological proof for this hypothesis, but it is interesting to note that the optical function of the diatom shell was already subject to speculations in the biological literature some time ago [Schmid 1996].

In addition to light guiding properties, the curved shape of the cell wall with the regular pore pattern leads to a light focusing effect that can be explained by the superposition of scattered waves in the structure. For *Coscinodiscus wailesii*, a focus length of about 100 μm for red light was obtained [De Stefano et al. 2007]. Wavelength-dependent studies show that this focusing effect is not present for ultraviolet wavelengths [De Tommasi et al. 2010], corroborating the idea that the cell wall has a protective function against ultraviolet radiation. Yamanaka et al. [2008] investigated a freshwater species, *Melosira* sp., and found an increased apparent absorption in the blue spectral region that was attributed to group velocity effects in the photonic crystal structure.

Diffraction is also important, and the sixfold symmetry of the valve of *Coscinodiscus* sp. gives rise to sixfold diffraction patterns, with decreasing efficiencies from longer to shorter wavelengths [Noyes et al. 2008].

Biosilica in diatoms not only acts as a passive optical element but also shows light emission. Isolated diatom frustules show blue fluorescence, but the exact color varies among different reports investigating different species [Orellana et al. 2004, Butcher et al. 2005, De Stefano et al. 2005, Qin et al. 2008, Mazumder et al. 2010, Kucki and Fuhrmann-Lieker 2011]. The origin of this fluorescence is not clear but can in principle arise either from the organic matrix that may contain heterocyclic moieties [Ingalls et al. 2010] or from surface states in mesoporous silica [Chiodini et al. 2000, Shieh et al. 2004]. De Stefano and coworkers [2005] proposed an optical chemical sensor based on the autofluorescence since intensity and peak position of the luminescence are found to be sensitive for gases and organic vapors.

In later sections, artificial hybrid materials based on diatom frustules are discussed.

2.3.5 Sponge Fibers

The most significant production of biosilica in the animal kingdom is the spicules of sponges. The phylum Porifera (sponges) is divided into three classes, of which one, the Calcarea, forms calcium carbonate skeletons, whereas the other two, Hexactinellida and Demospongiae, rely on silica structures as building blocks. Their skeleton consist of silica rods, the spicules, with lengths ranging from millimeters to several meters [Wang et al. 2009]. Two proteins, silicatein [Cha et al. 1999] and silintaphin-1 [Müller et al. 2009b], perform a similar role as silaffins and polyamines in the biomineralization of diatoms. Silicatein and the scaffold protein silintaphin-1

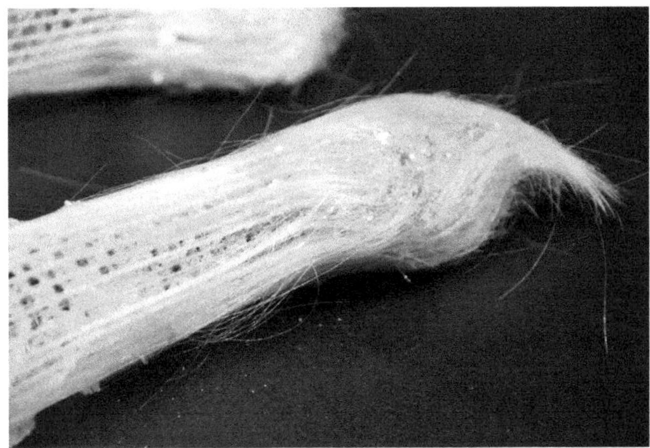

FIGURE 2.13
The silica mesh of the Venus flower basket *Euplectella aspergillum* (Museum Zeche Zollverein, Essen, Germany).

are located within the axial canal of the spicules and form the crystallization core for silica precipitation. Also in the case of sponges, the silica consists of sintered sol nanoparticles. For the demosponge *Tethya aurantia*, a mean particle diameter of 74 nm was measured [Weaver et al. 2003]. The structural hierarchy in sponge biosilica from the nanoscale to the macroscale was pointed out by Aizenberg and coworkers [2005] discussing the example of the hexactinellid genus *Euplectella* sp.

The function of sponge spicules as optical waveguides was described by Cattaneo-Vietti et al. [1996]. They measured light transmission through the spicules of the hexactinellid sponge *Rossella racovitzae* and demonstrated a large light-capturing surface by the cross-shaped apex of the spicules. Aizenberg and coworkers investigated the fiber optical properties of the beautiful Venus flower basket (*Euplectella aspergillum*, Figure 2.13) in detail [Sundar et al. 2003, Aizenberg et al. 2004]. They were able to show by interferometric refractive index profiling that the fiber consists of a core with high refractive index (1.46), surrounded by a cylinder with low refractive index, decreasing to 1.425, and an outer cladding of progressively increasing refractive index (Figure 2.14). The core has a thickness of 1–2 μm allowing only a few waveguide modes to propagate.

The biological function of the optical fiber property is not clear; however, symbiotic algae seem to benefit from it. It is known, for instance, that shade-adapted diatoms, green algae, and cyanobacteria live within Antarctic demosponges and may capture light propagating in the spicule [Gaino and Sarà 1994, Cattaneo-Vietti et al. 1996, Cerrano et al. 2000]. By using photosensitive paper, Brümmer and coworkers [2008] were able to locate the regions within deeper tissues of *Tethya aurantium* to which light is guided, corroborating the hypothesis that phototropic symbionts may extract that guided light.

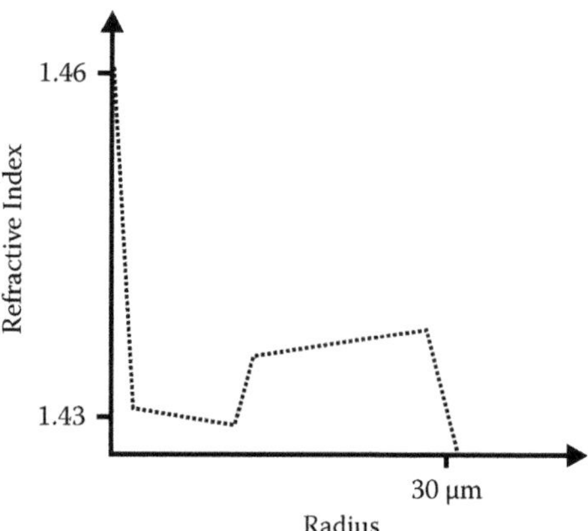

FIGURE 2.14
Simplified refractive index profile of the silica fiber of *Euplectella aspergillum*. (According to the data from Aizenberg, J., and Hendler, G., *J. Mater. Chem.*, 14:2066–2072, 2004.)

Very interesting is the wavelength dependence of light transmission through a sponge spicule. Müller and coworkers [2006] showed that only light between 615 and 1,310 nm can pass through the spicules of the hexactinellid *Hyalonema sieboldii*, determining a transmission efficiency of 60% after a path length of 5 cm. Similar results were obtained for the giant basal spicules of *Monorhapsis* sp. [Müller et al. 2009a]. Speculation about a possible photoreception system is corroborated by the fact that at least in the demosponge *Suberites domuncula*, the gene for the bioluminescence generating enzyme luciferase was detected [Müller et al. 2009b]. Speculation about a possible photoreception system are is corroborated by the fact that at least in the demosponge *Suberites domuncula*, the gene for the bioluminescence generating enzyme luciferase was detected. This may be a singular case; future research, however, will show whether there is a connection between the generation of light and its transmission in sponges.

Recently, also the emission properties of sponge biosilica have been investigated for photonic applications. It was found that sponge spicules show long-living fluorescence with a maximum at 770 nm and strong optical nonlinear behavior when pumped with femtosecond laser pulses [Kul'chin 2008, 2009, 2011].

2.4 Functionalizing Biological Structures

The efforts to add additional features to natural biomaterials will now be described. The addition of artificial ingredients during the formation of the biominerals or the posttreatment of biological structures aims at new hybrid materials that are nevertheless closely related to the natural forms from which they are derived.

2.4.1 Replacing Silica

One method of influencing the composition of the biomaterial is the replacement of the inorganic constituents. For biosilica, this approach means replacing silica partially by another, chemically similar compound that is tolerated by the organism. It turned out that other oxoanions, if available as a source, can be incorporated into biosilica to a certain extent. Many investigators have used diatoms since they allow easy experiments due to their fast reproduction cycle. Thus, successful incorporation was demonstrated for aluminum oxide [Gehlen et al. 2002], germanium oxide [Chiappino et al. 1977, Jeffryes et al. 2008b], and titania [Jeffryes et al. 2008a]. Concentrations of 1.6 weight percent Ge and 2.3 weight percent Ti in SiO_2 [Jeffryes et al. 2008a] were achieved.

Interesting photoluminescent and electroluminescent spectra were reported in the case of germanium-inserted frustules [Jeffryes et al. 2008b]. For the construction of electroluminescent devices, diatom frustules covered by $HfSiO_4$ were placed between an indium tin oxide anode and an aluminum cathode. They exhibit an emission spectrum with several sharp peaks for which photonic crystal effects are discussed. However, such spectral features were not observed in the photoluminescence spectrum.

2.4.2 In Vivo Fluorochromation

In vivo fluorochromation is a functionalization method that has been applied for some time in biomineralization studies, and that is gaining impetus from a new materials science perspective. Fluorescent marker dyes accumulate in distinct cell compartments and are deposited eventually in the biomineral. Again, many experiments have been made with biosilica. In the case of diatoms, the marker dyes accumulate in acidic vesicles, especially the SDVs, and are incorporated into the silica cell wall (Figure 2.16 (a)). After isolation of the frustules, they remain in the cell wall unless the biocomposite is dissolved completely by fluoride treatment. Thus, the organic part of the biomineral is modified.

Fluorescent dyes that worked successfully in this respect are compiled in Figure 2.15. The first dye that was used for this purpose is Rhodamine 123 [Li et al. 1989]. A recent study in our lab shows that in fact many rhodamine dyes can be used, provided they have a free acid function [Kucki 2009, Kucki and Fuhrmann-Lieker 2011]. By varying the substitution pattern of the rhodamine chromophore, several emission colors, from green to red, were achieved. A dye addressing the blue spectral region that is not covered by rhodamines is the pH-sensitive oxazole dye PDMPO, also known as Lysosensor DND-160 [Shimizu et al. 2001]. Yet a different chromophore

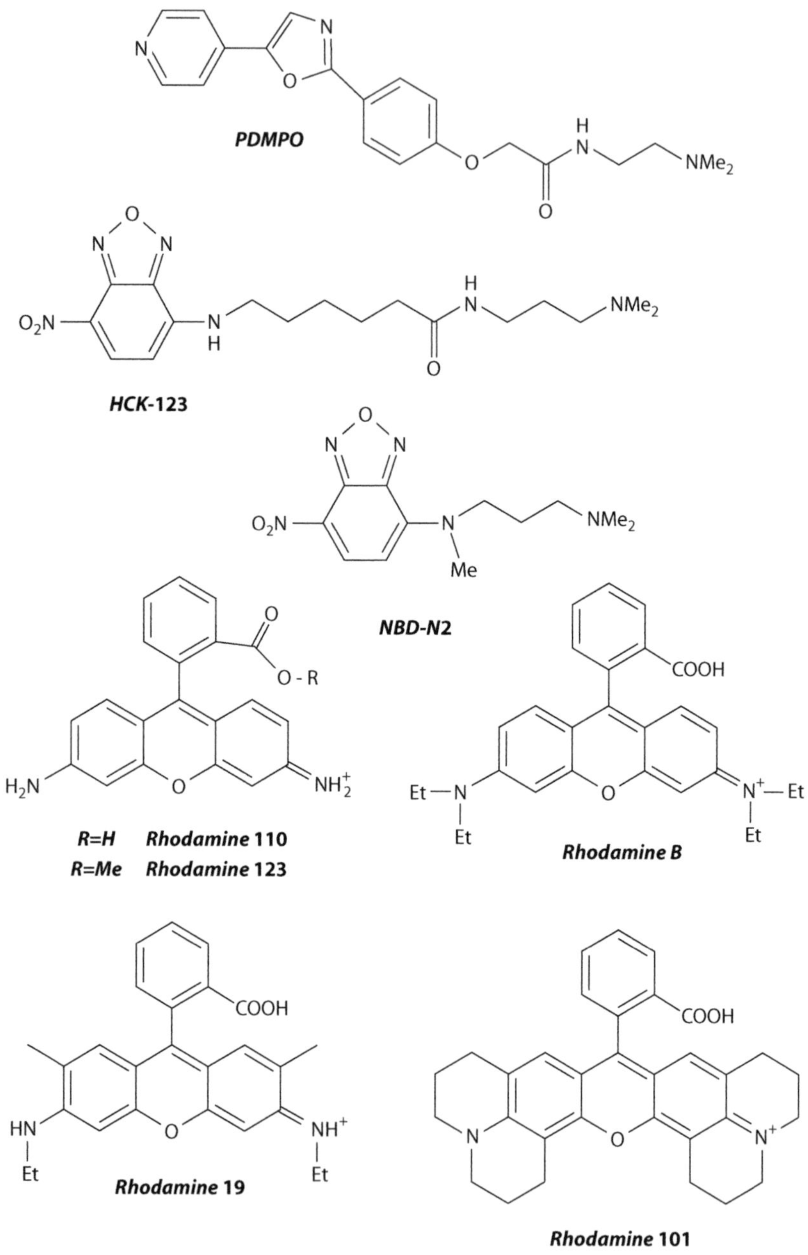

PDMPO

***HCK*-123**

NBD-N2

R=H ***Rhodamine* 110**
R=Me ***Rhodamine* 123**

Rhodamine B

***Rhodamine* 19**

***Rhodamine* 101**

FIGURE 2.15
Fluorescent dyes that have been applied as markers in biosilica, generating emissive biominerals. Examples include rhodamines, oxazole, and oxadiazole dyes.

class was introduced by Desclés and coworkers [2008] with the efficient dye HCK-123, a 1,2,5-oxadiazole dye. Based on this chromophore, derivatives such as NBD-N2 have been developed recently [Annenkov et al. 2010]. Those fluorescent biosilica structures may be useful for optical investigation of diatoms. They can be regarded as semiartificial biominerals since they are produced by the living organisms under special nutrient conditions.

2.4.3 Surface Functionalization

Instead of in vivo functionalization, a postmodification of isolated biominerals is also possible and may lead to new composite materials. The high surface area of mesostructured biominerals allows the attachment of functional compounds with high density, aiming at applications in heterogenous catalysis or—as considered in this book—in photonics. For the attachment of molecules to biosilica, silica surface chemistry is applied, and conveniently the functional units are attached via a linker. Rosi et al. [2004] functionalized isolated diatom frustules with 3-aminopropyltrimethoxysilane and coupled DNA nucleotides to the amino function. Then, gold nanoparticles functionalized with the complementary DNA sequence were attached, resulting in a gold-plated frustule. Because of the nanostructured pattern, interesting plasmonics effects of such metal-plated structures and possible applications in surface-enhanced Raman scattering [Payne et al. 2005] can be expected. In an alternative reaction sequence, rhodamine dyes were attached to the aminosilane by esterification first, and then coupled to the silica surface [Kucki et al. 2006]. Figure 2.16 (b) shows a fluorescence image of a shell fragment of *Coscinodiscus granii* functionalized by this method.

A slightly different approach was used by Y. Yu et al. [2010]. They sensitized the surface of frustules by treatment with trifluoroacetic acid and tin(II) chloride before electroless deposition of a thin silver nanoparticle layer, that in turn acted as catalyst for the final electroless deposition of a gold coating using formaldehyde and a Au(I) solution.

Applying atomic layer deposition (ALD) as a coating technique from the gas phase, Losic and coworkers [2006] succeeded in coating diatom frustules with TiO_2. Titanium(IV)chloride and water were used as precursors in the ALD process. From the reduction of the pore size in the cell wall, a coating thickness of approximately 23 nm after 1,000 deposition cycles was deduced. In these structures, optical confinement for waveguiding is significantly improved because of the high refractive index of titanium dioxide. The technique is also transferable to the synthesis of other coatings, such as Al_2O_3, ZrO_2, SnO_2, V_2O_5, ZnO, or TiN.

In another report, a coating with ZnO was prepared in a wet chemical route by soaking diatoms in a zinc acetate solution, followed by baking [Cai and Sandhage 2005]. At higher temperatures (1050°C) the porous coating consisting of ZnO nanoparticles can be transformed to Zn_2SiO_4 by reaction with the underlying silica.

(a)

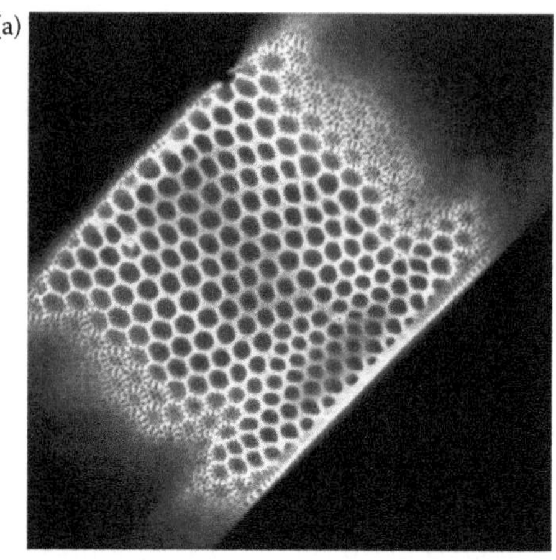

(b)

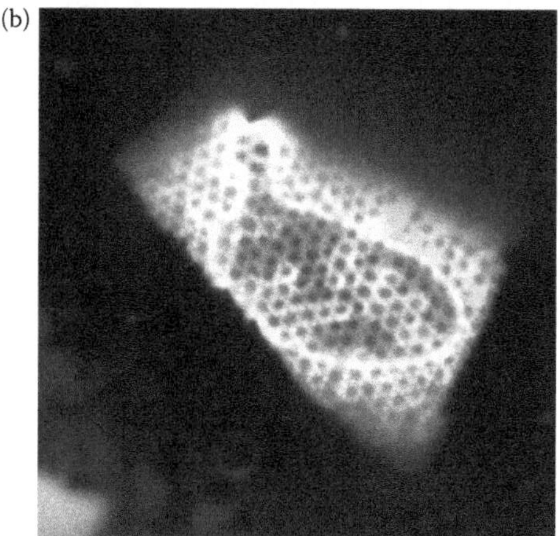

FIGURE 2.16 (See color insert.)
Confocal laser scanning microscope images of diatom shells (*Coscinodiscus granii*) with (a) an incorporated rhodamine dye after in vivo fluorochromation [Kucki 2009] (b) with a rhodamine dye (Rhodamine B) covalently attached to the silica surface.

2.4.4 Biomineral Replicas

By special shape-preserving displacement reactions that were developed in the past decades, biosilica can also be completely transformed into another material system. One example is the transformation of diatoms into MgO, with a complete removal of silicon out of the structure [Sandhage et al. 2002]. Diatom silica reacts with magnesium in the gas phase, following the reaction

$$SiO_2(s) + 2\,Mg \rightarrow 2\,MgO(s) + \{Si\}$$

where {Si} denotes silicon in a liquid alloy with magnesium. Energy-dispersive x-ray analysis showed that Si is not present in the transformed frustule that retained its original shape. Similarly thermodynamically allowed is the reaction of silica with other alkali and earth alkali metals, yielding Li_2O, CaO, or SrO. If metal fluorides are used as reactants, Si is removed as gaseous silicon fluoride, for example, in the reaction

$$SiO_2(s) + TiF_4(g) \rightarrow TiO_2(s) + SiF_4(g)$$

Analogously, ZrO_2, Al_2O_3, Fe_2O_3, Nb_2O_3, or Ta_2O_3 may be prepared from the corresponding fluoride precursors. TiO_2 was also used as the intermediate step toward a transformation of diatom frustules into $BaTiO_3$ and $SrTiO_3$ by reaction with molten $Ba(OH)_2$ or $Sr(OH)_2$, respectively [Dudley et al. 2006]. These perovskites are very interesting for photonic application because of their electrooptical and photorefractive properties.

If the transformation reaction is carefully controlled, silicon may be retained in the structure as a solid component. After removal of the oxide, as demonstrated for MgO, a photoluminescent porous nanocrystalline silicon replica can be obtained that may be interesting also for photonic applications [Bao et al. 2007].

In addition to photonics, diatom replicas may find applications in sensors, microfluidics, and other fields of materials science, but here the reader is referred to recent reviews in the literature [Losic et al. 2009, Yang et al. 2011].

Templating techniques have also been applied for calcium carbonate structures. The porous skeleton plates of the sea urchin *Heliocidaris erythograma* were transformed into a polymer replica by filling the network structure with a resin monomer. After hardening, the calcium carbonate was dissolved by acid, and the replica was filled with $CaCl_2$ and Na_2CO_3 solutions to precipitate $CaCO_3$ again. Surprisingly, a single crystal with the original complex morphology was formed in the interconnected channels [Park and Meldrum 2002]. A direct optical application was not given, but one can imagine the potential of this approach in the formation of nanostructured single-crystalline optical materials.

2.5 Biomimetic Approaches Toward Optical Materials

We finally turn to techniques in which the biological form is not used directly, but the biomineralization principles are applied in order to mimic the morphogenesis of special nanoarchitectures. Since the field is growing very rapidly, only a few examples are given, with special emphasis on the materials introduced above as important optical materials.

The morphology of crystals can be influenced by many additives, and very often the precipitation of inorganic materials under the influence of these additives is denominated as biomimetic (for examples, see Mann [2006] or Xu et al. [2007]). Here, the focus is especially, but not exclusively on structures formed under the influence of natural biopolymers or their simplified synthetic analogues as a kind of restriction of the term *biomimetic*.

2.5.1 Calcium Carbonate

The typical rhombohedral shape of $CaCO_3$ crystals changes to microspheres by addition of block copolypeptides that contain polyaspartate or polyglutamate blocks that also occur in nacre as a structure directing motif [Euliss et al. 2006]. Vaterite and calcite were detected in these biominerals. Hollow calcite crystals were obtained in micellar solutions containing citric acid and cetyl trimethylammonium bromide [Xiao and Yang 2010]. In the form of thin films, vaterite and aragonite were prepared on a chitosan matrix using polyaspartate as additive [Sugawara et al. 2006]. Additional functionality was added to calcium carbonate by using a conducting polymer, carboxylated polyaniline, as additive [Neira-Carrillo et al. 2008]. The new hybrid material exhibits fluorescent properties as well as electric conduction.

2.5.2 Silica

In contrast to calcium carbonate where anionic polymers are used as the precipitating agent, silica precipitation is induced by cationic polymers. Silaffines isolated from diatoms are able to precipitate silica within a few minutes as nanospheres [Kröger et al. 1999, 2002]. The same can be done by polyamines without the peptide backbone, in the presence of phosphate. These studies showed that the size of the resulting nanospheres is determined by the concentration of phosphate [Sumper et al. 2003]. The morphology changes from spherical to an open network if a polyamine-stabilized sol is used as a precursor instead of silicic acid [Sumper 2004].

The size and dispersity of nanospheres formed in a reaction mixture containing silicic acid, cetyl trimethylammonium bromide, poly(allylamine), and citrate was determined by Begum et al. [2010]. The poly(allylamine) was also tagged with the fluorescent dye fluorescein isothiocyanate, resulting in fluorescent biomimetic nanospheres.

An interesting photonic application of silica nanosphere precipitation was described by Brott et al. [2001]: if polycationic peptides are incorporated into a polymer diffraction grating that is exposed to silicic acid, an ordered array of silica nanospheres is deposited along the grooves of the hologram. The diffraction efficiency of the grating increases substantially, demonstrating a direct application of biomineralization in the production of photonic devices.

Other mesostructures than spheres are accessible by using appropriate organic components in silica deposition. With amphiphilic block copolymers, multilamellar vesicles, unilamellar nanofoams, and multilamellar vesicles with sponge-like walls can be obtained [M.H. Yu et al. 2010]. SiO_2 nanofibers were synthesized by grafting silicatein to pentafluorophenyl acrylate as backbone [Tahir et al. 2009]. Fibers with a thickness between 10 and 20 nm and an aspect ratio of 1,000 formed around this backbone by adding tetraethoxysilane as precursor.

Volkmer and coworkers [2003] were able to synthesize star-shaped shells with silica spines, mimicking the morphology of radiolarians. The structures grow at the interface of chlorocyclohexane droplets containing arachidic acid and an aqueous cetyl trimethylammonium bromide solution. Mineralization is strongly enhanced by adding orthotitanate precursors in addition to tetraalkoxysilane to the microemulsion reaction mixture.

2.5.3 Titanium Dioxide

The chemical similarity of the oxoanions of silicon and titanium in the formation of SiO_2 and TiO_2, as already mentioned in the previous example, is the starting point for developing biomimetic techniques for the synthesis of photonic materials that are not found in living nature. Indeed, it was demonstrated that silica-precipitating biomolecules are able to precipitate titanium dioxide as well. Silicatein filaments from *Tethya aurantia* were shown to mineralize Titanium(IV)bis(ammonium lactato)dihydroxide to titanium dioxide [Sumerel et al. 2003]. Sewell and Wright [2006] used a peptide from the diatom *Cylindrotheca fusiformis* and the same titanium precursor to synthesize titanium dioxide nanoparticles with a size of 50 ± 20 nm. The use of poly(lysin) instead of the natural peptide resulted in larger particles (140 ± 60 nm). Films of TiO_2 have been prepared on surfaces of poly(2-dimethylamino)ethylmethacylate as a synthetic analogue to polyamines [Yang et al. 2008].

2.5.4 Zinc Oxide

Zinc oxide nanostructures can be prepared artificially by various methods, but also biomimetic mineralization has been employed. Interestingly, peptides that normally do not take part in biomineralization have been used for that purpose, namely, silk fibroin peptides from the silk worm *Bombyx mori* [Huang et al. 2008, Yan et al. 2009]. Olive-like ZnO nanoparticles formed around these peptides after immersion in a zinc nitrate solution. Photoluminescence with peaks at 410, 470, and 530 nm was observed for these nanoparticles. Ring- and disc-like structures were obtained by using poly(acrylamide)s as additives [Peng et al. 2006].

2.5.5 Metal Sulfides and Metals

As mentioned above, chalogenides, especially II/VI semiconductor nanocrystals, and metal nanoparticles find widespread application in photonics because of the size dependence of their optical properties. A biomimetic control of CdS nanoparticle growth can be achieved by addition of phytochelatins [Bae and Mehra 1998, Liu et al. 2010]. Ag_2S nanoclusters with potential applications in imaging and photosensitizing were prepared using urease or cucurbit[7]uril as additive [Pejoux et al., 2010, de la Rica and Velders 2011]. Similarly, silver nanoparticles can be obtained in vitro with the aid of silver-binding peptides [Naik et al. 2002, Dong et al. 2005]. Details and further examples on the biomimetic synthesis of silver and gold nanoparticles can be found in the literature [Xie et al. 2010] and will not be treated here further.

2.6 Summary

Nature offers a unique collection of sculptured biominerals that exhibit interesting optical properties due to their micro- and nanostructure. In many cases, a true biological photonic function is not clear and remains a speculation until further investigations. However, by learning from nature, biomimetic strategies can be employed for the synthesis of hybrid biomineral composites that may consist of natural and artificial photonic materials. They may act as photonic crystals or optical waveguides, mirrors or electrooptical elements. The development of such microstructured photonic elements is just at the very beginning and may become a wider research area in the future, characterized by the unique combination of biological self-organization with microoptical engineering.

Acknowledgments

For providing new insights, interesting discussions, results, and photos, the author thanks all his coworkers.

Photo credits: Figure 2.2, E. Tatarov; Figures 2.5 and 2.13, I. Lieker; Figure 2.6, K. Pfaff; and Figures 2.11 and 2.16, M. Kucki, S. Landwehr, H. Rühling, and M. Maniak.

References

Addad, L., and Weiner, S. (1992). Control and design principles in biological mineralization. *Angew. Chem. Int. Edit. Engl.* 31:153–169.

Aizenberg, J., and Hendler, G. (2004). Designing efficient microlens arrays: lessons from nature. *J. Mater. Chem.* 14:2066–2072.

Aizenberg, J., Sundar, V.C., Yablon, A.D., Weaver, J.C., and Chen G. (2004). Biological glass fibres: correlation between optical and structural properties. *Proc. Natl. Acad. Sci. USA* 101:3358–3363.

Aizenberg, J., Tkachenko, A., Weiner, S., Addad, L., and Hendler, G. (2001). Calcitic microlenses as part of the photoreceptor in brittlestars. *Nature* 412:819–822.

Aizenberg, J., Weaver, J.C., Thanawala, M.S., Sundar, V.C., Morse, D.E., and Fratzl, P. (2005). Skeleton of *Euplectella* sp.: structural hierarchy from the nanoscale to the macroscale. *Science* 309:275–278.

Annenkov, V.V., Danilovtseva, E.N., Zelnsky, S.N., Basharina, T.N., Safonova, T.A., Korneva, E.S., Likhoshway, Y.V., and Grachev, M.A. (2010). Novel fluorescent dyes based on oligopropylamines for the in vivo staining of eukaryotic unicellular algae. *Analyt. Biochem.* 407:44–51.

Bae, W., and Mehra, R.K. (1998). Properties of glutathione- and phytochelatin- capped CdS bionanocrystallites. *J. Inorg. Biochem.* 69:33–43.

Bao, Z., Weatherspoon, M.R., Shian, S., Cai, Y., Graham, P.D., Alan, S.M., Ahmad, G., Dickerson, M.B., Church, B.C., Kang, Z., Abernathy III, H.W., Summers, C.J., Liu, M., and Sandhage, K.H. (2007). Chemical reduction of three-dimensional silica micro-assemblies into microporous silicon replicas. *Nature* 446:172–175.

Baur, W.H. (1956). Über die Verfeinerung der Kristallstrukturbestimmung einiger Vertreter des Rutiltyps: TiO2, SnO2, GeO2 und MgF2. *Acta Cryst.* 9:515–520.

Begum, G., Rana, R.K., Singh, S., and Satyanarayana, L. (2010). Bioinspired silification of functional materials: fluorescent monodisperse mesostructure silica nanospheres. *Chem. Mater.* 22:551–556.

Bevan, D.J.M., Rossmanith, E., Mylrea, D.K., Ness, S.E., Taylor, M.R., and Cuff, C. (2002). On the structure of aragonite—Lawrence Bragg revisited. *Acta Cryst.* B58:448–456.

Brott, L.L., Naik, R.R., Pikas, D.J., Kirkpatrick, S.M., Tomlin, D.W., Whitlock, P.W., Clarson, S.J., and Stone, M.O. (2001). Ultrafast holographic nanopatterning of biocatalytically formed silica. *Nature* 413:291–293.

Brümmer, F., Pfannkuchen, M., Baltz, A., Hauser, T., and Thiel, V. (2008). Light inside sponges. *J. Exp. Marine Biol. Ecol.* 367:61–64.

Brunner, E., Richthammer, P., Ehrlich, H., Paasch, S., Simon, P., Ueberlein, S., and van Pée, K.H. (2009). Chitin-based organic networks part of cell wall silica in the diatom *Thalassiosira pseudonana*. *Angew. Chem. Int. Ed.* 121:9724–9727.

Butcher, K.S.A., Ferris, J.M., Phillips, M.R., Wintrebert-Fouquet, M., Jong Wah, J.W., Jovanovic, N., Vyvermann, W., and Chepurnov, V.A. (2005). A luminescence study of porous diatoms. *Mater. Sci. Eng. C* 25:658–663.

Cai, Y., and Sandhage, K.H. (2005). Zn2SiO4-coated microparticles with biologically-controlled 3D shapes. *Phys. Stat. Sol. A* 202:R105–107.

Cattaneo-Vietti, R., Bavestello, G., Cerrano, C., Sarà, M., Benatti, U., Giovine, M., and Galno, E. (1996). Optical fibres in an Antarctic sponge. *Nature* 383:397–398.

Cerrano, C., Arillo, A., Bavestrello, G., Calcinai, B., Cattaneo-Vietti, R., Penna, A., Sarà, M., and Totti, C. (2000). Diatom invasion in the antarctic hexactellinid sponge *Scolymastra joubini*. *Polar Biol.* 23:441–444.

Cha, J.N., Shimizu, K., Zhou, Y. Christianssen, S.C., Chmelka, B.F., Stucky, G.D., and Morse, D.E. (1999). Silicatein filaments and subunits from a marine sponge direct the polymerization of silica and silicones in vitro. *Proc. Natl. Acad. Sci.* 96:361–365.

Checa, A., Ramírez-Rico, J., González-Segura, A., and Sánchez-Navas, A. (2009). Nacre and false nacre (foliated aragonite) in extant monoplacophorans (= *Tryblidiida*:Mollusca). *Die Naturwissenschaften* 96:111–122.

Chiappino, M.L., Azam, F., and Volcani, B.E. (1977). Effects of germanic acid on developing cell walls of diatoms. *Protoplasma* 93:191–204.

Chiodini, N., Meinardi, F., Morazzoni, F., Paleari, A., Scotti, R., and Di Matrino, D. (2000). Ultraviolet photoluminescence of porous silica. *Appl. Phys. Lett.* 76:3209–3211.

Clarkson, E., and Levi-Setti, R. (1975). Trilobite eyes and the optics of DesCartes and Huygens. *Nature* 254:663–667.

Clarkson, E., Levi-Setti, R., and Horváth, G. (2006). The eyes of trilobites: the oldest preserved visual system. *Arthropod Struct. Dev.* 35:247–259.

Dal Negro, A., and Ungaretti, L. (1971). Refinement of the crystal structure of aragonite. *Am. Mineral.* 56:768–772.

Davies, P., Dollimore, D., and Heal, G.R. (1978). Polymorph transition kinetics by DTA. *J. Therm. Anal.* 13:473–487.

de la Rica, R., and Velders, A.H. (2011). Biomimetic crystallization of Ag_2S canoclusters in nanopore assemblies. *J. Am. Chem. Soc.* 133:2875–2877.

De Stefano, L., Rendina, I., De Stefano, M., Bismuto, A., and Maddalena, P. (2005). Marine diatoms as optical chemical sensors. *Appl. Phys. Lett.* 87:233902–13.

De Stefano, L., Rea, I., Rendina, I., De Stefano, M., and Moretti, L. (2007). Lensless light focusing with the centric marine diatom *Coscinodiscus wailesii*. *Opt. Express* 15:18082–88.

De Tommasi, E., Rea, I., Mocella, V., Moretti, L., De Stefano, M., Rendina, I., and De Stefano, L. (2010). Multi-wavelength study of light transmitted through a single marine diatom. *Opt. Express* 18:12203–12.

Desclés, J., Vartanian, M., El Harrak, A., Quinet, M., Bremond, N., Sapriel, G., Bibette, J., and Lopez, P.J. (2008). New tools for labeling silica in living diatoms. *New Phytol.* 177:822–829.

Dong, Q., Su, H., and Zhang, D. (2005). In situ depositing silver nanoclusters on silk fibroin fibres supports by a novel biotemplate redox technique at room temperature. *J. Phys. Chem. B* 109:17429–34.

Dudley, S., Kalem, T., and Akinc, M. (2006). Conversion of SiO_2 diatom frustules to $BaTiO_3$ and $SrTiO_3$. *J. Am. Ceram. Soc.* 89:2434–2439.

Euliss, L.E., Bartl, M.H., and Stucky, G.D. (2006). Control of calcium carbonate crystallization utilizing amphiphilic block copolypeptides. *J. Cryst. Growth* 286:424–430.

Farrugia, L.J. (1997). Ortep-3 for Windows. *J. Appl. Cryst.* 30:565.

Fuhrmann, T., Landwehr, S., El Rharbi-Kucki, M., and Sumper, M. (2004). Diatoms as living photonic crystals. *Appl. Phys. B.* 78:257–260.

Fujishima, A., and Honda, K. (1972). Electrochemical photolysis of water at a semiconductor electrode. *Nature* 238:37–38.

Gaino, E., and Sarà, M. (1994). Siliceous spicules of *Tethya seychellensis* (*Porifera*) support the growth of green algae: a possible light conducting system. *Mar. Ecol. Prog. Ser.* 108:147–151.

Gál, J., Horváth, G., Clarkson, E.N.K., and Haiman, O. (2000). Image formation by bifocal lenses in a trilobite eye? *Vision Res.* 40:843–853.

Gehlen, M., Beck, L., Calas, G., Flank, A.-M., Van Bennekom, A.J., and Van Beusekom, J.E.E. (2002). Unraveling the atomic structure of biogenic silica: evidence of structural association of Al and Si in diatom frustules. *Geochim. Cosmochim. Acta* 66:1601–1609.

Graf, D.L. (1961). Crystallographic tables for the rhombohedral carbonates. *Am. Mineral.* 46:1283–1316.

Hecht, E. (1987). *Optics*. 2nd ed. Reading, MA: Addison Wesley.
Hellwege, K.-H., and Hellwege, A.M. (eds.). (1962). *Landolt-Boernstein—Zahlenwerte und Funktionen*. Vol. II-8, 6th ed. Springer, Berlin.
Horn, M., Schwerdtfeger, C.F., and Meagher, E.P. (1972). Refinement of the structure of anatase at several temperatures. *Z. Kristallogr.* 136:273–281.
Huang, Z., Yan, D., Yang, M., Liao, X., Kang, Y., Yin, G., Yao, Y., and Baoqing, H. (2008). Preparation and characterization of the biomineralized zinc oxide particles in spider silk peptides. *J. Colloid Interf. Sci.* 325:356–362.
Ingalls, A.E., Whitehead, K., and Bridoux, M.C. (2010). Tinted windows: the presence of UV absorbing compounds called mycosporine-like amino acids embedded in the frustules of marine diatoms. *Geochim. Cosmochim. Acta* 74:104–115.
Jeffryes, C., Gutu, T., Jiao, J., and Rorrer, G.L. (2008a). Metabolic insertion of nanostructured TiO_2 into the patterned biosilica of the diatom *Pinnularia* sp., by a two-stage bioreactor cultivation process. *ACS Nano* 2:2103–2112.
Jeffryes, C., Solanki, R., Rangineni, Y., Wang, W., Chang, C., and Rorrer, G.L. (2008b). Electroluminescence and photoluminescence from nanostructured diatom frustules containing metabolically inserted germanium. *Adv. Mater.* 20:2633–2637.
Johnson, S.G., Fan, S., Villeneuve, P.R., Joannopoulos, J.D., and Kolodziejski, L.A. (1999). Guided modes in photonic crystal slabs. *Phys. Rev. B* 60:5751–5758.
Jones, L.H.P., Milne, A.A., and Sanders, J.V. (1966). Tabashir: an opal of plant origin. *Science* 151:464–466.
Kokta, J. (1930). Physicochemical properties of opal and their relation to artificially prepared amorphous silica acids. *Rozpravy Ceske Akad. 2. Kl.* 40 (21):1–25.
Kröger, N. (2007). Prescribing diatom morphology: toward genetic engineering of biological nanomaterials. *Curr. Opin. Chem. Biol.* 11:662–669.
Kröger, N., Deutzmann, R., and Sumper, M. (1999). Polycationic peptides from diatom biosilica that direct silica nanosphere formation. *Science* 286:1129–1132.
Kröger, N., Lorenz, S., Brunner, E., and Sumper, M. (2002). Self-assembly of highly phosphorylated silaffins and their function in biosilica morphogenesis. *Science* 298:584–586.
Kröger, N., and Wetherbee, R. (2000). Pleuralins are involved in theca differentiation in the diatom *Cylindrotheca fusiformis*. *Protist* 151:263–273.
Kucki, M. (2009). Biological photonic crystals: diatoms. Dye functionalization of biological silica nanostructures. PhD thesis, University of Kassel, Kassel. urn:nbn:de:hebis:34–2009091430073/5/DissertationMelanieKucki.pdf.
Kucki, M., and Fuhrmann-Lieker, T. (2011). Staining diatoms with rhodamine dyes: control of emission colour in photonic biocomposites. *J. R. Soc. Interface*, doi:10.1098/rsif.2011.0424.
Kucki, M., Landwehr, S., Rühling, H., Maniak, M., and Fuhrmann-Lieker, T. (2006). Light-emitting photonic crystals—the bioengineering of metamaterials. *Proc. SPIE* 6182:6182S1–9.
Kul'chin, Y.N., Bezverbny, A.V., Bukin, O.A., Voznesensky, S.S., Golik, S.S., Mayor, A.Y., Shchipunov, Y.A., and Nagorny, I.G. (2011). Nonlinear optical properties of biomineral and biomimetical nanocomposite structures. *Laser Physics* 21:630–636.
Kul'chin, Y.N., Bukin, O.A., Voznesenskiy, S.S., Galkina, A.N., Gnedenkov, S.V., Drozdov, AL., Kuryavyi, V.G., Mal'tseva, T.L., Nagornyi, S.L., Sinebryukhov, A.I., and Cherednichenko, A.I. (2008). Optical fibres based on natural biological minerals—sea sponge spicules. *Quant. Electron.* 38:51–55.

Kul'chin, Y.N., Voznesenski, S.S., Bukin, O.A., Bezverbnyi, A.V., Drozdov, A.L., Nagomy, I.G., and Galkina, A.N. (2009). Spicules of glass sponges as a new type of self-organizing natural photonic crystal. *Opt. Spectrosc.* 107:442–447.

Lee, M.R. Torney, C., and Owen, A.W. (2007). Magnesium-rich intralensar structures in schizochroal trilobite eyes. *Paleontology* 50:1031–1037.

Li, C.-W., Chu, S., and Lee, M. (1989). Characterizing the silica deposition vesicle of diatoms. *Protoplasma* 151:158–163.

Liu, F., Kang, S.H., Lee, Y.-I., Choa, Y.-H., Mulchandani, A., Myung, N.V., and Chen, W. (2010). Enzyme mediated synthesis of phytochelatin-capped CdS nanocrystals. *Appl. Phys. Lett.* 97:123703–13.

Losic, D., Mitchell, J.G., and Voelcker, N.H. (2009). Diatomaceous lessons in nanotechnology and advanced materials. *Adv. Mater.* 21:2947–2958.

Losic, D., Triani, G., Evans, P.J. Atanacio, A., Mitchell, J.G., and Voelcker, N.H. (2006). Controlled pore structure modification of diatoms by atomic layer deposition of TiO_2. *J. Mater. Chem.* 16:4029–4034.

Lowenstam, H.A., and Weiner, S. (1989). *On biomineralization*. Oxford University Press, New York.

Mann, S. (2006). *Biomineralization. Principles and concepts in bioinorganic materials chemistry*. Oxford University Press, Oxford.

Mazumder, N., Gogoi, A., Kalita, R.D., Ahmed, G.A., Buragohain, A.K., and Choudhury, A. (2010). Luminescence studies of fresh water diatom frustules. *Indian J. Phys.* 84:665–669.

Metzler, R.A., Abrecht, M., Olabisi, R.M., Ariosa, D., Johnson, C.J., Frazer, B.H., Coppersmith, S.N., and Gilbert, P.U.P.A. (2007). Columnar nacre architecture and possible formation mechanism. *Phys. Rev. Lett.* 98:268102–14.

Müller, W.E.G., Kasueske, M., Wang, X., Schröder, H.C., Wang, Y., Pisignano, D., and Wiens, M. (2009a). Luciferase a light source for the silica-based optical waveguides (spicules) in the demosponge *Suberites domuncula*. *Cell. Mol. Life Sci.* 66:537–552.

Müller, W.E.G., Wang, X., Cui, F.-Z., Jochum, K.P., Tremel, W., Bill, J., Schröder, H.C., Natalio, F., Schlossmacher U., and Wiens, M. (2009b). Sponge spicules as blueprints for the biofabrication of inorganic-organic composites and biomaterials. *Appl. Microbiol. Biotechnol.* 83:397–413.

Müller, W.E.G., Wendt, K., Geppert, C., Wiens, M., Reiber, A., and Schröder, H.C. (2006). Novel photoreception system in sponges? Unique transmission properties of the stalk spicules from the hexactinellid *Hyalonema sieboldii*. *Biosensors Bioelectron.* 21:1149–1155.

Naik, R.R. Stringer, S.J., Agarwal, G., Jones, S.E., and Stone, M.O. (2002). Biomimetic synthesis and patterning of silver nanoparticles. *Nat. Mater.* 1:169–172.

Navrotsky, A. (2003). Energetics of nanoparticle oxides: interplay between surface energy and polymorphism. *Geochem. Trans.* 4:34–37.

Neira-Carrillo, A., Acevedo, D.F., Miras, M.C., Barbero, C.A., Gebauer, D., Coelfen, H., and Arias, J. (2008). Influence of conducting polymers based on carboxylated polyaniline on in vitro $CaCO_3$ crystallization. *Langmuir* 24:12496–12507.

Noll, F., Sumper, M., and Hampp, N. (2002). Nanostructure of diatom silica surfaces and of biomimetic analogues. *Nano Lett.* 2:91–95.

Notomi, N., Suzuki, H., and Tamamura, T. (2001). Directional lasing oscillation of two-dimensional organic photonic crystal lasers. *Appl. Phys. Lett.* 78:1325–1327.

Noyes, J., Sumper, M., and Vukusic, P. (2008). Light manipulation in a marine diatom. *J. Mater. Res.* 23:3229–3235.
O'Regan, B., and Graetzel, M. (1991). A low-cost, high-efficiency solar cell based on dye-sensitised colloidal TiO_2 films. *Nature* 353:737–740.
Orellana, M.V., Petersen, T.W., and van den Engh, G. (2004). UV-excited blue autofluorescence of pseudo-*Nitzschia* multiseries (Bacillariophyceae). *J. Phycol.* 40:705–710.
Ostwald, W. (1897). Studies upon the forming and changing solid bodies. *Z. Physikal. Chemie.* 22:289–330.
Park, R.J., and Meldrum, F.C. (2002). Synthesis of single crystals of calcite with complex morphologies. *Adv. Mater.* 14:1167–1169.
Park, Y.R., and Kim, K.J. (2005). Structural and optical properties of rutile and anatase TiO_2 thin films: effects of cobalt doping. *Thin Solid Films* 484:34–38.
Parkinson, J., and Gordon, R. (1999). Beyond micromachining: the potential of diatoms. *Tibtech* (Elsevier) 17:190–196.
Payne, E.K., Rosi, N.L., Xue, C., and Mirkin, C. (2005). Sacrificial biological templates for the formation of nanostructured metallic microshells. *Angew. Chem. Int. Ed.* 44:5064–167.
Pejoux, C., de la Rica, R., and Matsui, H. (2010). Biomimetic crystallization of sulfide semiconductor nanoparticles in aqueous solution. *Small* 6:999–1002.
Peng, Y., Xu, A.W., Deng, B., Antonietti, M., and Coelfen, H. (2006). Polymer-controlled crystallization of zinc oxide hexagonal nanorings and disks. *J. Phys. Chem. B* 110:2988–2993.
Picket-Heaps, J., Schmid, A.M., and Edgar, L.A. (1990). The cell biology of diatom valve formation. *Prog. Phycol. Res.* 7:1–168.
Qin, T., Gutu, T., Jiao, J., Chang, C.-H., and Rorrer, G.L. (2008). Photoluminescence of silica nanostructures from bioreactor culture of marine diatom *Nitzschia frustulum*. *J. Nanosci. Nanotechnol.* 8:2392–2398.
Ramachandran, G.N. (1947). Birefringence of crystals and its temperature-variation. Part I. Calcite and aragonite. *Proc. Indian Acad. Sci.* 26:77–92.
Roessler, U. (ed.). (1999). *Landolt-Boernstein—New Series*. Vol. III 41B. Springer, Berlin.
Rosi, N.L., Thaxton, C.S., and Mirkin, C.A. (2004). Control of nanoparticle assembly by using DNA-modified diatom templates. *Angew. Chem. Int. Ed.* 43:5500–5503.
Sandhage, H.H., Dickerson, M.B., Husemann, P.M., Caranna, M.A., Clifton, J.D., Bull, T.A., Heibel, T.J., Overton, W.R., and Schönwälder, M.E.A. (2002). Novel, bioclastic route to self-assembled, 3D, chemically tailored meso/nanostructures: shape-preserving reactive conversion of biosilica (diatom) microshells. *Adv. Mater.* 14:429–433.
Schaefer, K., and Lax, E. (eds.). (1961). *Landolt-Boernstein—Zahlenwerte und Funktionen*. Vol. II-4, 6th ed. Springer, Berlin.
Schmid, A.M. (1996). Available from http://www.indiana.edu/~diatom/silica.dis (accessed September 14, 2011).
Schmid, A.M., and Schulz, D. (1979). Wall morphogenesis in diatoms: deposition of silica by cytoplasmic vesicles. *Protoplasma* 100:267–288.
Sewell, S.L., and Wright, D.W. (2006). Biomimetic synthesis of titanium dioxide utilizing the R5 peptide derived from *Cylindrotheca fusiformis*. *Chem. Mater.* 18:3108–3113.
Shieh, J., Cho, A., Lai, Y., Dai, B., Pan, F., and Chao, K. (2004). Stable blue luminescence from mesoporous silica films. *Electrochem. Solid-State Lett.* 7:G319–322.

Shimizu, K., Del Amo, Y., Brzezinski, M.A., Stucky, G.D., and Morse, D.E. (2001). A novel fluorescent silica tracer for biological silification studies. *Chem. Biol.* 8:1051–1060.

Simpson, T.L., and Volcani, B.E. (eds.). (1981). *Silicon and siliceous structures in biological systems*. Springer, New York.

Speiser, D.I., Ernisse, D.J., and Johnson, S. (2011). A chiton uses aragonite lenses to form images. *Curr. Biol.* 21:665–670.

Stowasser, A., Rapaport A., Layne, J.E., Mogan, R.C., and Buschbek, E.K. (2010). Biological bifocal lenses with image separation. *Curr. Biol.* 20:1482–1486.

Sugawara, A., Oichi, A., Suzuki, H., Shigesato, Y., Kogure, T., and Kato, T. (2006). Assembled structures of nanocrystals in polymer/calcium carbonate thin-film composites formed by the cooperation of chitosan and poly(aspartate). *J. Polym. Sci. A Polym. Chem.* 44:5153–5160.

Sumerel, J.L., Yang, W., Kisailus, D., Weaver, J.C., Choi, J.H., and Morse, D.E. (2003). Biocatalytically templated synthesis of titanium dioxide. *Chem. Mater.* 12:4804–4809.

Sumper, M. (2002). A phase separation model for the nanopatterning of diatom biosilica. *Science* 295:2430–2433.

Sumper, M. (2004). Biomimetic patterning of silica by long-chain polyamines. *Angew. Chem. Int. Ed.* 43:2251–2254.

Sumper, M., and Brunner, E. (2006). Learning from diatoms: nature's tool for the production of nanostructured silica. *Adv. Funct. Mater.* 16:17–27.

Sumper, M., Lorenz, S., and Brunner, E. (2003). Biomimetic control of size in the polyamine-directed formation of silica nanospheres. *Angew. Chem. Int. Ed.* 42:5192–5195.

Sundar, V.C., Yablon, A.D., Grazul, J.L., Ilan, M., and Aizenberg, J. (2003). Fibre-optical features of a glass sponge. *Nature* 424:899–900.

Tahir, M.N., Natalio, F., Berger, R., Barz, M., Theato, P., Schroeder, H.C., Mueller, W.E.G., and Tremel, W. (2009). Growth of fibrous aggregates of silica nanoparticles: fibre growth by mimicking biogenic silica patterning process. *Soft Matter* 5:3657–3662.

Tesson, B., and Hildebrand M. (2010). Dynamics of silica cell wall morphogenesis in the diatom *Cyclotella cryptica*: substructure formation and the role of microfilaments. *J. Struct. Biol.* 169:62–74.

Towe, K.M. (1973). Trilobite eyes: calcified lenses in vivo. *Science* 179:1008–1009.

Turnbull, A.G. (1973). A thermochemical study of vaterite. *Geochim. Cosmochim. Acta* 37:1593–1601.

Volkmer, D., Tugulu, S., Fricke, M., and Nielsen, T. (2003). Morphosynthesis of star-shaped titania-silica shells. *Angew. Chem. Int. Ed.* 42:58–61.

Vrieling, E.G., Gieskes, W.W.C., and Beelen, T.P.M. (1999). Silicon deposition in diatoms: control by the pH inside the silicon deposition vesicle. *J. Phycol.* 35:548–559.

Vrieling, E.G., Beelen, T.P.M., van Santen, R.A., and Gieskes, W.W.C. (2002). Mesophases of (bio)polymer-silica-particles inspire a model for silica biomineralization in diatoms. *Angew. Chem. Int. Ed.* 41:1543–1546.

Wang, X., Schroeder, H.C., and Mueller, W.E.G. (2009). Giant siliceous spicules from the deep-sea glass sponge *Monoraphis chuni*. *Intern. Rev. Cell Mol. Biol.* 273:69–115.

Weaver, J.C., Pietrasanta, L.I., Hedin, N., Chmelka, B.F., Hansma, P.K., and Morse, D.E. (2003). Nanostructural features of demosponge biosilica. *J. Struct. Biol.* 144:271–281.

Weiner, S., and Traub, W. (1984). Macromolecules in mollusc shells and their functions in biomineralization. *Philos. Trans. R. Soc. B* 304:425–434.

Wenzl, S., Hett, R., Richthammer, P., and Sumper, M. (2008). Silacidines: highly acidic phosphopeptides from diatom shells assist silica precipitation in vitro. *Angew. Chem. Int. Ed.* 47:1729–1732.

Wetherbee, R., Crawford, S., and Mulvaney, P. (2000). The nanostructure and development of diatom biosilica. In Bäuerlein, E. (ed.), *Biomineralization*. Wiley-VCH, Weinheim. pp. 177–194.

Wise, S.W. (1970). Microarchitecture and mode of formation of nacre (mother of pearl) in pelecypods, gastropods and cephalopods. *Eclogae Geol. Helv.* 63:775–797.

Xiao, J., and Yang, S. (2010). Hollow calcite crystals with complex morphologies formed from amorphous precursors and regulated by surfactant micellar structures. *Cryst. Eng. Commun.* 12:3296–3304.

Xie, J., Tan, Y.N., and Lee, J.Y. (2010). Biological and biomimetic synthesis of metal nanomaterials. In Kumar, C. (ed.), *Biomimetic and bioinspired nanomaterials*. Wiley-VCH, Weinheim. pp. 251–282.

Xu, A.-W., Ma, Y., and Coelfen, H. (2007). Biomimetic mineralization. *J. Mater. Chem.* 17:415–449.

Yamanaka, S., Yano, R., Usami, H., Hayashida, N., Ohguchi, M., Takeda, H., and Yoshino, K. (2008). Optical properties of diatom silica frustule with special reference to blue light. *J. Appl. Phys.* 103:074701–15.

Yan, D., Yin, G., Huang, Z., Yang, M., Liao, X., Kang, Y., Yao, Y., Hao, B., and Han, D. (2009). Characterization and bacterial response of zinc oxide particles prepared by a biomineralization process. *J. Phys. Chem. B* 113:6047–6053.

Yang, S.H., Kang, K., and Choi, I.S. (2008). Biomimetic approach to the formation of titanium dioxide thin films by using poly(2-(dimethylamino)ethylmethacrylate). *Chem. Asian J.* 3:2097–2104.

Yang, W., Lopez, P.J., and Rosengarten, G. (2011). Diatoms: self assembled silica nanostructures, and templates for bio/chemical sensors and biomimetic membranes. *Analyst* 135:42–53.

Yu, M.H., Yuan, P., Zhang, J., Wang, H.N., Zhang, Y., Hu, Y.F., Wang, Y.H., and Yu, C.Z. (2010). A bioinspired route to various siliceous vesicular structures. *J. Nanosci. Nanotech.* 10:612–615.

Yu, Y., Addai-Mensah, J., and Losic, D. (2010). Synthesis of self-supporting gold microstructures with three-dimensional morphologies by direct replication of diatom templates. *Langmuir* 26:14068–72.

Zu, P., Tang, Z.K., Wong, G.K.L., Kawasaki, M., Ohtomo, A., Koinuma, H., and Segawa, Y. (1997). Ultraviolet spontaneous and stimulated emission from ZnO microcrystallite thin films at room temperature. *Solid State Commun.* 103:459–463.

3

Biomimetics of Optical Nanostructures

Andrew R. Parker, Torben Lenau, and Akira Saito

CONTENTS

3.1 Engineering of Optical Nanostructures 56
3.1.1 Engineering of Antireflectors 57
3.1.2 Engineering of Iridescent Devices 59
3.1.3 Cell Culture 63
3.1.4 Diatoms and Coccolithophores 64
3.1.5 Iridoviruses 68
3.1.6 The Mechanisms of Natural Engineering and Future Research 69
References 70
3.2 Nature-Inspired Structural Color Applications 72
3.2.1 Shell Structures in Insects 73
3.2.2 Light Reflection Principles in Nature 74
3.2.2.1 Narrowband Reflection in Nature 76
3.2.2.2 Narrowband Reflection Applications 79
3.2.3 Metallic Appearance in Nature 82
3.2.3.1 Multilayer Metallic Reflectors in Beetles 83
3.2.3.2 Helical Bouligand Metallic Reflectance in Beetles 85
3.2.3.3 Metallic Appearance in Fishes 87
3.2.4 Metallic Reflection Using Multilayer Materials and Coatings 88
References 94
3.3 Fabrication of *Morpho* Butterfly-Specific Structural Color Aiming at Industrial Applications 96
3.3.1 Principles of the *Morpho* Color 97
3.3.2 Reproduction of *Morpho* Color in Textiles 99
3.3.3 Fabrication of the Nanosized Color-Producing Part 100
3.3.4 Reproduction with Discrete Multilayers 101
3.3.5 Mass Production for Applications 106
3.3.6 Control of Optical Properties (Angular Dispersion) 108
3.3.7 Control of Optical Properties (Spectra) 109
3.3.8 Recent Progress for Industrial Applications 109
3.3.9 Summary and Outlook 111
References 113

3.1 Engineering of Optical Nanostructures

Andrew R. Parker

Three centuries of research, beginning with Hooke and Newton, have revealed a diversity of optical devices at the nanoscale (or at least the submicron scale) in nature [Parker 2000]. These include structures that cause random scattering, 2D diffraction gratings, 1D multilayer reflectors, and 3D liquid crystals (Figure 3.1 (a)–(d)). In 2001 the first photonic crystal was identified as such in

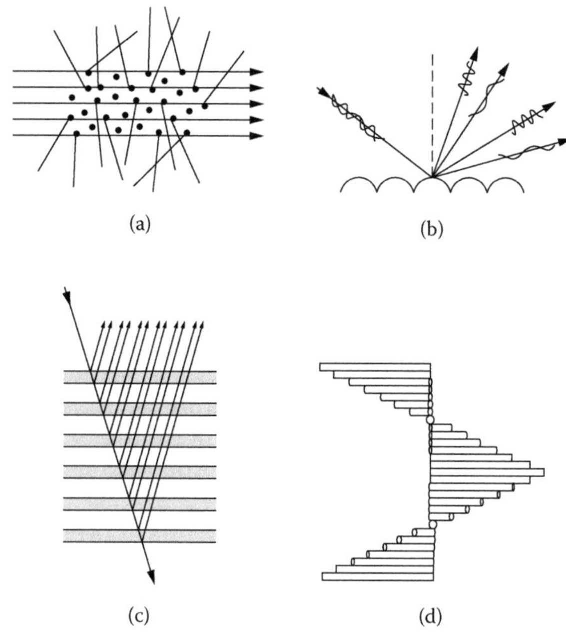

FIGURE 3.1
Summary of the main types of optical reflectors found in nature; a-d where a light ray is (generally) reflected only once within the system (i.e. they adhere to the single scattering, or First Born, approximation), and E and F where each light ray is (generally) reflected multiple times within the system. (a) An irregular array of elements that scatter incident light into random directions. The scattered (or reflected) rays do not interfere. (b) A diffraction grating, a surface structure, from which incident light is diffracted into specific angular directions resulting in a spatial separation of its angular wavelength/colour components. Each corrugation is about 500 nm wide. Diffracted rays superimpose either constructively or destructively. (c) A multilayer reflector, composed of thin (ca. 100 nm thick) layers of alternating refractive index, where light rays reflected from each interface in the system superimpose either constructively or destructively. (d) A 'liquid crystal' composed of nano-fibres arranged in layers, where the nano-fibres of one layer lie parallel to each other yet are orientated slightly differently to those of adjacent layers. Hence spiral patterns can be distinguished within the structure. The height of the section shown here – one 'period' of the system – is around 200 nm.

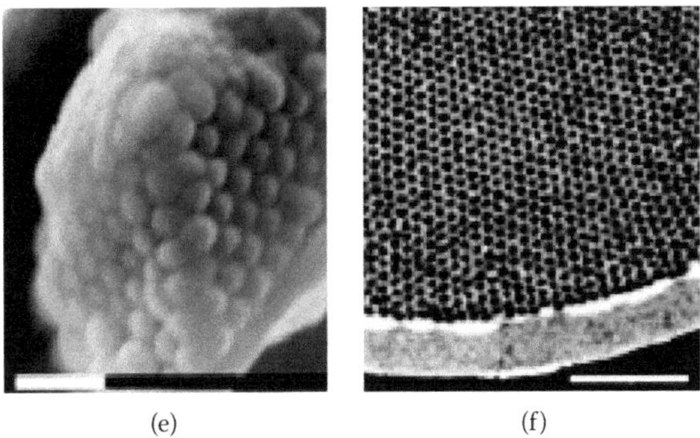

(e) (f)

FIGURE 3.1 ***(Continued)***
Summary of the main types of optical reflectors found in nature; a-d where a light ray is (generally) reflected only once within the system (i.e. they adhere to the single scattering, or First Born, approximation), and E and F where each light ray is (generally) reflected multiple times within the system. (e) Scanning electron micrograph of the "opal" structure – a close-packed array of sub-micron spheres (a "3D photonic crystal") – found within a single scale of the weevil *Metapocyrtus* sp.; scale bar = 1 μm. (f) Transmission electron micrograph of a section through a hair (neuroseta) of the sea mouse *Aphrodita* sp. (Polychaeta), showing a cross-section through a stack of sub-micron tubes (a "2D photonic crystal"); scale bar = 5 μm.

animals [Parker et al. 2001], and since then the scientific effort in this subject has accelerated. Now we know of a variety of 2D and 3D photonic crystals in nature (e.g., Figure 3.1 (e) and (f)), including some designs not encountered previously in physics.

Biomimetics is the extraction of good design from nature. Some optical biomimetic successes have resulted from the use of conventional (and constantly advancing) engineering methods to make direct analogues of the reflectors and antireflectors found in nature. However, recent collaborations between biologists, physicists, engineers, chemists, and material scientists have ventured beyond merely mimicking in the laboratory what happens in nature, leading to a thriving new area of research involving biomimetics via cell culture. Here, the nanoengineering efficiency of living cells is harnessed, and nanostructures such as diatom "shells" can be made for commercial applications via culturing the cells themselves.

3.1.1 Engineering of Antireflectors

Some insects benefit from antireflective surfaces, either on their eyes to see under low-light conditions, or on their wings to reduce surface reflections in transparent (camouflaged) areas. Antireflective surfaces therefore occur on the corneas of moth and butterfly eyes [Miller et al. 1966] and on the transparent wings of hawkmoths [Yoshida et al. 1997]. These consist of nodules, with rounded tips, arranged in a hexagonal array with a periodicity of around

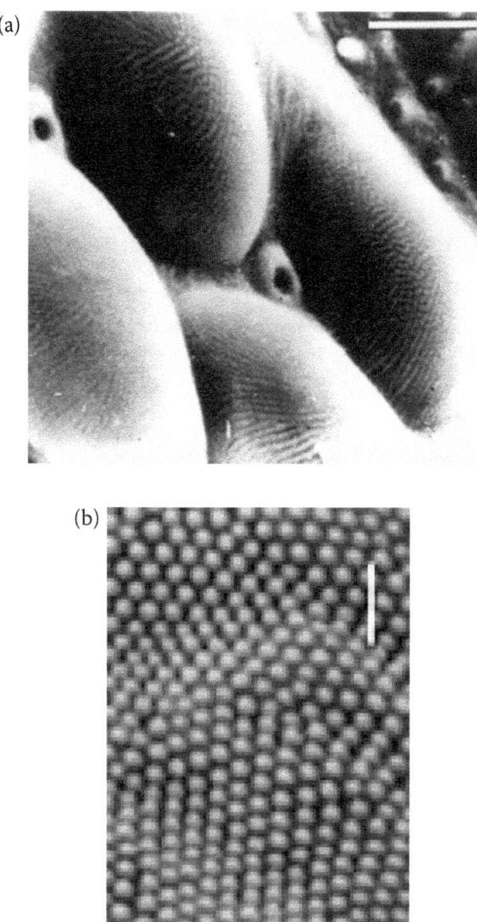

FIGURE 3.2
Scanning electron micrographs of antireflective surfaces. (a) Fly-eye antireflector (ridges on four facets) on a 45 million year old dolichopodid fly's eye. Micrograph by P. Mierzejewski, reproduced with permission. and (b) moth-eye antireflective surfaces.

240 nm (Figure 3.2 (a)). Effectively they introduce a gradual refractive index profile at an interface between chitin (a polysaccharide, often embedded in a proteinaceous matrix with a refractive index of 1.54) and air, and hence reduce reflectivity by a factor of 10.

This moth-eye structure was first reproduced at its correct scale by crossing three gratings at 120° using lithographic techniques, and employed as antireflective surfaces on glass windows in Scandinavia [Gale 1989]. Here, plastic sheets bearing the antireflector were attached to each interior

(c)
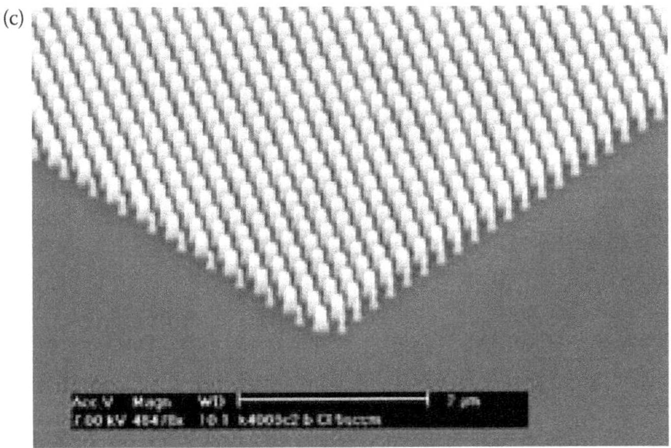

FIGURE 3.2 (*Continued*)
Scanning electron micrographs of antireflective surfaces. (c) Moth-eye mimic fabricated using ion-beam etching. Micrograph by S.A. Boden and D.M. Bagnall, reproduced with permission. Scale bars = 3 μm (a), 1 μm (b), 2 μm (c).

surface of triple-glazed windows using refractive index-matching glue to provide a significant difference in reflectivity. Today the moth-eye structure can be made with extreme accuracy using e-beam etching [Boden and Bagnall 2006], and is also employed commercially on solid plastic and other lenses.

A different form of antireflective device, in the form of a sinusoidal grating of 250 nm periodicity, was discovered on the cornea of a 45 million-year-old fly preserved in amber [Parker et al. 1998] (Figure 3.2 (b)). This is particularly useful where light is incident at a range of angles (within a single plane, perpendicular to the grating grooves), as demonstrated by a model made in photoresist using lithographic methods [Parker et al. 1998]. Consequently it has been employed on the surfaces of solar panels, providing a 10% increase in energy capture through reducing the reflected portion of sunlight [Beale 1999]. Again, this device is embossed onto plastic sheets using holographic techniques.

3.1.2 Engineering of Iridescent Devices

Many birds, insects (particularly butterflies and beetles), fishes, and lesser known marine animals display iridescent (changing color with angle) or "metallic" colored effects resulting from photonic nanostructures. These appear comparatively brighter than the effects of pigments and often function in animals to attract the attention of a potential mate or to startle a predator. An obvious application for such visually attractive and optically sophisticated devices is within the anticounterfeiting industry. For secrecy reasons, work in this area cannot be described, although devices are sought at different levels of sophistication, from effects that are discernable by the eye to fine-scale optical characteristics (polarization and angular properties, for example) that can be read only by specialized detectors. However, new research aims to exploit these devices in the cosmetics, paint, printing/ink, and clothing industries. They are even being tested in art to provide a sophisticated color change effect.

Original work on exploiting nature's reflectors involved copying the design but not the size, where reflectors were scaled up to target longer wavelengths. For example, rapid prototyping was employed to manufacture a microwave analogue of a *Morpho* butterfly scale that is suitable for reflection in the 10–30 GHz region. Here layer thicknesses would be in the order of 1 mm rather than 100 nm, as in the butterfly, but the device could be employed as an antenna with broad radiation characteristics, or as an antireflection coating for radar. However, today techniques are available to manufacture nature's reflectors at their true size.

Nanostructures causing iridescence include photonic crystal fibers, opal and inverse opal, and unusually sculpted 3D architectures. Photonic crystals are ordered, often complex, subwavelength (nano) lattices that can control the propagation of light at the single wave scale in the manner that atomic crystals control electrons [Yablonovich 1999]. Examples include opal (a hexagonal or cubic close packed array of 250 nm spheres) and inverse opal (a hexagonal array of similar sized holes in a solid matrix). Hummingbird feather barbs contain ultra-thin layers with variations in porosity that cause their iridescent effects, due to the alternating nanoporous/fully dense ultrastructure [Cohen et al. 2004]. Such layers have been mimicked using aqueous-based layering techniques. The greatest diversity of 3D architectures can be found in butterfly scales, which can include microribs with nanoridges, concave multilayered pits, blazed gratings, and randomly punctate nanolayers [Ghiradella 1989, Berthier 2005]. The cuticles of many beetles contain structurally chiral films that produce iridescent effects with circular or elliptical polarization properties [De Silva et al. 2005]. These have been replicated in titania for specialized coatings, where a mimetic sample can be compared with the model beetle and an accurate variation in spectra with angle is observed (Figure 3.3). The titania mimic can be nanoengineered for a wide range of resonant wavelengths; the lowest so far is a pitch of 60 nm for a circular Bragg resonance at 220 nm in a Sc_2O_3 film (Ian Hodgkinson, personal communication).

Biomimetic work on the photonic crystal fibers of the *Aphrodita* sea mouse is underway. The sea mouse contains spines (tubes) with walls packed with smaller tubes of 500 nm, with varying internal diameters (400–450 nm). These provide a bandgap in the red region and are to be manufactured via an extrusion technique. Larger glass tubes packed together in the proportion of the spine's nanotubes will be heated and pulled through a drawing

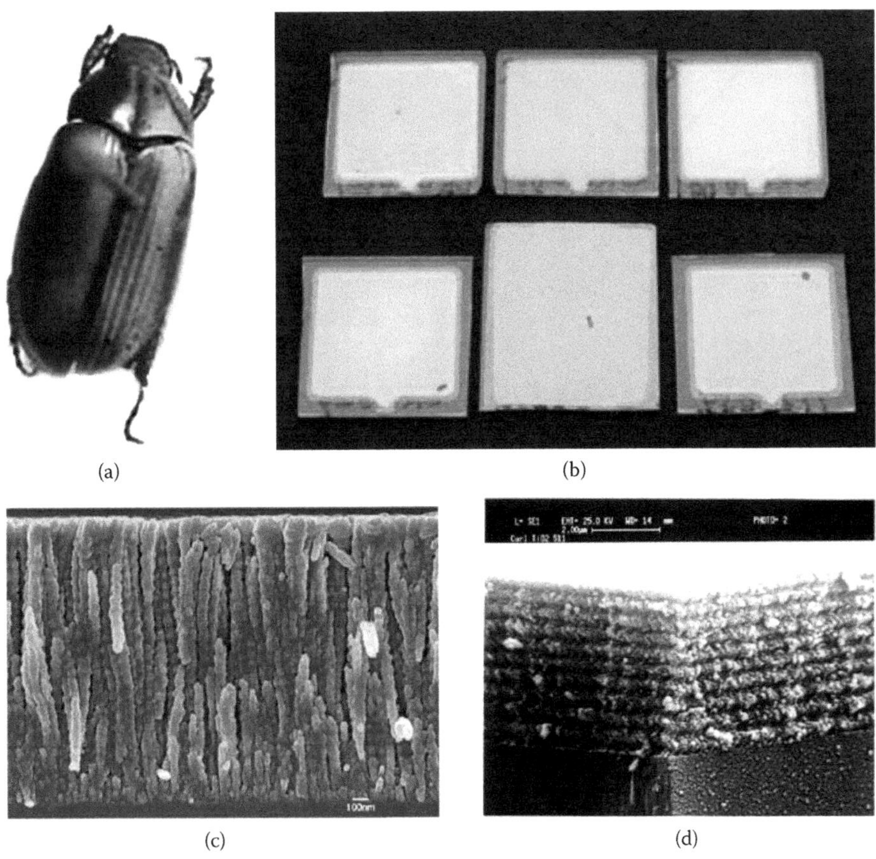

FIGURE 3.3 (See color insert.)
(a) A Manuka (scarab) beetle with (b) titania mimetic films of slightly different pitches. (c) Scanning electron micrograph of the chiral reflector in the beetle's cuticle. (d) Scanning electron micrograph of the titania mimetic film. Images by L. DeSilva and I. Hodgkinson, reproduced with permission.

tower until they reach the correct dimensions. The sea mouse fiber mimetics will be tested for standard PCF applications (e.g., in telecommunications), but also for anticounterfeiting structures readable by a detector.

Analogues of the famous blue *Morpho* butterfly (Figure 3.4 (a)) scales have been manufactured [Kinoshita et al. 2002, Watanabe et al. 2005]. Originally, corners were cut. Where the *Morpho* wing contained two layers of scales—one to generate color (a quarter-wave stack) and another above it to scatter the light—the model copied only the principle. The substrate was roughened at the nanoscale, and coated with 80 nm thick layers alternating in refractive index [Kinoshita et al. 2002]. Therefore the device retained a quarter-wave stack centered in the blue region, but incorporated a degree of randomness to generate

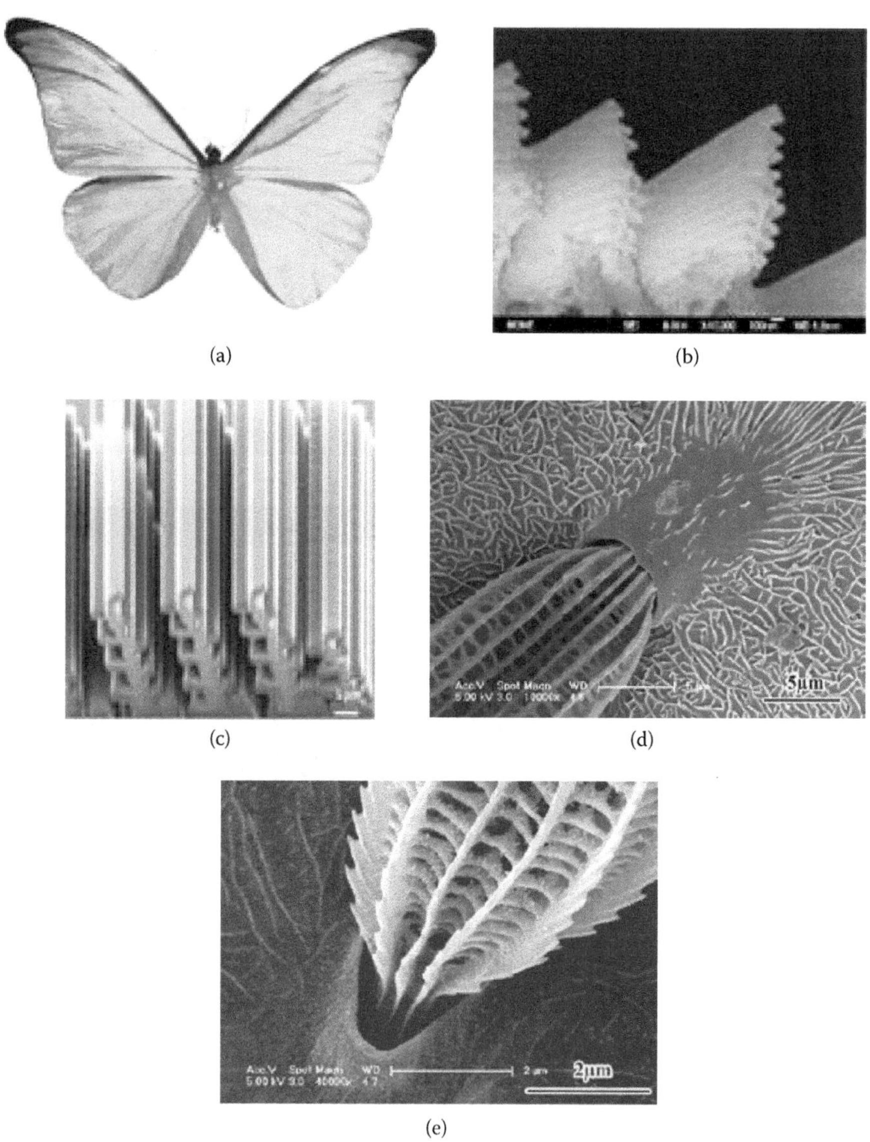

FIGURE 3.4
(a) A *Morpho* butterfly with (b) a scanning electron micrograph of the structure causing the blue reflector in its scales. (c) A scanning electron micrograph of the FIB-CVD fabricated mimic. A Ga^+ ion beam (beam diameter 7 nm at 0.4pA; 30kV), held perpendicular to the surface, was used to etch a precursor of phenanthrene ($C_{14}H_{10}$). Both give a wavelength peak at around 440 nm and at the same angle (30°). (d) Scanning electron micrograph of the base of a scale of the butterfly *Ideopsis similes*. (e) Scanning electron micrograph of a ZnO replica of the same part of the scale in (d). (a)–(c) by K. Watanabe, and (d) and (e) by W. Zhang, all reproduced with permission of the authors. Scale bars: (b) and (c) = 100 nm, (d) = 5 μm, (e) = 2 μm.

scattering. The engineered device closely matched the butterfly wing—the color observed changed only slightly with a changing angle over 180°, an effect difficult to achieve and useful for a broad-angle optical filter without dyes.

A new approach to making the 2D "Christmas tree" structure (a vertical, elongated ridge with several layers of 70 nm thick side branches; Figure 3.4 (b)) has been achieved using focused-ion-beam chemical vapor deposition (FIB-CVD) [Watanabe et al. 2005]. By combining the lateral growth mode with ion beam scanning, the Christmas tree structures were made accurately (Figure 3.4 (c)). However, this method is not ideal for low-cost mass production of 2D and 3D nanostructures, and therefore the ion-beam-etched Christmas trees are currently limited to high-cost items, including nano- or microsized filters (such as pixels in a display screen or a filter). Recently further corners have been cut in manufacturing the complex nanostructures found in many butterfly scales, involving the replication of the scales in ZnO, using the scales themselves as templates [Zhang et al. 2006] (Figure 3.4 d and e).

3.1.3 Cell Culture

Sometimes nature's optical nanostructures have such an elaborate architecture at such a small (nano) scale that we simply cannot copy them using current engineering techniques. Additionally, sometimes they can be made as individual reflectors (as for the *Morpho* structure), but the effort is so great that commercial-scale manufacture would never be cost-effective.

An alternative approach to making nature's reflectors is to exploit an aspect other than design—that the animals or plants can make them efficiently. Therefore, we can let nature manufacture the devices for us via cell culture techniques. Animal cells are in the order of 10 μm in size and plant cells up to about 100 μm, and hence suitable for nanostructure production. The success of cell culture depends on the species and on the type of cell from that species. Insect cells, for instance, can be cultured at room temperature, whereas an incubator is required for mammalian cells. Cell culture is not a straightforward method, however, since a culture medium must be established to which the cells adhere, before they can be induced to develop to the stage where they make their photonic devices.

Current work in this area centers on butterfly scales. The cells that make the scales are identified in chrysalises, dissected, and plated out. Then the individual cells are separated, kept alive in culture, and prompted to manufacture scales through the addition of growth hormones. Currently we have cultured blue *Morpho* butterfly scales in the lab that have identical optical and structural characteristics to natural scales. The cultured scales could be embedded in a polymer or mixed into a paint, where they may float to the surface and self-align. Further work, however, is required to increase the level of scale production and to harvest the scales from laboratory equipment in appropriate ways. A far simpler task emerges where the iridescent organism is single-celled.

3.1.4 Diatoms and Coccolithophores

Diatoms are unicellular photosynthetic microorganisms. The cell wall is called the frustule and is made of the polysaccharide pectin impregnated with silica. The frustule contains pores (Figure 3.5 (a)–(c)) and slits that give the protoplasm access to the external environment. There are more than 100,000 different species of diatoms, generally 20–200 μm in diameter or length, but some can be up to 2 mm long. Diatoms have been proposed to build photonic devices directly in 3D [Fuhrmann et al. 2004]. The biological function of the optical property (Figure 3.5 (d)) is at present unknown, but may affect light collection by the diatom. This type of photonic device can be made in silicon using a deep photochemical etching technique (initially developed by Lehmann [1993]) (e.g., Figure 3.5 (e)). However, there is a new potential here since diatoms carry the added advantage of exponential growth in numbers—each individual can give rise to 100 million descendants in a month.

Unlike most manufacturing processes, diatoms achieve a high degree of complexity and hierarchical structure under mild physiological conditions. Importantly, the size of the pores does not scale with the size of the cell, thus maintaining the pattern. The presence of these pores in the silica cell wall of the diatom *Coscinodiscus granii* means that the frustule can be regarded as a photonic crystal slab waveguide [Fuhrmann et al. 2004]. Furthermore, Fuhrmann et al. presented models to show that light may be coupled into the waveguide and give photonic resonances in the visible spectral range.

The silica surface of the diatom is amenable to simple chemical functionalization (e.g., Figure 3.6 (a)–(c)). An interesting example of this uses a DNA-modified diatom template for the control of nanoparticle assembly [Rosi et al. 2004]. Gold particles were coated with DNA complementary to that bound to the surface of the diatom. Subsequently, the gold particles were bound to the diatom surface via the sequence-specific DNA interaction. Using this method up to seven layers were added, showing how a hierarchical structure could be built onto the template.

Porous silicon is known to luminesce in the visible region of the spectrum when irradiated with ultraviolet light [Cullis et al. 1997]. This photoluminescence (PL) emission from the silica skeleton of diatoms was exploited in the production of an optical gas sensor [De Stefano et al. 2005]. It was shown that the PL of *Thalassiosira rotula* is strongly dependent on the surrounding environment. Both the optical intensity and peaks are affected by gases and organic vapors. Depending on the electronegativity and polarizing ability, some substances quench the luminescence, while others effectively enhance it. In the presence of the gaseous substances NO_2, acetone, and ethanol, the photoluminescence was quenched. This was because these substances attract electrons from the silica skeleton of the diatoms and hence quench the PL. Nucleophiles, such as xylene and pyridine, which donate electrons, had the opposite effect, and increased PL intensity almost

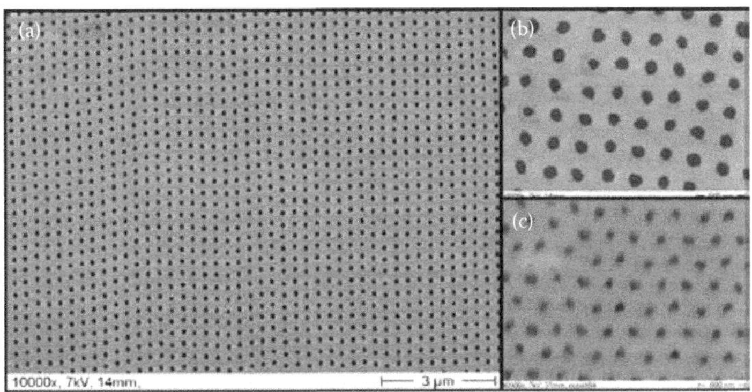

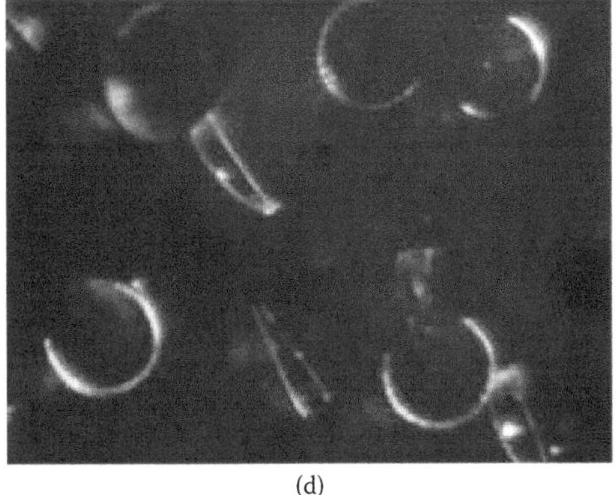

(d)

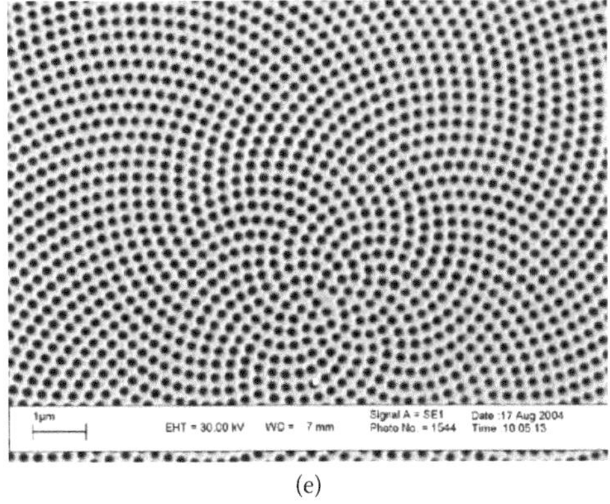

(e)

FIGURE 3.5
(a–c) Scanning electron micrographs of the intercalary band of the frustule from two species of diatoms, showing the square array of pores from *C. granii* ((a) and (b)) and the hexagonal arrays of pores from *C. wailesii* (c). These periodic arrays are proposed to act as photonic crystal waveguides. (d) Iridescence of the *C. granii* girdle bands. (e) Southampton University mimic of a diatom frustule (patented for photonic crystal applications); scanning electron micrograph (by G. Parker, reproduced with permission).

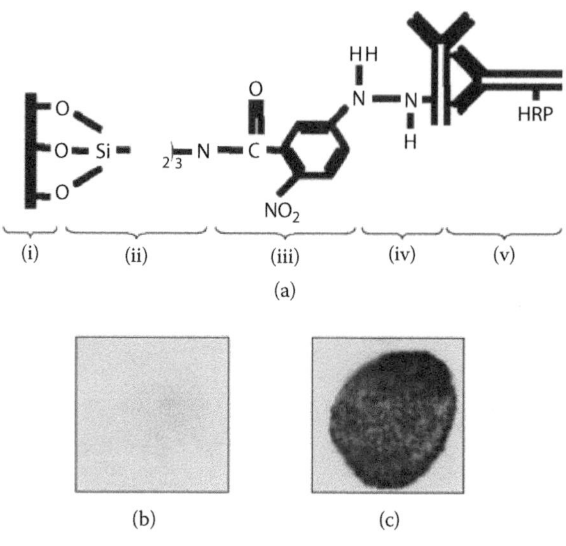

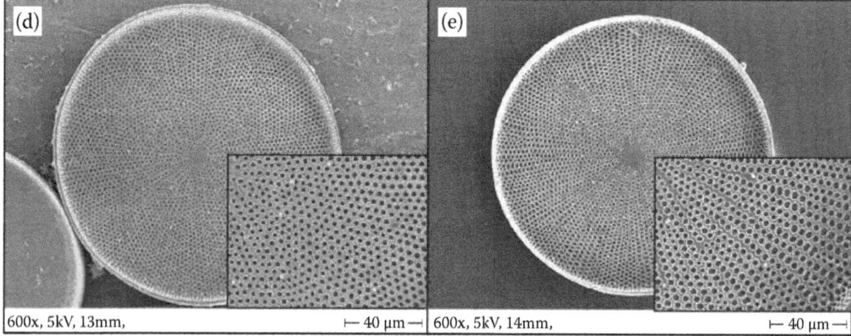

FIGURE 3.6

Modification of natural photonic devices. (a)–(c) Diatom surface modification. The surface of the diatom was silanized, then treated with a heterobifunctional cross-linker, followed by attachment of an antibody via a primary amine group. (a) (i) Diatom exterior surface (ii) APS (iii) ANB-NOS (iv) primary antibody (v) secondary antibody with HRP conjugate. Diatoms treated with primary and secondary antibody with (b) no surface modification (c) after surface modification. (d), (e) Scanning electron micrographs showing the pore pattern of the diatom *C. wailesii* (d) and after growth in the presence of nickel sulphate (e). Note the enlargement of pores, and hence change in optical properties, in (e).

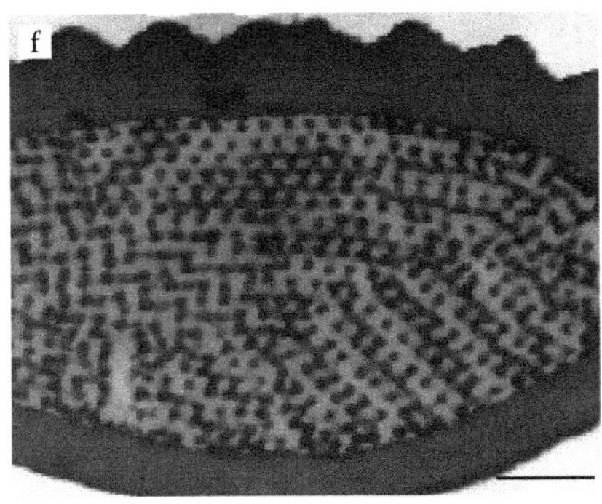

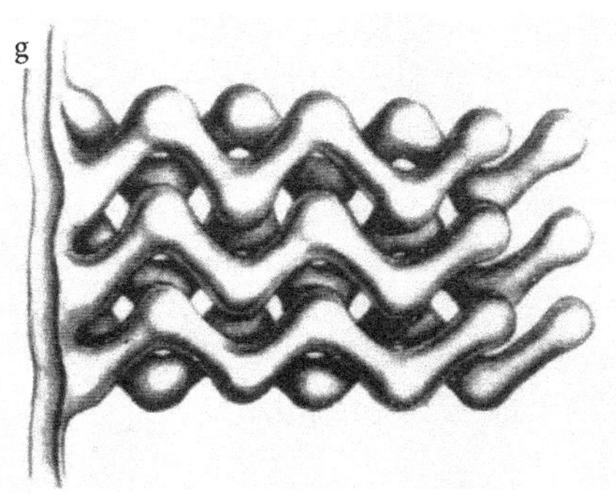

FIGURE 3.6 (*Continued*)
Modification of natural photonic devices. (f) 'Photonic crystal' of the weevil *Metapocyrtus* sp., section through a scale, transmission electron micrograph; scale bar: 1 µm (see Parker[33]). (g) A comparatively enlarged diagrammatic example of cell membrane architecture: tubular christae in mitochondria from the chloride cell of sardine larvae (from Threadgold, L.T. *The ultrastructure of the animal cell.* Pregamon press, Oxford, 313 pp. (1967)). Evidence suggests that pre-existing internal cell structures play a role in the manufacture of natural nanostructures; if these can be altered then so will the nanostructure made by the cell.

10 times. Both quenching and enhancements were reversible as soon as the atmosphere was replaced by air.

The silica inherent to diatoms does not provide the optimum chemistry/refractive index for many applications. Sandhage et al. [2002] have devised an inorganic molecular conversion reaction that preserves the size, shape, and morphology of the diatom while changing its composition. They perfected a gas/silica displacement reaction to convert biologically derived silica structures such as frustules into new compositions. Magnesium was shown to convert SiO_2 diatoms by a vapor phase reaction at 900°C to MgO of identical shape and structure, with a liquid Mg_2Si by-product. Similarly, when diatoms were exposed to titanium fluoride gas the titanium displaced the silicon, yielding a diatom structure made up entirely of titanium dioxide, a material used in some commercial solar cells.

An alternative route to silica replacement hijacks that native route for silica deposition in vivo. Rorrer et al. [2004] sought to incorporate elements such as germanium into the frustule, a semiconductor material that has interesting properties that could be of value in optoelectronics, photonics, thin-film displays, solar cells, and a wide range of electronic devices. Using a two-stage cultivation process the photosynthetic marine diatom *Nitzschia frustulum* was shown to assimilate soluble germanium and fabricate Si-Ge oxide nanostructured composite materials.

Porous glasses impregnated with organic dye molecules are promising solid media for tunable lasers and nonlinear optical devices, luminescent solar concentrators, gas sensors, and active waveguides. Biogenic porous silica has an open sponge-like structure and its surface is naturally OH terminated. Hildebrand and Palenik [2003] have shown that Rhodamine B and 6G are able to stain diatom silica in vivo, and determined that the dye treatment could survive the harsh acid treatment needed to remove the surface organic layer from the silica frustule.

Now attention is beginning to turn additionally to coccolithophores—single-celled marine algae, also abundant in marine environments. Here, the cell secretes calcitic photonic crystal frustules that, like diatoms, can take a diversity of forms, including complex 3D architectures at the nano- and microscales.

3.1.5 Iridoviruses

Viruses are infectious particles made up of the viral genome packaged inside a protein capsid. The iridovirus family comprises a diverse array of large (120–300 nm in diameter) viruses with icosahedral symmetry. The viruses replicate in the cytoplasm of insect cells. Within the infected cell the virus particles produce a paracrystalline array that causes Bragg refraction of light. This property has largely been considered aesthetic to date, but the research group of Vernon Ward (New Zealand), in collaboration with the Biomaterials Laboratory at Wright-Patterson Air Force Base, is using iridoviruses to create biophotonic crystals. These can be used for the control of light, with this laboratory undertaking large-scale virus production and purification as well as targeting manipulation of the surface of iridoviruses for altered crystal properties. These can provide a structural platform for a broad range of optical technologies, ranging from sensors to waveguides.

Virus nanoparticles, specifically *Chilo* and *Wiseana* invertebrate iridoviruses, have been used as building blocks for iridescent nanoparticle assemblies. Here, virus particles were assembled in vitro, yielding films and monoliths with optical iridescence arising from multiple Bragg scattering from close-packed crystalline structures of the iridoviruses. Bulk viral assemblies were prepared by centrifugation, followed by the addition of glutaraldehyde, a cross-linking agent. Long-range assemblies were prepared by employing a cell design that forced virus assembly within a confined geometry followed by cross-linking. In addition, virus particles were used as core substrates in the fabrication of metallodielectric nanostructures. These comprise a dielectric core surrounded by a metallic shell. More specifically, a gold shell was assembled around the viral core by attaching small gold nanoparticles to the virus surface using inherent chemical functionality of the protein capsid [Radloff et al. 2005]. These gold nanoparticles then acted as nucleation sites for electroless deposition of gold ions from solution. Such nanoshells could be manufactured in large quantities, and provide cores with a narrower size distribution and smaller diameters (below 80 nm) than currently used for silica. These investigations demonstrated that direct harvesting of biological structures, rather than biochemical modification of protein sequences, is a viable route to create unique, optically active materials.

3.1.6 The Mechanisms of Natural Engineering and Future Research

Where cell culture is concerned it is enough to know that cells *do* make optical nanostructures, which can be farmed appropriately. However, in the future an alternative may be to emulate the natural engineering processes ourselves, through reacting the same concentrations of chemicals under the same environmental conditions, and possibly substituting analogous nano- or macromachinery.

To date, the process best studied is the silica cell wall formation in diatoms. The valves are formed by the controlled precipitation of silica within a specialized membrane vesicle called the silica deposition vesicle (SDV). Once inside the SDV, silicic acid is converted into silica particles, each measuring approximately 50 nm in diameter. These then aggregate to form larger blocks of material. Silica deposition is molded into a pattern by the presence of organelles such as mitochondria spaced at regular intervals along the cytoplasmic side of the SDV [Schmid 1994]. These organelles are thought to physically restrict the targeting of silica from the cytoplasm, to ensure laying down of a correctly patterned structure. This process is very fast, presumably due to optimal reaction conditions for the synthesis of amorphous solid silica. Tight structural control results in the final species-specific, intricate exoskeleton morphology.

The mechanism whereby diatoms use intracellular components to dictate the final pattern of the frustule may provide a route for directed evolution. Alterations in the cytoplasmic morphology of *Skeletonema costatum* have been observed in cells grown in sublethal concentrations of mercury and zinc [Smith 1983], resulting in swollen organelles, dilated membranes, and vacuolated cytoplasm. Frustule abnormalities have also been reported in *Nitzschia liebethrutti* grown in the presence of mercury and tin [Saboski 1977]. Both metals resulted in a reduction in the length-to-width ratios of the diatoms, fused pores, and a reduction in the number of pores per frustule. These abnormalities were thought to arise from enzyme disruption either at the silica deposition site or at the nuclear level. We grew *C. wailesii* in sublethal concentrations of nickel and observed an increase in the size of the pores (Figure 3.6 (d) and (e)), and a change in the phospholuminescent properties of the frustule. Here, the diatom can be "made to measure" for distinct applications such as stimuli-specific sensors.

Further, *trans*-Golgi-derived vesicles are known to manufacture the coccolithophore 3D "photonic crystals" [Corstjens and Gonzales 2004]. So the organelles within the cell appear to have exact control of (photonic) crystal growth ($CaCO_2$ in the coccolithophores) and packing (SiO_3 in the diatoms) [Klaveness and Guillard 1975, Klaveness and Paasche 1979]. Indeed, Ghiradella [1989] suggested that the employment of preexisting, intracellular structures lay behind the development of some butterfly scales, and Overton [1966] reported the action of microtubules and microfibrils during butterfly scale morphogenesis. Further evidence has been found to suggest that these mechanisms, involving the use of molds and nanomachinery (e.g., Figure 3.6 (f) and (g)), reoccur with unrelated species, indicating that the basic eukaryote (containing a nucleus) cell can make complex photonic nanostructures with minimal genetic mutation [Parker 2006].

For further information on the topic of evolution of optical devices in nature, including those found in fossils, or when they first appeared on earth, the reader is referred to the literature [Parker 2003, 2005].

The ultimate goal in the field of optical biomimetics, therefore, could be to replicate such machinery and provide conditions under which, if the correct ingredients are supplied, the optical nanostructures will self-assemble with precision.

References

Beale, B. (1999). Fly eye on the prize. *The Bulletin*, 46–48 (May 25).

Berthier, S. (2005). *Les coulers des papillons ou l'imperative beauté. Proprietes optiques des ailes de papillons*. Springer, Paris.

Boden, S.A., and Bagnall, D.M. (2006). Biomimetic subwavelength surfaces for near-zero reflection sunrise to sunset. Presented at Proceedings of 4th World Conference on Photovoltaic Energy, Conversion, Waikoloa, HI.

Cohen, R.E., Zhai, L., Nolte, A., and Rubner, M.F. (2004). pH gated porosity transitions of polyelectrolyte multilayers in confined geometries and their applications as tunable Bragg reflectors. *Macromolecules* 37:6113–6123.

Corstjens, P.L.A.M., and Gonzales, E.L. (2004). Effects of nitrogen and phosphorus availability on the expression of the coccolith-vesicle v-ATPase (subunit C) of *Pleurochrysis* (Haptophyta). *J. Phycol.* 40:82–87.

Cullis, A.G., Canham, L.T., and Calcott, P.D.J. (1997). The structural and luminescence properties of porous silicon. *J. Appl. Phys.* 82:909–965.

De Silva, L., Hodgkinson, I., Murray, P., Wu, Q., Arnold, M., Leader, J., and Mcnaughton, A. (2005). Natural and nanoengineered chiral reflectors: structural colour of manuka beetles and titania coatings. *Electromagnetics* 25:391–408.

De Stefano, L., Rendina, I., De Stefano, M., Bismuto, A., and Maddalena, P. (2005). Marine diatoms as optical chemical sensors. *Appl. Phys. Lett.* 87:233902.

Fuhrmann, T., Lanwehr, S., El Rharbi-Kucki, M., and Sumper, M. (2004). Diatoms as living photonic crystals. *Appl. Phys. B* 78:257–260.

Gale, M. (1989). Diffraction, beauty and commerce. *Phys. World* 2:24–28.

Ghiradella, H. (1989). Structure and development of iridescent butterfly scales: lattices and laminae. *J. Morph.* 202:69–88.

Hildebrand, M., and Palenik, B. (2003). *Investigation into the optical properties of nanostructured silica from diatoms*. Grant report. Available from http://www.stormingmedia.us/26/2663/A266314.html (accessed January 15, 2012).

Kinoshita, S., Yoshioka, S., Fujii, Y., and Okamoto, N. (2002). Effect of macroscopic structure in iridescent color of the peacock feathers. *Forma* 17:103.

Klaveness, D., and Guillard, R.R.L. (1975). The requirement for silicon in *Synura petersenii* (Chrysophyceae). *J. Phycol.* 11:349–355.

Klaveness, D., and Paasche, E. (1979). Physiology of coccolithophorids. In Levandowsky, M., Hutner, S.H., and Lwoff, A. (eds.) *Biochemistry and physiology of protozoa*. 2nd ed., vol. 1. Academic Press, New York, pp. 191–213.

Lehmann, V. (1993). On the origin of electrochemical oscillations at silicon electrodes. *J. Electrochem. Soc.* 143:1313–1318.

Miller, W.H., Moller, A.R., and Bernhard, C.G. (1966). The corneal nipple array. In Bernhard, C.G. (ed.), *The functional organisation of the compound eye*, 21–33. Pergamon Press, Oxford.

Overton, J. (1966). Microtubules and microfibrils in morphogenesis of the scale cells of *Ephestia kuhniella*. *J. Cell Biol.* 29:293–305.

Parker, A.R. (2000). 515 million years of structural colour. *J. Opt. A* 2:R15–28.

Parker, A.R. (2003). *In the blink of an eye*. Simon & Schuster, London/Perseus Press, Cambridge, MA.

Parker, A.R. (2005). A geological history of reflecting optics. *J. R. Soc. Lond. Interface* 2:1–17.

Parker, A.R. (2006). Conservative photonic crystals imply indirect transcription from genotype to phenotype. *Rec. Res. Develop. Entomol.* 5:1–10.

Parker, A.R., Hegedus, Z., and Watts, R.A. (1998). Solar-absorber type antireflector on the eye of an Eocene fly (45Ma). *Proc. R. Soc. Lond. B* 265:811–815.

Parker, A.R., McPhedran, R.C., McKenzie, D.R., Botten, L.C., and Nicorovici, N.-A.P. (2001). Aphrodite's iridescence. *Nature* 409:36–37.

Radloff, C., Vaia, R.A., Brunton, J., Bouwer, G.T., and Ward, V.K. (2005). Metal nanoshell assembly on a virus bioscaffold. *Nano Lett.* 5:1187–1191.

Rorrer, G.L., Chang, C.H., Liu, S.H., Jeffryes, C., Jiao, J., and Hedberg, J.A. (2004). Biosynthesis of silicon-germanium oxide nanocomposites by the marine diatom *Nitzschia frustulum*. *J. Nanosci. Nanotechnol.* 5:41–49.

Rosi, N.L., Thaxton, C.S., and Mirkin, C.A. (2004). Control of nanoparticle assembly by using DNA-modified diatom templates. *Angew Chem. Int. Ed.* 43:5500–5503.

Saboski, E. (1977). Effects of mercury and tin on frustular ultrastructure of the marine diatom *Nitzschia liebethrutti*. *Water Air Soil Pollut.* 8:461–466.

Sandhage, K.H., Dickerson, M.B., Huseman, P.M., Caranna, M.A., Clifton, J.D., Bull, T.A., Heibel, T.J., Overton, W.R., and Schoenwaelder, M.E.A. (2002). Novel, bioclastic route to self-assembled, 3D, chemically tailored meso/nanostructures: shape-preserving reactive conversion of biosilica (diatom) microshells. *Adv. Mater.* 14:429–433.

Schmid, A.M.M. (1994). Aspects of morphogenesis and function of diatom cell walls with implications for taxonomy. *Protoplasma* 181:43–60.

Smith, M.A. (1983). The effect of heavy metals on the cytoplasmic fine structure of *Skeletonema costatum* (Bacillariophyta). *Protoplasma* 116:14–23.

Watanabe, K., Hoshino, T., Kanda, K., Haruyama, Y., and Matsui, S. (2005). Brilliant blue observation from a Morpho-butterfly-scale quasi-structure. *Jpn. J. Appl. Phys.* 44:L48–L50.

Yablonovitch, E. (1999). Liquid versus photonic crystals. *Nature* 401:539–541.

Yoshida, A., Motoyama, M., Kosaku, A., and Miyamoto, K. (1997). Antireflective nanoprotuberance array in the transparent wing of a hawkmoth *Cephanodes hylas*. *Zool. Sci.* 14:737–741.

Zhang, W., Zhang, D., Fan, T., Ding, J., Gu, J., Guo, Q., and Ogawa, H. (2006). Bio-mimetic zinc oxide replica with structural color using butterfly (*Ideopsis similis*) wings as templates. *Bioinspir. Biomim.* 1:89–95.

3.2 Nature-Inspired Structural Color Applications

Torben Lenau

The photonic structures that cause structural colors in nature are fascinating and can inspire new man-made counterparts in industrial products. Artificial structures that are inspired by the ones found in nature can be advantageous compared to the traditional dyes and pigments that absorb certain wavelengths and to industrial metal coating. Structural colors show interesting properties, including many that are crucial for functional materials such as durability, environmental sustainability, and electric and magnetic insulation. They are visually aesthetic and decorative, exhibiting bright colors, color shift, and metallic appearance.

This section describes photonic principles found in nature that cause structural colors and how they can be mimicked by various artificial techniques, including from surface plating, coating, physical treatment, and others.

A natural question is why we should use structural colors. There are several reasons:

1. They give more bright colors—especially in the short-wavelength part of the visible spectrum.
2. The colors are more long lasting since a physical structure generally is more stable than pigments (a chemical compound) and therefore do not fade. The *Philocteanus rubroaureus* beetle in Figure 3.7 is a brilliant example of the durability of structural colors. The beetle was collected in the 18th century and is therefore more than 200 years old, and it still exhibits bright vibrant colors.

FIGURE 3.7 (See color figure at http://www.crcpress.com/product/isbn/9781439877463) *Philocteanus rubroaureus* (Coleoptera Buprestidae) beetle collected in the 18th century illustrates the durability of the structural colors. Scale bar = 10 mm. (The beetle is kindly provided from the collections of the Zoological Museum, Copenhagen University. Graphics by Michael Barfoed.)

3. Tunable color structures can be designed to specific needs, making reflection or transmission of selected wavelengths possible.
4. The color additives normally used in polymers can be replaced by nanoscale structures in the polymer itself, making recycling more feasible.
5. Bright and shiny metallic appearance can be achieved without using metal, which gives the possibility of electrical and thermal insulation, and improves recycling and disposal properties.
6. Permanent and reversible color change is feasible by changing the physical dimensions in the nanoscale structures.

3.2.1 Shell Structures in Insects

All insects have an outer stiff shell called the cuticle that surrounds its body. The structural colors in insects are produced in delicate structures that are an integral part of the cuticle. The cuticle is produced layer by layer from various materials segregated by the underlying living epidermal cells. The segregation is modified over time, causing changes in thickness, chemical composition, and optical properties. Chitin is a dominant component in many of the layers, but the cuticle is a composite structure also including many other materials, such as various proteins, melanin, lipids, and others. The many layers in the beetle cuticle are normally described from their position in different sections, as illustrated in Figure 3.8. From the outside to the inwards the overall sections include the epicuticle, the exocuticle, and the endocuticle [Lenau and Barfoed 2008, Neville 1975].

Insects change their cuticle on a regular basis. This is the way they grow. The process of cuticle shift is called ecdysis. Just before the ecdysis a new

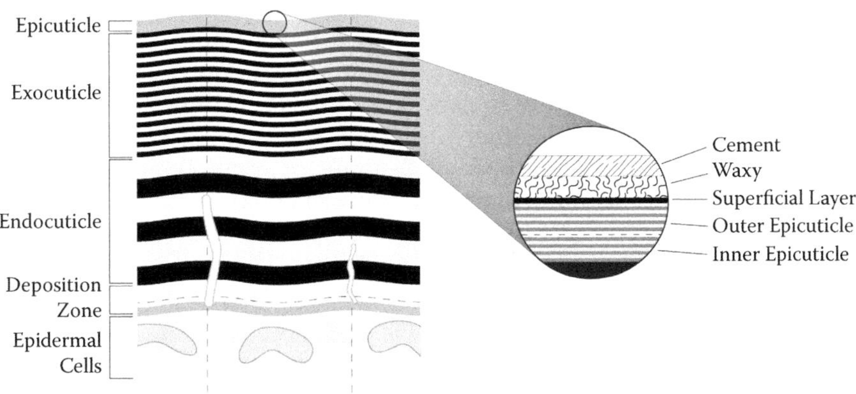

FIGURE 3.8
Enlarged cross section of a beetle cuticle showing the most important layers. (Graphics by Thomas Pil Winkel.)

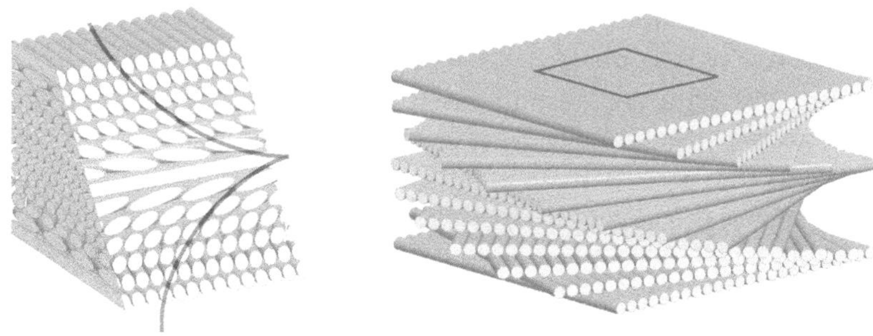

FIGURE 3.9
Model illustrating helical Bouligand structure. The left picture illustrates the arcs seen in micrographs of oblique cut sections. The dark square in the picture at the right shows where the cut-out is made. (Based on Neville, A.C., *Biology of the Arthropod Cuticle*, Springer-Verlag, Berlin, 1975, p. 178. Graphics by Thomas Nissen.)

epicuticle is formed right over the epidermal cells. After the ecdysis, the new epicuticle becomes the outermost layer. The underside of this becomes the starting point of the new exocuticle. The material for the formation of the cuticle is made by the epidermal cells and released into the fluid of the lower endocuticle. It is assumed that all subsequent processes to a 3D architecture occur passively by self-assembly. The epidermal cells produce the new material as tiny nanofibrils that form layers on the underside of the cuticle where the microfibrils are orientated in parallel. Each successive layer is slightly twisted, and the result is a helical structure, as shown in Figure 3.9.

3.2.2 Light Reflection Principles in Nature

One of the principles used in nature to selectively reflect colored light is thin-layer interference. The interference occurs when the incoming light is reflected from two parallel surfaces in a transparent material with a refractive index different from the surrounding material. The wavelength equal to four times the optical thickness of the layer is strengthened by constructive interference, while other wavelengths are suppressed by destructive interference. The optical thickness is found by multiplying the physical layer thickness with its refractive index. Since materials in nature only have small differences in the refractive index, most pronounced reflections come from multiple reflections in so-called quarter-wave stacks, with alternating layers having high and low refractive indices, as illustrated in Figure 3.10. Reflectors can have more than 200 layers.

For an ideal multilayer stack in air the reflectivity can be estimated from Equation (3.1). The reflectivity is the ratio between the reflected and the incident light intensity. n_a and n_b are the refractive indices for the two different types of layers in the stack, and p the number of double layers [Tilley 2000]. Typical refractive indices in insect cuticula are 1.5–1.6, which is also the typical value

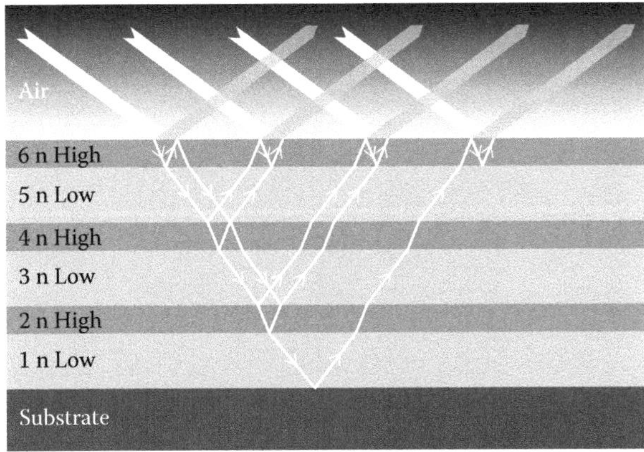

FIGURE 3.10
Reflection by thin-layer interference in three double layers. The reflected light has a wavelength equal to four times the optical thickness of each of the transparent thin layers. (Graphics by Tomas Benzon.)

for many commonly used polymers [Parker-TexLoc 2009]. A reflectivity of 90% can be achieved with as few as seven double layers if $n_a = 1.5$ and $n_b = 2.0$. When the refractive index difference decreases, the number of layers has to increase to give the same reflectivity (29 layers for $n_b = 1.6$). One way of achieving lower refractive index is to add very small air inclusions, like air bubbles in the material. Using the Bruggemann effective medium approximation as described by Stefano et al. [2007], it can be estimated that a refractive index of 1.25 can be obtained if half the material is replaced by small air cavities.

$$R = \left[1 - \left(\frac{n_a}{n_b} \right)^{2\rho} \bigg/ 1 + \left(\frac{n_a}{n_b} \right)^{2\rho} \right] \tag{3.1}$$

Another principle for light reflection is found in the so-called helicoidal Bouligand structure [Neville 1975, Giraud-Guille et al. 2003]. The reflection is also caused by interference, but from a single birefringent material. From a biomimetic point of view it can be an advantage to use a single substance instead of controlled formation of two different types of layers. The birefringent material is in the form of very small fibrils that are oriented in the same direction within a layer. The fibrils are birefringent, meaning that the refractive index is different along and perpendicular to the fibril axis. The fibrils are chiral and therefore orient themselves with a twist in successive layers. The distance between layers for which the direction of the fibers has taken a full 360° rotation, as shown in Figure 3.11, is called the pitch. The reflector will reflect wavelengths that have an optical thickness equal to half a pitch. Incident light will meet a changing refractive index as it passes down

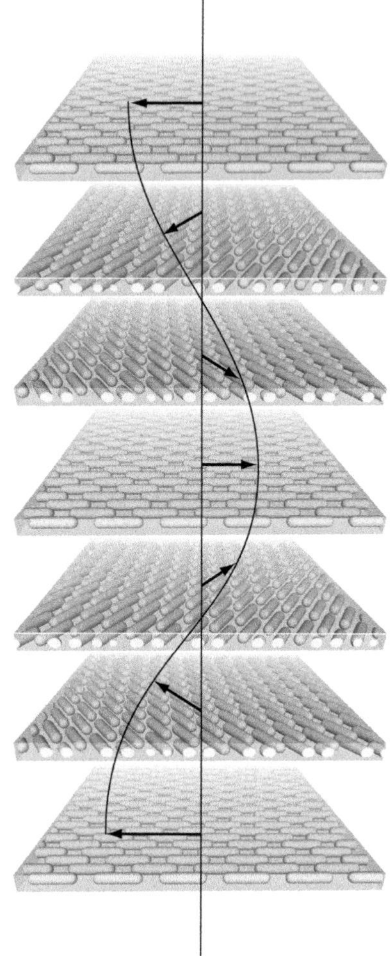

FIGURE 3.11
Reflection in helicoidal Bouligand structures. The birefringent fibrils in ajacent layers are slightly twisted. The fibrils in the figure form a full 360° rotation from the lower to the top layer. This distance is called the pitch. (Graphics by Thomas Nissen.)

through the structure, causing reflection of circular polarized light through interference. As described earlier, most insects have Bouligand structures in their cuticle, but only a limited number of insects—in particular scarabid beetles—use birefringence to give optical effects.

3.2.2.1 Narrowband Reflection in Nature

Colors are used for many different purposes in nature, ranging from identification of food sources, over search for prey, to sexual selection. In order to understand the amazing colors in nature and their function, it is important to remember that most animals, birds, and insects see color in a different way than what we humans do. We only see a limited part of the color spectrum, namely, from violet (400 nm) to red (750 nm), which we interpret from the combined sense signals from three different types of cone-shaped photoreceptors using trichromatic color vision. Many animals, however, see other parts of the light spectrum. Insects and birds often have vision sensitive to ultraviolet light in the region of 300–400 nm [Bennett et al. 1996].

Kestrels (*Falco tinnuncullus*) are able to detect vole scent marks from urine and feces when UV light is present [Viitala et al. 1995]. The vole scent marks reflect UV light differently than the surroundings and can therefore be identified by the kestrels. Female starlings (*Sturnus vulgaris*) rank males differently when UV wavelengths are present or absent [Bennett et al. 1997]. Again, the reflection in the UV part of the light spectrum plays a significant role—this time for the sexual selection. Parrots (*Psittaciformes*) also use UV colors for mate selection, but unlike many other birds, they live for many years [Carvalho et al. 2011]. This is surprising since those organisms that use UV light typically reproduce quickly and live for only a shorter period. UV light is much more energetic than the light of visual wavelengths, and organic matter, in particular in the eyes, is therefore more prone to damage. Carvalho and colleagues [2011] report that parrots have UV-sensitive pigments in their photoreceptors, but also that the eyes can handle the damaging effect from the UV light.

Some marine organisms, namely, silica diatoms, protect themselves to the exposure from UV light by scattering the light [Raven and Waite 2004]. Diatoms are unicellular algae with silica exoskeletons called frustules that consist of two overlapping thecae. The frustule looks like a small pillbox or set of petri dishes. Each theca has a periodic pattern, which is assumed to scatter the light most effectively in the short-wavelength part of the light spectrum [Stefano et al. 2007], as can be seen in Figure 3.12.

Stefano and colleagues [2007] use a finite-element model to demonstrate the UV shielding properties in diatoms, and measurements by Noyes and colleagues [2008] confirm that thecae show much lower transmission for blue than for red wavelengths. The examined diatom frustules have a size of around 100–350 μm in diameter [Noyes et al. 2008, Fuhrmann et al. 2004], and the thecae consist of two silica layers. The outer layer has pores with a size around 250 nm and lattice constant around 400 nm, and the inner layer holes with sizes around 1.3 μm and lattice constant around 2 μm [Noyes et al. 2008].

Several insects and birds exhibit bright reflections caused by structural colors. One of the most well-known phenomena is the bright blue reflection from the *Morpho* butterflies [Kinoshita et al. 2002, Vukusic et al. 1999]. The *Heterorrhina* beetles exhibit a uniform green color. Neville [1975] describes how the exocuticle has a lamelar structure with alternating layers, which

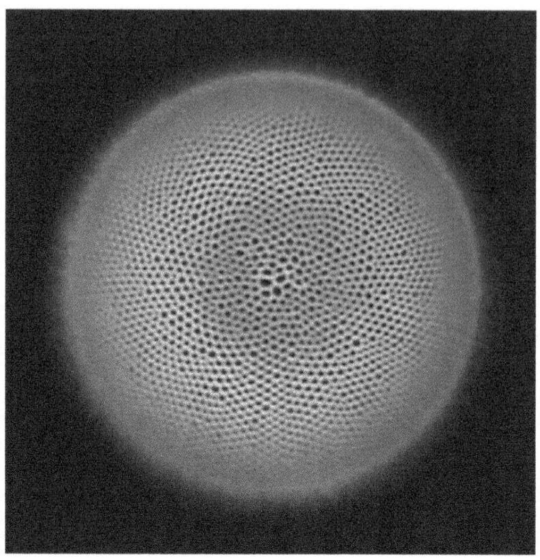

FIGURE 3.12
Light micrograph of a diatom of the *Coscinodiscus* species. Diameter is 650 μm. (Photograph with kind permission from Nina Lundholm, Copenhagen University.)

functions as a multilayer interference reflector. This is supported by the observation that a piece of plasticized elytral cuticle (part of the wing) swelled and changed its color to red. When suitable pressure was applied to the surface, it produced a blue color in the middle of the depression, with colors grading through the spectrum.

Many of the blowflies (Calliphoridae), like the green-golden *Lucilia caesar,* also have remarkable colors and metallic reflecting surfaces.

Reversible color shifts are seen in several species. One of the most well known is the chameleon (Chamaeleonidae) color change used for social signaling and for camouflage [Stuart-Fox et al. 2006, 2008, Prusten 2008]. The mechanism in the chameleon skin responsible for the color change is a combination of yellow pigment particles, scattering particles that produce blue colors due to Tyndall scattering, and branched cells with melamin granules. By dispersing or concentrating the melamin granules the amount of light absorption can be controlled and hence change the reflected color.

Color shift is also seen in beetles. According to Neville [1977] *Aspidomorpha tecta* turns yellow before any metallic colors appear 3 to 4 days after the final ecdysis. When the living beetle is provoked parts of its surface changes color from gold to red in about 2 min. The reflector in *A. tecta* is placed in the endocuticle, i.e., the lower part of the cuticula that has not been hardened. This is also the reason why the beetle loses its golden metallic color when it dies.

3.2.2.2 Narrowband Reflection Applications

Structural color reflectors often use the thin-layer interference principle in either single or multiple layers. Single-layer interference surfaces are found in a number of places. Kikuti and colleagues [2004] describe three different chemical and electrochemical surface treatments of stainless steel resulting in single-layer metal-oxide films with interference colors. Layer thicknesses were measured to be between 70 and 400 nm. Colors span most of the visible spectrum depending on the coating thickness. Apart from the coloring effect the treatment with thin coatings improves the pitting corrosion resistance. A similar experiment was carried out at the Technical University of Denmark with good results, as shown in Figure 3.13.

Similar oxide films can be seen in Japanese colored stainless steel cutlery [Nakano 2011], as shown in Figure 3.14, and in stainless steel facade elements [Euro-inox 2011]. The phenomenon is also seen—unwanted—on stainless steel cooking equipment caused by oxidation from the gas fire, as shown in Figure 3.15. The advantage of these surfaces is that they are hard and lasting—the thin oxide film does not fade. Among disadvantages is their sensitivity to other coatings, which can change their optical properties. For example, fingerprints change the reflected color.

An iridescent textile using interference structures is inspired by the *Morpho* butterfly [Kinoshita and Yoshioka 2005, Iohara et al. 2000]. The textile filaments are made of polyester and have a flattened shape and a thickness of 15–17 μm, in which 61 layers of polyamide and polyester with a thickness of 70–90 nm are incorporated. The reflector stack is covered with a thicker coating of polyester. The reflectors can be aligned in the textile due to the flat shape of the filament resulting in better reflectivity. The textile

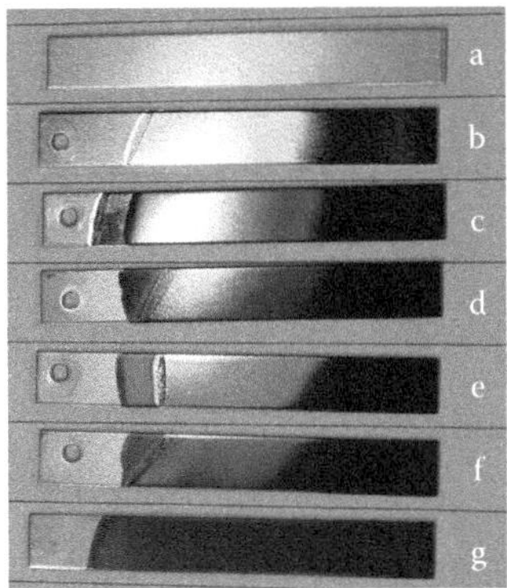

FIGURE 3.13 (See color insert.)
Stainless steel samples electropolished and coated with a single transparent layer resulting in interference colors. (a) Raw 304 stainless steel, (b) electropolished, (c)–(g) electropolished and coated with increased layer thickness. (Photograph courtesy of Per Møller and Torben Lenau, Technical University of Denmark.)

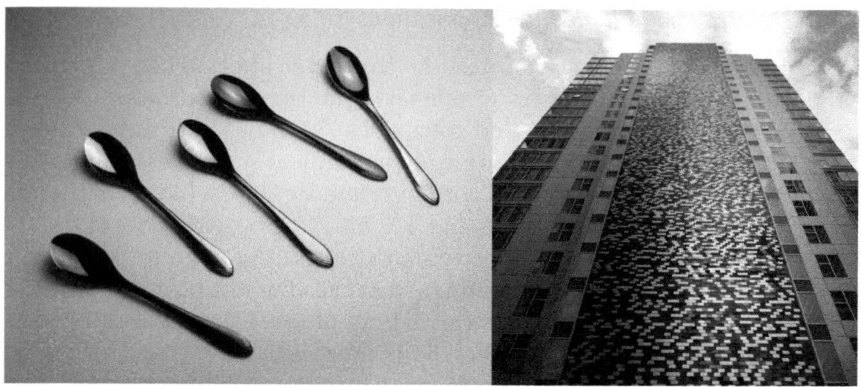

FIGURE 3.14 (See color figure at http://www.crcpress.com/product/isbn/9781439877463)
Stainless steel spoons with interference colors from a thin transparent oxide layers. (Courtesy of Nakano Acl.) The tiles in the building at right are pictured courtesy of Millennium Tiles LLC. The building was designed by Ruffcorn Hinthorne Mott and Stine Architects in Seattle.

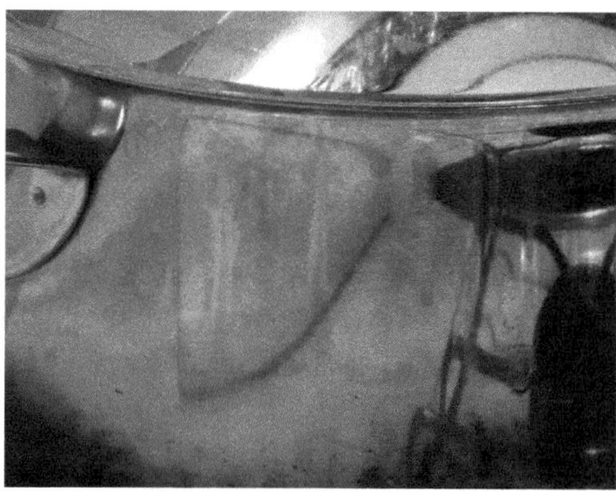

FIGURE 3.15 (See color figure at http://www.crcpress.com/product/isbn/9781439877463)
A stainless steel cooking pot showing interference colors.

reflects a single metallic color that changes depending on the viewing angle. Colors from red to blue can be obtained by adjusting the layer thickness. Applications considered are textiles for apparel, curtains, car cloth, and short-cut fibers for car paint (Figure 3.16).

A possible application of UV-reflecting particles based on structural colors is the identification of polymers. This is analogous to how kestrels identify the UV-reflecting vole trails. The particles could work like a name tag that can be identified by spectrometers. The particle should be mixed into the polymer similarly as traditional pigments or added as a surface coating. Each type of polymer would reflect a unique pattern of UV wavelengths, i.e., have a specific UV color. A spectrometer will be able to identify polymers that to the human eye would look identical. If polymers were marked in this way the

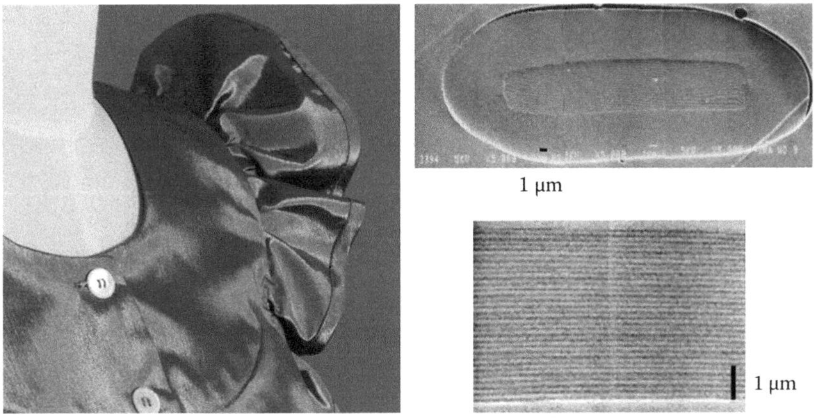

FIGURE 3.16
Shown left is a woman's dress made from a textile blend (60% silk and 40% Morphotex). (Design and picture courtesy of Donna Sgro.) The Morphotex fiber exhibits interference colors caused by a multilayer reflector in the core of the fiber. Right are shown two scanning microscopic images of the cross section of a flattened polyester fiber mimicking the *Morpho* butterfly. The fiber includes 61 alternate layers of nylon 6 and polyester with a thickness of 70–90 nm. (From Kinoshita, S., and Yoshioka, S., *ChemPhysChem*, 6:1443–1459, 2005. Photographs with kind permission from Teijin Fibers Ltd.)

FIGURE 3.17
Symbols for type identification of polymers for packaging materials. (From American Chemistry Council, Plastic Packaging Resins, available from http://plastics.americanchemistry.com/Education-Resources/Plastics-101/Plastics-Resin-Codes-PDF.pdf, accessed October 30, 2011.)

automatic sorting of scrap plastic becomes feasible, making recycling easier. Today plastic packaging products are marked during the molding with a symbol that identifies the polymer type as shown in Figure 3.17. However, the symbol can only be seen from one side of the product and the automatic recognition can therefore be difficult.

Cathell and Schauer [2007] describe how color-based metal-ion detectors can be made using very thin layers of alginate, a polysaccharide. Alginate is isolated from algea and spin-coated onto silicon wafers. The thin layer of alginate is water stable after cross-linking with calcium. The wafers have a distinct structural color that changes when exposed to different metal-ions.

Heinlein and Kasch [1999] describe a commercial available paint pigment that has similar optical properties as a rose beetle. A similar paint is shown in Figure 3.18. The pigment has a pronounced iridescence, and the resulting

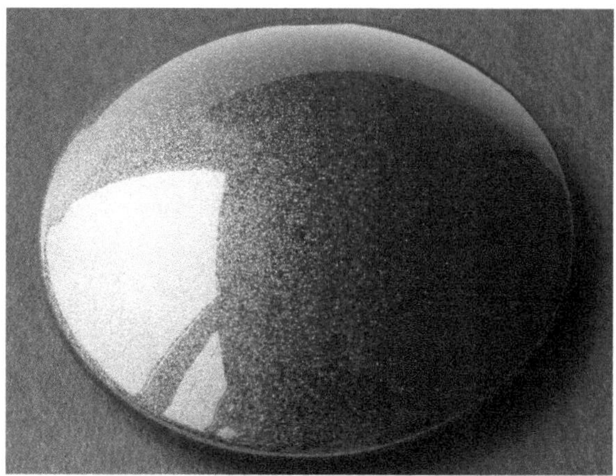

FIGURE 3.18 (See color figure at http://www.crcpress.com/product/isbn/9781439877463)
Computer rendering illustrating the effect in iridescent paint based on liquid crystals. (Graphics by Tomas Benzon.)

paint will therefore have a different color, e.g., change from red to green, depending on the viewing angle. The pigments are small platelets made from an organosilicone material and are typically about 35 μm wide. The material is a nematic liquid crystal doped with a chiral agent resulting in a helicoidal-layered structure. The pigments are transparent and the resulting colors are interference colors. The pigments must be applied to a dark light absorbing background for the color change effect to be distinct.

Srinivasarao [1999] described how they produced a cholesteric liquid crystal material similar to the beetle structure. This was done by mixing a commercial nematic liquid crystal with a chiral dopant. By changing the amount of dopant the pitch of the helicoidal structure could be controlled, and hence the color of the reflected light. The purpose was to convey optical measurements on reflectance spectra on structures similar to those found in beetles.

Parnell and colleagues [2011] reported to have made self-assembled copolymer blends that can be tuned to reflect any color within the visible region, as shown in Figure 3.19. They use two different polystyrene-polyisoprene (PS-PI) diblock polymers that are dissolved in a given ratio in the solvent *o*-xylene and placed between thin glass coverslips with 80 μm spacers. The ratio between the two diblocks determines the resulting reflected color. The self-assembly into alternating thin layers is achieved by using manual oscillatory shear.

3.2.3 Metallic Appearance in Nature

Metallic reflectance is seen in several species. Two of the well-studied phenomena are the reflection in beetles and that in fish. Metallic or mirror-like appearance is achieved in nature by reflecting all (or most)

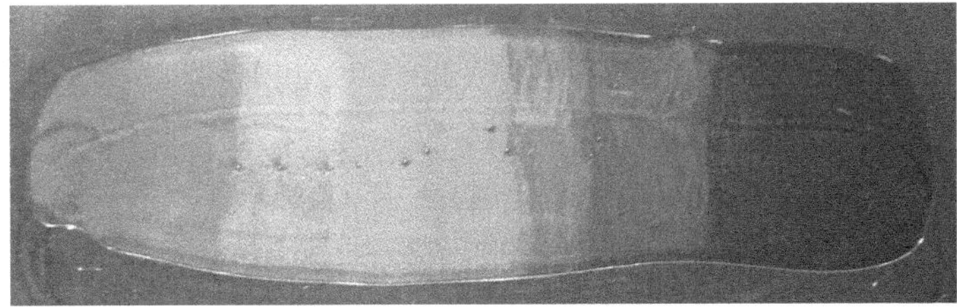

FIGURE 3.19 (See color insert.)
Range of colors that can be made by mixing two block copolymers in varying proportions. (With kind permission from Dr. Andrew Parnell, Sheffield University.)

of the visible wavelengths from the incoming light. This can be done using interference in thin-layer systems or helical Bouligand structures.

A natural question is why the beetles have such conspicuous appearance, which seems far from the camouflage look of other insects. The iridescent and metallic colored beetles are most common in the tropical rain forests of Southeast Asia, South America, and Australia [Beckmann 2011]. These forests are characterized of being very dark at the forest floor. It seems that there is an intense fight for light, materialized in elaborate leaf ornamentation wherever light strikes. Within these forests light only comes in narrow and short lasting specs. This might be the exact reason why it is biologically successful to have high reflecting colors. One technique to avoid predators is "flash coloration." When a reflective beetle moves between light and dark, the result is a high visual contrast that confuses the predator. Of course it also helps to reflect the surroundings [Schmidt 2006]. The coloration of other beetles typically allows them "to disappear into their environment" by matching dominant color patterns [Beckmann 2011].

A typical property for an interference reflector is the iridescence where the reflected color changes with the viewing angle. This is typically not the case for the metallic interference reflectors. A possible explanation is that the reflector covers a broader band than the visible spectrum. When the viewing angle changes from normal to grazing angles, substructures reflecting in the near-infrared region become red, while red structures turn to yellow, and so forth. But since we only see the visible part of the spectrum we will not detect a shift in color.

3.2.3.1 Multilayer Metallic Reflectors in Beetles

Metallic reflection can be achieved using a multilayer reflector having a pairwise set of layers with a high and a low refractive index. Each pair of layers reflects a different wavelength by interference since it has a different thickness. The basic chitinous substance is the same in all layers (apart from the top epicuticle), and the difference in refractive index is achieved through different hardening (called tanning) processes. A golden *Anoplagnathus parvulus*

(Figure 3.20) has a so-called chirped stack of layers with pairwise thicker layers in the top of the elytra (the cover wings) and thinner layers in the bottom [Parker et al. 1998]. The difference in layer thickness should result in the reflection of many different wavelengths giving the golden metallic appearance.

The golden *Aspidomorpha tecta* (Figure 3.21) also has a chirped stack of layers, but it is placed in the endocuticle just above the epidermis (the layer-producing tissue) [Hinton 1973, Neville 1977]. Figure 3.22 shows a micrograph where 44 × 2 interference layers can be seen. *A. tecta* loses its golden metallic appearance when it dies. The reason is that the endocuticle with the reflector in the living beetle is still soft and has not been hardened like the exocuticle placed closer to the surface of the cuticle. When the beetle dies the endocuticle

FIGURE 3.20 (See color insert.)
The golden beetle *Anoplagnathus parvulus.* (With kind permission from Robert Perger from the Coleop-Terra Organization.)

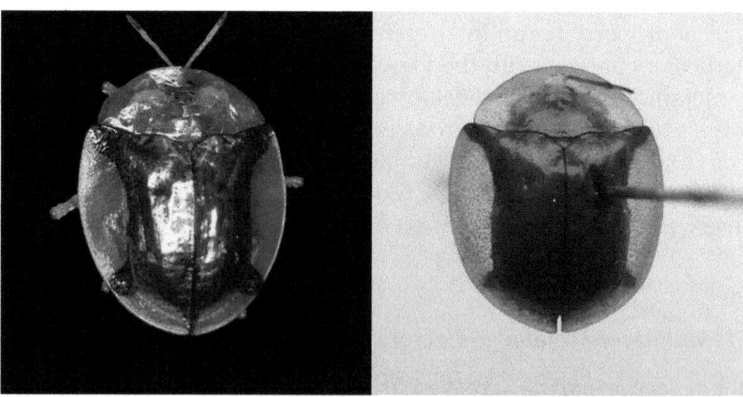

FIGURE 3.21 (See color insert.)
The *Aspidomorpha tecta* beetle at left is alive and has a golden appearance, while the museum specimen at the right has lost the golden color. The beetles are about 10 mm long. (Photograph courtesy of the Aquazoo/Löbbecke-Museum in Düsseldorf and the Zoological Museum at Copenhagen University.)

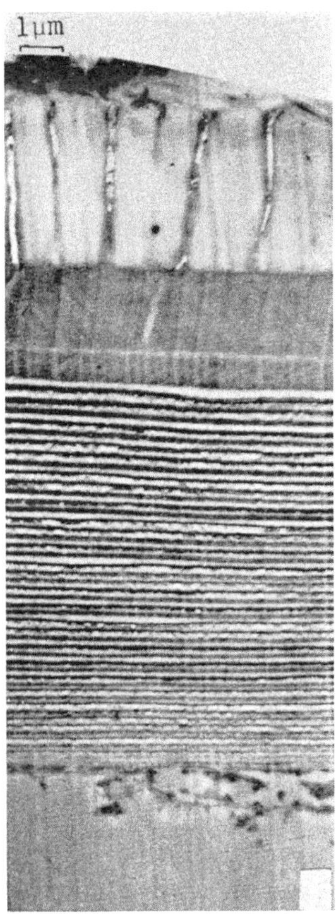

FIGURE 3.22
Electron micrograph of a vertical cut through the cuticle of *Aspidomorpha tecta* with a golden appearance. (From Hinton, H.E., *Proc. Br. Ent. Nat. Hist. Soc.*, 6:43–54, 1973. With kind permission from the *British Journal of Entomology Proceedings*.)

dries and the optical properties are changed [Neville 1977]. The reason that the reflection is golden and not silvery (as some other beetles actually are) is that the reflection of long-wavelength red colors is more pronounced than other wavelengths. Actually, the very young *A. tecta* shows a red color just after molting. After some days during which more layers are being produced the color changes to gold metallic. The gold color can also be achieved through red and yellow pigment tinting of a silver reflector.

3.2.3.2 Helical Bouligand Metallic Reflectance in Beetles

Another way to achieve metallic appearance is the so-called helical Bouligand structures found in the golden scarabid beetle *Plusiotis resplendens* shown

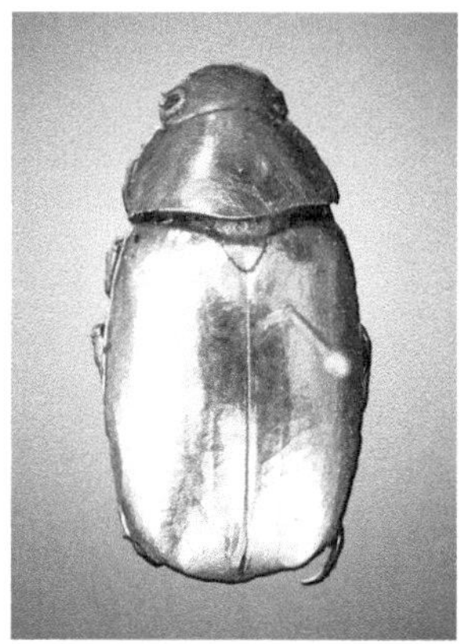

FIGURE 3.23 (See color insert.)
The golden beetle *Plusiotis resplendens*. (The beetle is kindly provided from the collections of the Zoological Museum, Copenhagen University.)

in Figure 3.23. The helicoidal structures can be seen in the cuticle of many insects as described earlier, but only a few beetles like the *Plusiotis* have the right combination of optical properties that causes this peculiar reflectance. The structures are particularly interesting from a biomimetic point of view since they are made from a single substance. In other multilayer structures at least two different substances are required.

The metallic appearance in the helical structure is caused by different sizes of the pitch down through the structure. This is analogous to the chirped multilayer structures where different layer thickness causes the broadband reflectance. The intensity of a given color increases with the number of pitches reflecting the wavelength of the color [Parker 2000, Hodgkinson et al. 2004, Neville 1975]. The helical structure also changes the polarization of the light. Many Scarabaeidae beetles reflect left circular polarized light, which makes them easy to detect in collections using a circular polarizing filter. Since the reflector only reflects the left circular polarized light and transmits the right circular, the theoretical maximum reflectance is 50%. However, *Plusiotis resplendens* is remarkable since it reflects both left and right circular polarized light, thereby enhancing the reflectivity [Caveney 1971, Neville 1975]. Figure 3.24 shows an electron micrograph of an oblique cut through the optically active reflecting layer in the outer exocuticle of *Plusiotis resplendens*. The phenomenon is explained in the following way. A layer of chitin fibrils (U) is sandwiched between two helicoidal structures (h1, h2). The thickness of the whole optical system is measured to 22 μm. Left circular polarized light (LCP) is reflected from the h1 system, whereas right circular polarized light (RCP) goes through h1 and is retarded about half a wavelength through U, which turns it left circular. This then gets reflected in h2, changed again through U to RCP, and transmitted through h1 to the surface. The whole construction thus reflects both LCP and RCP.

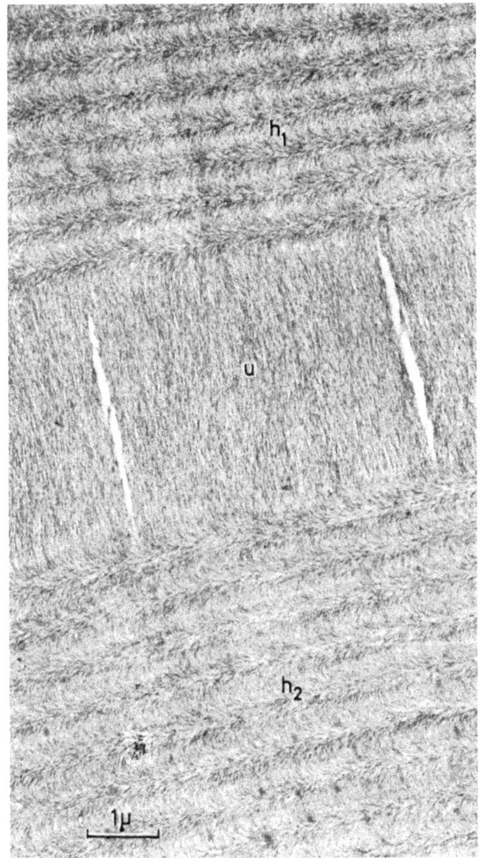

FIGURE 3.24
Electron micrograph of an oblique section cut through the optically active reflecting layer in the outer exocuticle of *Plusiotis resplendens*. (From Caveney, S., *Proc. R. Soc. Lond. B*, 178:205–225, 1971. With kind permission from the Royal Society of London.)

3.2.3.3 Metallic Appearance in Fishes

Many fish like herring, salmon, and mackerel have silvery reflecting areas on the sides, as shown in Figure 3.25 [Denton 1971, Denton and Land 1971, Fujii 2000]. The purpose of the reflectors is camouflage. When seen from the side the reflecting areas make sure that the fish will reflect light with an intensity equal to its surroundings. According to Denton [1971], it is the experience of divers that silvery fishes are very difficult to distinguish when seen from the side. Seen from above the background is typically dark, which is also the appearance of the topside of the fish. The bottom side of the fish has the poorest camouflage properties, since most fish will be clearly seen from below as silhouettes against the sunlight coming from above.

FIGURE 3.25 (See color figure at http://www.crcpress.com/product/isbn/9781439877463) Herring (*Clupea harengus*) with dark top and metallic reflecting sides.

Denton describes the reflection mechanism in the sprat *Clupea sprattus* and the herring *C. harengus* [Denton 1971, Denton and Land 1971]. The silverlike reflectors are found in the fish scales where small platelets function as multilayer reflectors that reflect light selectively through interference [Denton 1971]. The platelets each have about five sets of layers with alternating high and low refractive indices, as shown in Figure 3.26. The high index layers in the platelets are called guanine crystals and have a refractive index of around 1.8 [Fujii 1993]. They consist of the nitrogenous compound guanine ($C_5H_5N_5O$) and the related hypoxanthine ($C_5H_4N_4O$). The low index layers are cytoplasm with a refractive index of 1.33. The platelets have a size of approximately 6 × 24 μm and the thickness of the guanine crystals is about 100 nm [Vukusic 2003]. Platelets with thicker layers of cytoplasm reflect the longer wavelengths in the red end of the visible spectrum, while platelets with thinner layers reflect the shorter wavelength. Since the scales have many platelets a mixture of many different colors is reflected [Fujii 2000]. The scale therefore functions as a kind of chaotic arranged chirped multilayer reflector.

3.2.4 Metallic Reflection Using Multilayer Materials and Coatings

Polymer products with a metallic appearance are usually made by coating with a thin layer of metal by electrochemical, chemical, or vacuum-evaporation methods. The metal reflectance can be used for decorative purposes, like on water taps, automobile panels, jewelry, and household equipment, and functional purposes, like light reflectors in lamps and solar collectors. A number of advantages could be gained if the coatings on the polymer products were also made from polymers. Since polymers, in contrast to metals, are very poor conductors of heat and electricity, they are well suited for applications in hazard-proof electrical products or well-tempered handling surfaces. Other advantages are the improved recycling and disposal properties.

Weber and colleagues [2000] describe the development of flat coextruded multilayer polymer films that function as high reflective mirrors. Figure 3.27 shows a picture of such a reflecting polymer film. The films are commercially

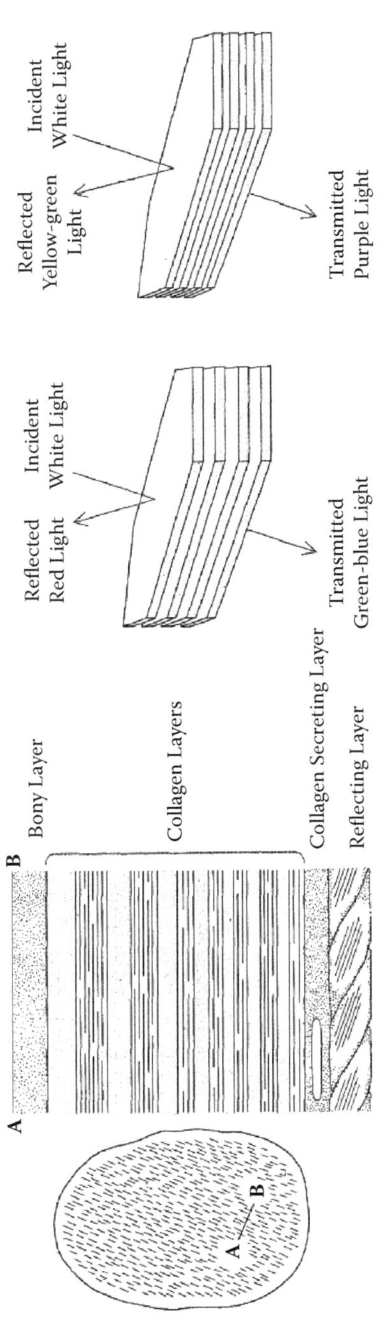

FIGURE 3.26
Fish scale from a juvenile sprat (left) showing the different layers. Length of AB is about 1 mm. The reflecting layer at the bottom holds small platelets, each with about five so-called crystals (right), i.e., sheets of high refractive guanin (shown as dark layers) separated by low index cytoplasm (shown as white layers). The thickness of the cytoplasm layers determines the reflected color, and the mixture of platelets with different thicknesses causes the broadband metallic reflectance. (Redrawn from Denton, E., *Sci. Am.*, 224:65–72, 1971.)

FIGURE 3.27
High reflective polymer film made from thin layers of polymer films. (With kind permission from 3M Corporation.)

available and can be used for piping visible light over long distances as well as decorative products and sunlight reflector applications. The mirrors can be made largely insensitive to polarization effects by controlling the Brewster angle of the p-polarized light, i.e., light with its electric field in the plane of incidence. An effect is that high reflectivity can be achieved for all reflection angles. This is done using so-called giant birefrefringent optics (GBO). The principle is that at least one of the materials in the multilayer stack is birefringent; i.e., the refractive index (n) in the plane of the layer (xy-direction) is different from the refractive index perpendicular to the plane (z-direction). The effect is achieved by choosing two materials that have the same index of refraction in the z-direction, but a different index of refraction in the layer plane. The materials used are poly(methyl methacrylate) (PMMA), having $n_z = n_{xy} = 1.5$, and a birefringent polyester having $n_z = 1.5$ and $n_{xy} = 1.8$. High reflectivity is achieved by having a large number of layers, and the broadband reflectivity is reached by using two blocks of layers with different layer thicknesses. The films are very efficient, but the manufacturing method restricts applications to flat and single curved surfaces. However, if such reflecting polymers could be applied to double-curved and free-formed shapes a wide range of applications becomes possible.

Lenau et al. [2009] describe an experiment in which a multilayer polymer coating can be made by simply dipping an object in two different polymer solutions. After a range of different polymers and solvents were tried, the selected polymers were polystyrene (PS) with a refractive index of 1.59 and polyvinyl acetate (PVAC) with a refractive index of 1.47. PS dissolves in tetrahydroforan (THF) and PVAC dissolves in acetone, and the two solvents do not dissolve the other polymer. The solvents are very volatile, causing very uneven coatings, but uniform layer thicknesses were achieved by carrying out the dipping in closed containers where the air was saturated with solvent. In this way only the surface tension forces were important for

FIGURE 3.28 (See color figure at http://www.crcpress.com/product/isbn/9781439877463)
Reflections from a double layer of PVAC (1.25%) and PS (1%) on glass substrate made by dipping in closed containers.

controlling the layer thickness. The goal was to make an ideal quarter-wave stack, and this meant that the desired layer thicknesses for reflecting 550 nm yellow light were 86 nm for PS and 94 nm for PVAC. Ordinary microscope glasses were used as a substrate. By changing the polymer concentration in the solvent the layer thickness could be controlled, and it was possible to get weak visible color reflections, as shown in Figure 3.28. Layer thicknesses were measured using a nanoscan atomic force microscope. It was not possible to apply more than two layers. Even though the polymers only dissolved in one solvent, the solvent used for the third layer affected the underlying layers—they cracked and made small bubbles. However, the experiment shows that it is possible to apply uniform layers of the right optical thickness using dipping. The challenge is to find a suitable combination of polymers and solvents for more layers.

De Silva and colleagues [2005] describe the making of an artificial optical equivalent to a beetle-like Bouligand reflector, which they refer to as a birefringent and chiral material. They mimicked the manuka scarab beetle *Pyronota festiva* that has green or red colors and reflects circular polarized light. The chiral inorganic titania material was made with a so-called serial bideposition apparatus, where the material is vacuum deposited layer by layer onto a stepwise rotating substrate.

The optical properties in Bouligand structures in beetle shells are very similar to those found in cholesteric liquid crystals (CLCs); however, a major difference is that the beetle shells are solid in contrast to the liquid CLC. For both structures the reflected part of the light is circular polarized in either a left-handed or a right-handed direction, which means that a reflector with a single-handedness only reflects up to half the light [Makow 1980].

In order to explore if metal reflections can be achieved Lenau and colleagues [2009] experimented with two types of CLC: a nematic LC Merck

MLC 6608 doped with either a left- or a right-handed dopant (MLC 6247 and MLC 6248) to get CLC. The birefringence for 6608 is $\Delta n = 0.083$ at $\lambda = 589$ nm [Rao et al. 2009]. The compounds were positioned between two glass plates and measurements were made using a halogen light source and an Elliott photospectrometer. By changing the dopants, different reflected colors were obtained ranging from red to yellow, green, and blue (see Figure 3.29). The width of the reflected spectra was around 60 nm, and the reflectivity was from 25 to 35%. Up to 50% reflectivity is theoretically possible for reflection of single-handed polarized light. Srinivasarao [1999] reports reflections of about 40%. Makow [1980] superimposes a left- and right-handed CLC and measures 85% reflectance. This is the same principle used by the *Plusiotis resplendens* beetle. Jeong and colleagues [2006] report that they have combined two layers of different handed CLC achieving more than 90% reflectivity. Lenau and colleagues [2009] also tried to superimpose a right- and left-handed CLC, as shown in Figure 3.30. The reflections from the samples seem larger from where the two samples overlap. Measurements of the samples in Figure 3.29 confirm that the combination of left- and right-handed CLC reduces the transmittance compared to two layers of left-handed CLC; i.e., they have a higher reflectance.

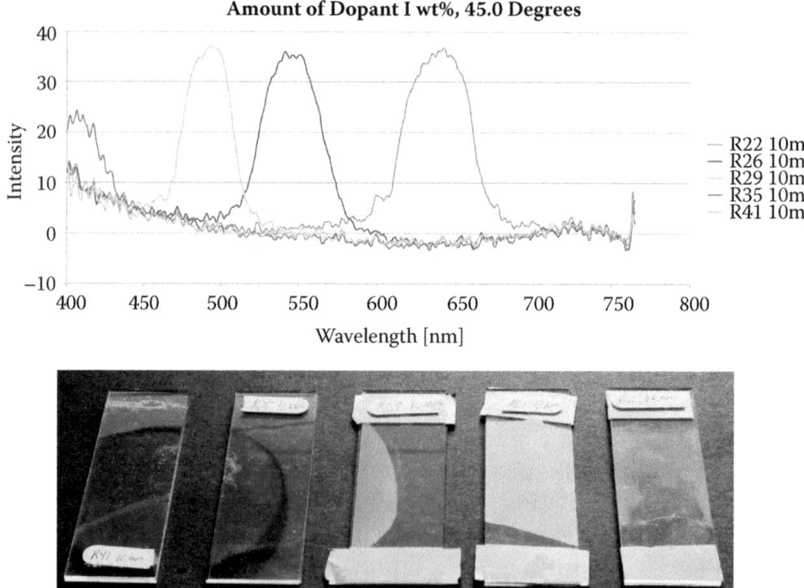

FIGURE 3.29 (See color insert.)
Reflectance spectra at 45° incidence and observation angle for Merck MLC 6608 doped with (from right to left) 22% (red curve), 26% (yellow curve with black spots), 29% (green curve), and 35% (blue curve) Merck MLC 6248. Measurements are normalized regarding light source and glass reflections. The picture shows slightly "colder" colors due to a different angle of incidence.

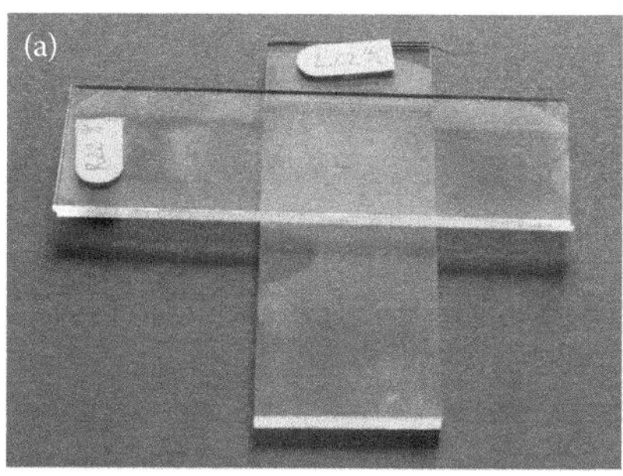

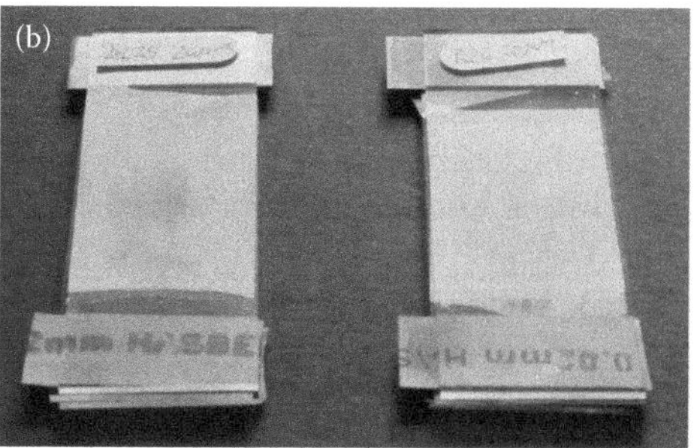

FIGURE 3.30 (See color insert.)
(a) A left-handed and a right-handed CLC placed between glass plates. Apparently the reflection looks larger where the two samples are superimposed. Samples have 22% dopant. (b) The samples are placed between three glasses. The left one has 2 × 25% left-handed dopant and the one to the right, which looks brighter, has 25% left-handed and 26% right-handed dopant.

Colorless metallic sheen requires that the whole visual spectrum be reflected. The question is if this can be done with CLC. Most papers in this field indicate that a CLC has a fairly low bandwidth—typically less than 100 nm. Mitov et al. [1999] describe the relationship between bandwidth $\Delta\lambda$ and birefringence Δn (the difference between the ordinary and the extraordinary refractive index) as $\Delta\lambda = p\ \Delta n$, where p is the pitch. Δn is normally limited to 0.3 for colorless organic materials, which sets the limit to 100 nm for $\Delta\lambda$. The solution to more broadband and metallic reflection could be to superimpose more than one CLC in order to cover the visual spectrum. Mitov and colleagues [1999] reported to have made a broadband reflector with a metallic aspect. They used an oligomer type CLC and two glass plates: one doped to become blue $\lambda = 445$ nm and another to become red $\lambda = 710$ nm. A sandwich cell is made with 20 μm spacers, and the sample is heated to 85°C, causing diffusion, and then quenched to room temperature. In this way the helix structure is maintained. The result is about 50% reflection over most of the visible spectrum, causing a colorless metallic appearance.

The viscous characteristic in CLC is an advantage when a movement of the material is desired, e.g., for temperature sensors. But surfaces in contact with humans should be wear resistant and must not stick to the fingers. Different methods for freezing or curing the CLC can therefore be applied.

One method is quenching, i.e., quick cooling from a higher temperature to a solid state at room temperature, which was independently reported by Palffy-Muhoray [1998] and Tamaoki et al. [1997]. They describe how a CLC that is cholesteric in the temperature range 87–115°C can be heated and quenched, resulting in a range of different colors. The material can be used for rewritable full-color media.

Another method is described by Zapotocky and colleagues [1999]. They describe a method by which colloidal particles with a size of about 1 μm can be used to stabilize the liquid CLC so it behaves like a solid.

References

American Chemistry Council. (2011). Plastic packaging resins. Available from http://plastics.americanchemistry.com/Education-Resources/Plastics-101/Plastics-Resin-Codes-PDF.pdf (accessed October 30, 2011).

Beckmann, P. (2011). Color and texture of beetles. Available from http://www.living-jewels.com/color.htm (accessed October 23, 2011).

Bennett, A.T.D., Cuthill, I.C., Partridge, J.C., and Lunau, K. (1997). Ultraviolet plumage colors predict mate preferences in starlings. *Proc. Natl. Acad. Sci. USA* 94:8618–8621.

Bennett, A.T.D., Cuthill, I.C., Partridge, J.C., and Maier, E.J. (1996). Ultraviolet vision and mate choice in zebra finches. *Nature*, 380(6573):433–435.

Carvalho, L.S., Knott, B., Berg, M.L., Bennett, A.T.D., and Hunt, D.M. (2011). Ultraviolet-sensitive vision in long-lived birds. *Proc. R. Soc. B*, 278:107–114.

Cathell, M.D., and Schauer, C.L. (2007). Structurally colored thin films of Ca^{2+}-cross-linked alginate. *Biomacromolecules*, 8:33–41.

Caveney, S. (1971). Cutile reflectivity and optical activity in scarab beetles: the role of uric acid. *Proc. R. Soc. Lond. B*, 178:205–225.

Denton, E. (1971). Reflectors in fishes. *Sci. Am.*, 224:65–72.

Denton, E.J., and Land, M.F. (1971). Mechanism of reflexion in silvery layers of fish and cephalopods. *Proc. R. Soc. Lond. B*, 178:43–61.

De Silva, L., Hodgkinson, I., Murray, P., Wu, Q.H., and Arnold, M. (2005). Natural and nanoengineered chiral reflectors: structural color of manuka beetles and titania coatings. *Electromagnetics*, 25:391–408.

Euro-inox. (2011). Guide to stainless steel finishes. Available from http://www.euro-inox.org/pdf/build/Finishes02_EN.pdf (accessed October 23, 2011).

Fuhrmann, T., Landwehr, S., el Rharbi-Kucki, M., and Sumper M. (2004). Diatoms as living photonic crystals. *Appl. Phys. B*, 78:257–260.

Fujii, R. (1993). Coloration and chromatophores. In Evans, D.H. (ed.), *The physiology of fishes*, 535–62. Boca Raton, FL: CRC Press.

Fujii, R. (2000). The regulation of motile activity in fish chromatophores. *Pigment Cell Res.*, 13:300–319.

Giraud-Guille, M.M., Besseau, L., and Martin, R. (2003). Liquid crystalline assemblies of collagen in bone and in vitro systems. *J. Biomechanics*, 36:1571–1579.

Heinlein, J., and Kasch, M. (1999). Pigments offer color effects matched only by nature. *Paint Coatings Ind.*, 15:58–61.

Hinton, H.E. (1973). Some recent works on the colours of insects and their likely significance. *Proc. Br. Ent. Nat. Hist. Soc.*, 6:43–54.

Hodgkinson, I.J., De Silva, L., Murray, P., Wu, Q.H., and Arnold, M. (2004). Modeling optical reflectance from chiral micro-mirrors embedded in manuka beetles. *Proc. SPIE*, 5509:15–23.

Iohara, K., Yoshimura, M., Tabata, H., and Shimizu, S. (2000). Structurally colored fibers. *Chem. Fibers Int.*, 50:38–39.

Jeong, S.M., Sonoyama, K., Takanishi, Y., Ishikawa, K., Takezoe, H., Nishimura, S., Suzaki, G., and Song, M.H. (2006). Optical cavity with a double-layered CLC mirror and its prospective application to solid state laser. *Appl. Phys. Lett.*, 89:241116.

Kikuti, E., Conrrado, R., Bocchi, N., Biaggio, S.R., and Rocha-Filho, R.C. (2004). Chemical and electrochemical coloration of stainless steel and pitting corrosion resistance studies. *J. Braz. Chem. Soc.*, 15:472–480.

Kinoshita, S., and Yoshioka, S. (2005). Structural colors in nature. A role of regularity and irregularity in the structure. *ChemPhysChem*, 6:1443–1459.

Kinoshita, S., Yoshioka, S., and Kawagoe K. (2002). Mechanisms of structural colour in the Morpho butterfly: cooperation of regularity and irregularity in an iridescent scale. *Proc. R. Soc. Lond. B*, 269:141714–21.

Lenau, T., Aggerbeck, M., and Nielsen S. (2009). Approaches to mimic the metallic sheen in beetles. *Proc. SPIE*, 7401:740107.

Lenau, T., and Barfoed, M. (2008). Colours and metallic sheen in beetle shells—a biomimetic search for material structuring principles causing light interference. *J. Adv. Eng. Mater.*, 10:299–314.

Makow, D.M. (1980). Peak reflectance and color gamut of superimposed left- and right-handed cholesteric liquid crystals. *Appl. Opt.*, 19:1274–1277.

Mitov, M., Boudet, A., and Sopéna, P. (1999). From selective to wide-band light reflection: a simple thermal diffusion in a glassy cholesteric liquid crystal. *Eur. Phys. J. B*, 8:327–330.

Nakano, N. (2011). Stainless steel coasters and spoons. Available from http://www.nakano-acl.co.jp (accessed October 23, 2011).

Neville, A.C. (1975). *Biology of the arthropod cuticle*. Springer-Verlag, Berlin.

Neville, A.C. (1977). Metallic gold and silver colours in some insect cuticles. *J. Insect Physiol.*, 23:1267–1274.

Noyes, J., Sumper, M., and Vukusic, P. (2008). Light manipulation in a marine diatom. *J. Mater. Res.*, 23:3229–3235.

Palffy-Muhoray, P. (1998). New designs in cholesteric colour. *Nature*, 391:745–747.

Parker, A.R. (2000). 515 million years of structural colour. *J. Opt. A Pure Appl. Opt.*, 2:R15–28.

Parker, A., McKenzie, D.R., and Large, M.C.J. (1998). Multilayer reflectors in animals using green and gold beetles as contrasting examples. *J. Exp. Biol.*, 201:1307–1313.

Parker-TexLoc. (2009). The TexLoc closet, hard to find information. Available from http://www.texloc.com/closet/cl_refractiveindex.html (accessed March 2009).

Parnell, A.J., Pryke, A., Mykhaylyk, O.O., Howse J.R., Adawi, A.M., Terrill, A.M., and Fairclough, J.P.A. (2011). Continuously tuneable optical filters from self-assembled block copolymer blends. *Soft Matter*, 7:3721–3725.

Prusten, M. (2008). High dynamic range image rendering of color in chameleons camouflage using optical thin films. *Proc. SPIE*, 7057:705709.

Rao, L., Gauza, S., and Wu, S.-T. (2009). Low temperature effects on the response time of liquid crystal displays., *Appl. Phys. Lett.*, 94:071112.

Raven, J.A., and Waite, A.M. (2004). The evolution of silicification in diatoms: inescapable sinking and sinking as escape? *New Phytol.*, 162:45–61.

Schmidt, J. (2006). Prey acquisition and predator avoidance by tiger beetles (Coleoptera: Cicindelidae). Available from http://www.colostate.edu/Depts/Entomology/courses/en507/papers_2001/schmidt.htm (accessed December 3, 2006).

Srinivasarao, M. (1999). Nano-optics in the biological world: beetles, butterflies, birds, and moths. *Chem. Rev*, 99:1935–1961.

Stefano, L.D., Stefano, M.D., Maddalena, P., Moretti, L., Rea, I., Mocella, V., and Rendina, I. (2007). Playing with light in diatoms: small water organisms with a natural photonic crystal structure. *Proc. SPIE*, 6593:659313.
Stuart-Fox, D., Moussalli, A., and Whiting, M.J. (2008). Predator-specific camouflage in chameleons. *Biol. Lett.*, 4:326–329.
Stuart-Fox, D., Whiting, M.J., and Moussalli, A. (2006). Camouflage and colour change: antipredator responses to bird and snake predators across multiple populations in a dwarf chameleon. *Biol. J. Linnean Soc.*, 88:437–446.
Tamaoki, N., Parfenov, A.V., Masaki, A., and Matsuda, H. (1997). Rewritable full-color recording on a thin solid film of a cholesteric low-molecular-weight compound. *Adv. Mater.*, 9:1102–5.
Tilley, R. (2000). *Colour and optical properties of materials*. Cardiff: John Wiley & Sons.
Viitala, J., Korplmäki, E., Palokangas, P., and Koivula, M. (1995). Attraction of kestrels to vole scent marks visible in ultraviolet light. *Nature*, 373:425–427.
Vukusic, P. (2003). Natural coatings. In Kaiser, N., and Pulker, H.K. (eds.), *Optical interference coatings*, 1–34. Springer, New York.
Vukusic, P., Sambles, J.R., Lawrence, C.R., and Wootton, R.J. (1999). Quantified interference and diffraction in single Morpho butterfly scales. *Proc. Biol. Sci.*, 266:1403–1411.
Weber M.F., Stover, C.A., Gilbert, L.R., Nevitt, T.J., and Ouderkirk, A.J. (2000). Giant birefringent optics in multilayer polymer mirrors. *Science*, 287:2451–2456.
Zapotocky, M., Ramos, L., Poulin, P., Lubensky, T.C., and Weitz, D.A. (1999). Particle-stabilized defect gel in cholesteric liquid crystals. *Science*, 283:209–212.

3.3 Fabrication of *Morpho* Butterfly-Specific Structural Color Aiming at Industrial Applications

Akira Saito

Colors are produced from various principles such as optical absorption (e.g., pigment), emission (e.g., light-emitting diode (LED)), interference (e.g., bubbles of soap or rainbow coloration on a compact disc), or scattering (e.g., blue sky or red sunset). The structural color is a type of coloration originating from microstructure variation at a length scale comparable to the optical wavelength. It is found in nature, for example, in pearls, jewel beetles, peacock tails, and fishes [Kinoshita and Yoshioka 2005, Berthier 2007]. This coloration is generally accompanied with a brilliant metallic luster, and has long attracted scientific interest [Hooke 1665, Newton 1730, Michelson 1911, Rayleigh 1918, Mason 1927, Anderson and Richards 1942, Ghiradella 1998]. Recently, a variety of mechanisms have been found in the nature. Some of them are related to photonic crystals [Parker et al. 2001, Ingram and Parker 2008, Saranathan et al. 2010, Liu et al. 2009, Biro and Vigneron 2011, Vukusic et al. 2003], which are a new trend in photonics.

The structural color has a variety of potential applications, because of its long-term resistance to discoloration by chemical change; furthermore,

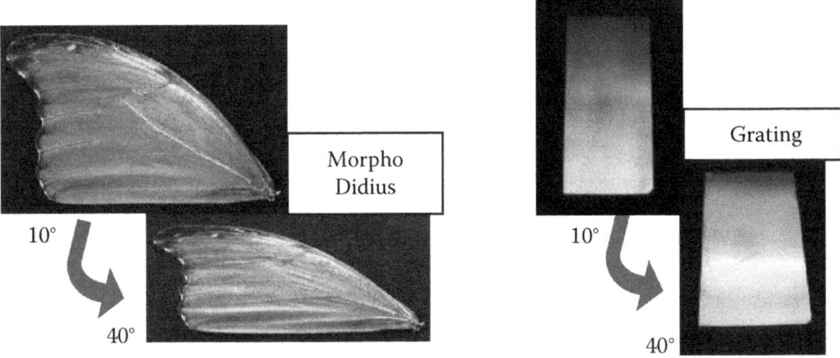

FIGURE 3.31 (See color insert.)
Angular dependences of the coloration on the viewing angle: (left) *Morpho* butterfly and (right) grating.

brilliant colors cannot be reproduced by pigments, and pigment-free coloration is preferable from an ecological viewpoint [Saito 2005]. This application-wise approach can be combined well with the biomimetic one that attracts interest because of the environmentally friendly aspect of green technology, as summarized in this chapter. Some artificial synthesis methods of structural colors have already been realized and might find industrial applications.

Many physical coloration technologies are based on thin-film or opal-like structures, which are tunable and have recently been studied for applications [Wang et al. 2011, Fudouzi 2011], but have an angular dependence of color. On the other hand, the *Morpho* butterfly has both high reflectivity (>60%) and a single color in wide angular range (>40° from the normal), which cannot be explained by interference phenomena solely (Figure 3.31). In this section, I review the biomimetic material design and fabrication techniques of *Morpho*-like colors by presenting the ideas and examples aiming toward practical applications.

3.3.1 Principles of the *Morpho* Color

The brilliant blue color of some *Morpho* butterflies has long been an important research subject [Kinoshita and Yoshioka 2005, Berthier 2007, Michelson 1911, Rayleigh 1918, Mason 1927, Anderson and Richards 1942]. The color is produced by the wing scales, which are composed of nearly transparent cuticle proteins. The principle of this phenomenon has been referred to as grating or multilayer, which also explains the high reflectivity of the blue coloration. This blue luster is not affected by chemical change and lasts for more than 100 years, and the structural color can be found even in fossils [Parker 2000].

However, the *Morpho* blue reflection spectrum cannot be explained by grating or multilayer—the color appears blue in too wide of an angular

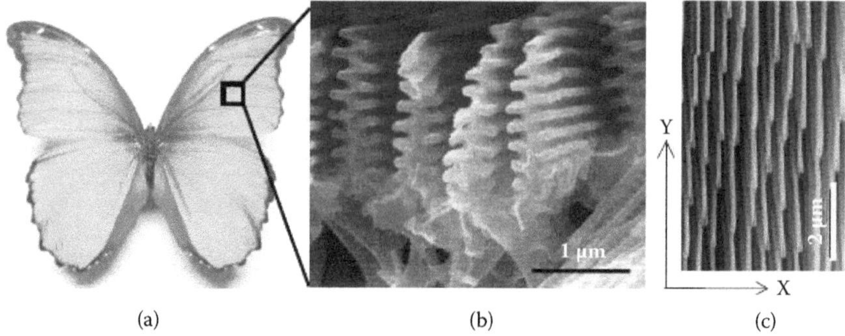

FIGURE 3.32
(a) Photograph of a *Morpho* butterfly (*Morpho didius*), and SEM images of the wing microstructure in cross-sectional (b) and top (c) views.

range. The uniformity of the color in such a wide angular range contradicts the characteristics of the interference effect. The lacking multicoloration has recently been explained by the discrete multilayer model [Kinoshita et al. 2002]. This model contains both ordered (regular) and random (irregular) structures, which form a specific architecture.

Figure 3.32 (a) shows the wing structure a typical *Morpho* butterfly (*Morpho didius*). A cross-sectional scanning electron microscopy (SEM) image (Figure 3.32 (b)) reveals a fine and complex 3D structure, with the ridges seen in a top view (Figure 3.32 (c)). This structure is characterized by one-dimensionally arrayed shelves, each composed of a multilayer. From these features, a structural model can be constructed, as shown in Figure 3.33 in cross-sectional and top views. This model can be summarized as follows [Saito et al. 2004, Saito 2005]:

1. The blue coloration is produced by interference in a single shelf, which is composed of alternating layers of high and low refractive index materials.
2. The blue color is diffracted into a wide angular range because the width of each shelf (~300 nm) is less than the wavelength of visible light.
3. The arrangement of individual domains has randomness in height (feature 3 in Figure 3.33 (a)) and in-plane structure (Figure 3.33 (b)); this randomness suppressed the multicolor (rainbow) interference and produces the speckle-like luster. The randomness in height is within one wavelength.
4. The close proximity of neighboring domains results in high reflectivity—if the gap is much wider than the wavelength, the incident light reaches the bottom of the multilayer and is absorbed or transmitted through the substrate. For high reflectivity the gap must be narrower than one wavelength (~450 nm).

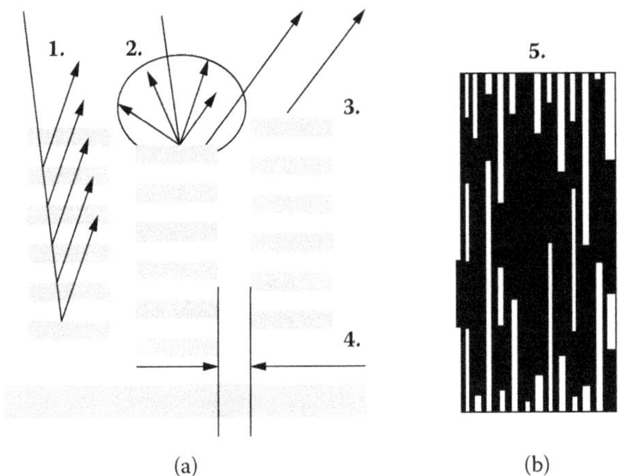

FIGURE 3.33
Schematic of the *Morpho* blue wing microstructure in (a) side and (b) top view: (1) interference in a single multilayer (shelf), (2) diffraction on narrow structure, (3) randomness in height suppresses the multicolor, (4) narrow gap results in high reflectivity, and (5) quasi-1D structure contains both the in-plane randomness and line shapes.

5. As shown in Figure 3.32 (c), the pattern extends along the Y-direction, but forms a random array of discrete linear sections. This quasi-1D anisotropy plays an important role in generating the high reflectivity in a limited angular range from the Y-direction. Otherwise, if the pattern is isotropic, the reflection is scattered two-dimensionally, decreasing critically at any viewing angle. The randomness also suppresses the multicolor interference, as mentioned above. As shown in Figure 3.33 (b), this quasi-1D pattern can be approximated as rectangular units of different heights distributed randomly with an interval of ~2 μm in the Y-direction.

3.3.2 Reproduction of *Morpho* Color in Textiles

One of the most well-known examples of direct biomimetic applications is adding luster to fibers. The color of the resulting textile material is derived from the multilayer interference of the *Morpho* butterfly, and this is the first application of the *Morpho* butterfly's coloration. Tabata fabricated a noncircular structurally colored fiber by conjugated melt spinning (Figure 3.34). The fiber has a sectional structure containing a stack of alternating layers of two polymers with different refractive indices. The unique optical characteristics of these dye-free, structurally colored fibers based on biomimetics and their application are discussed in the literature [Tabata 2005].

On the other hand, from the viewpoint of the optical principle, the small difference of the refractive indices between the two multilayer components needs a large number of layers (>50 layers) to realize a high reflectivity, and

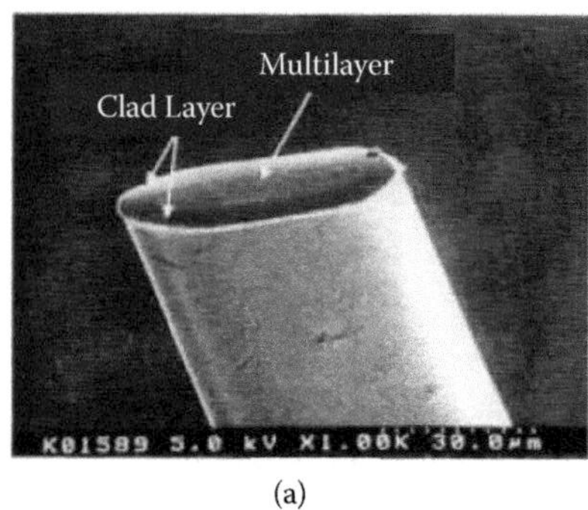

(a)

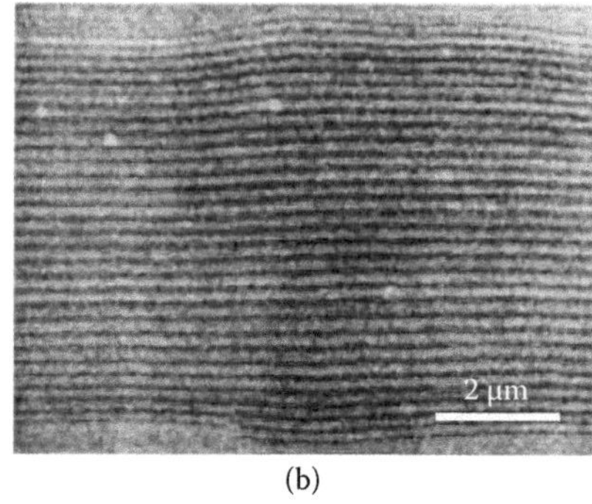

(b)

FIGURE 3.34
SEM images of a typical colored fiber (a) and its cross section (b). (From Tabata, H., in Kinoshita, S., and Yoshioka, S. (eds.), *Structural Colors in Biological Systems*, Osaka University Press, Osaka, 2005. With permission. Copyright © 2005, Osaka University Press.)

also such an ordered multilayer needs more disorder in the structure to prevent the angular dependence of the coloration.

3.3.3 Fabrication of the Nanosized Color-Producing Part

The principles discussed above originate from the disorder in the arrangement of the shelf structures, and the optical properties of a single shelf have been studied using a synthetic structure. Watanabe et al. [2005] fabricated a *Morpho* butterfly scale using focused-ion-beam chemical vapor deposition (FIB-CVD) and observed brilliant blue reflection from this quasi-structure in an optical microscope. They measured the reflection spectra from real *Morpho* butterfly scales and from the quasi-structure with a photonic multichannel spectral analyzer system, and found that they are similar [Watanabe et al. 2005].

Another approach to reproduce the specific coloration of the butterfly is to fabricate directly a replica of the butterfly scale using atomic layer deposition (ALD) [Huang et al. 2006], differently from the mimetic approach. Specifically, the structure of a *Morpho* butterfly wing scale was examined and replicated through low-temperature ALD of a uniform Al_2O_3 coating. An inverted structure was achieved by removing the butterfly wing template at high temperature, thereby forming a polycrystalline Al_2O_3 shell structure with precisely controlled thickness. The alumina replica not only copied the wing morphology, but also inherited the optical properties, such as the existence of a photonic bandgap. Reflection peaks in the violet/blue range were detected for both the original wings and their replica, while a simple alumina coating shifted the reflection peak to longer wavelengths because of the change of periodicity and refractive index. The alumina replicas also exhibited similar functional structures as waveguide and beam splitter, which may be used as the building blocks for photonic integrated circuits with high reproducibility and lower fabrication cost compared to traditional lithography techniques. Further research is being conducted in this direction [Liu et al. 2011].

3.3.4 Reproduction with Discrete Multilayers

The discussed above synthetic techniques may have potential limitations for industrial applications—FIB is costly and time-consuming, whereas ALD has a limited control and is hardly suited for mass production because it uses the mold made from the real butterfly scale.

For realistic applications, the design of the material and fabrication process should be simplified. Thus, while studying the principles on the discrete multilayer discussed above, we identified the details and principles of the *Morpho* blue butterfly wing structure. We emulated it by a dielectric multilayered nanostructure on stepped quartz, which was fabricated by electron beam (EB) lithography and dry etching, a simple and conventional technique in the semiconductor industry [Saito et al. 2004].

Specifically, a multilayer composed of alternating layers of high and low refractive index materials (Figure 3.35 (b)) was deposited on a nanopatterned surface shown in Figure 3.35 (a). The parameters of the nanopattern, i.e., the gap, width, randomness, and the quasi-one-dimensionality, were set using the model system of Figure 3.33. The most important step is then to engrave the substrate surface (Figure 3.35 (a)) by controlling the gap and width (set by parameter W in Figure 3.35 (b)), which simultaneously include randomness and quasi-1D anisotropy.

We composed a surface pattern containing randomly distributed rectangular units of 300 × (2,000 ± σ) nm (Figure 3.35 (a)), where σ = 500 nm is the standard deviation of the Gaussian distribution (i.e., W in Figure 3.35 (b) was set at 300 nm). The depth of the pattern (D in Figure 3.35 (b)) was set at

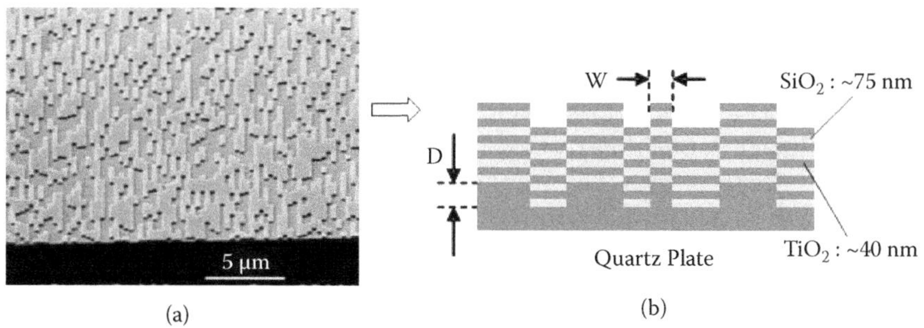

FIGURE 3.35
(a) SEM image of a quasi-1D pattern fabricated by conventional electron beam lithography on a quartz substrate before multilayer deposition. (b) Schematic of the multilayer fabricated by depositing TiO_2 and SiO_2 layers on the structure (a) to mimic the mechanisms presented in Figure 3.33.

FIGURE 3.36
SEM images on the experimental results of the reproduction processes. (a) Emulated master quartz plate before deposition of the TiO_2/SiO_2 multilayers. (b) After deposition of the TiO_2/SiO_2 multilayers on the master quartz plate.

110 nm to prevent the normal reflection of 440 nm light (blue color) from the multilayer. This depth condition helps the blue light scatter over a wide range of angles. To produce the desired surface pattern, conventional EB lithography and dry etching were applied to quartz substrates (Figure 3.35 (a)). A computer program was used to produce a random number to set the distance between the rectangular units.

The process was finalized by step-by-step electron beam-assisted deposition of seven bilayers of TiO_2 (high refractive index layer, ~40 nm thickness) and SiO_2 (low refractive index layer, ~75 nm thickness) (Figure 3.35 (b)). This simple process allowed reproducing the *Morpho* blue wing structure. Metal oxides are the best materials for the multilayer deposition because of their wide range of refractive indices and because their thickness can be accurately controlled. In all experiments, the material was designed to have the reflection maximum at 450 nm.

Figure 3.36 shows the SEM images in the top view before and after deposition of thin TiO_2/SiO_2 layers on the mold quartz substrate (Figure 3.36 (a) and (b)). The direct observation of nanopatterned surfaces before and after multilayer deposition reveals clearly that a 900 nm thick deposit with an etching depth of only 110 nm does not lose the discreteness of the substrate, and does not form a continuous flat multilayer film. The decrease of sharpness at the edge of the patterns after the multilayer deposition, which is shown in Figure 3.36 (b), is a result of the "step coverage" effect. This undesirable effect is caused by the deposition of TiO_2 and SiO_2 onto the vertical wall of the patterns. Although this effect does not affect seriously the optical properties, the decrease of this step coverage is one of the tasks in the future.

Figure 3.37 shows the cross-sectional SEM images of the fabricated multilayer structure on the original quartz plate (Figure 3.36 (a)). The images

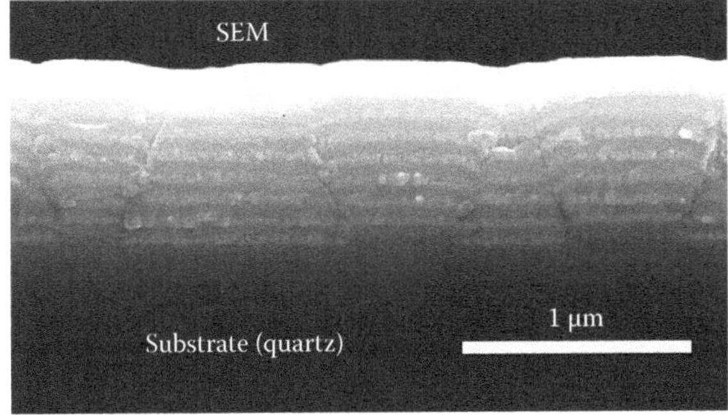

FIGURE 3.37
Cross-sectional SEM images of the reproduced *Morpho* structure. TiO_2/SiO_2 multilayers on a master quartz plate. SEM image of the replicated *Morpho* structure in cross-sectional view. A discrete multilayer is formed.

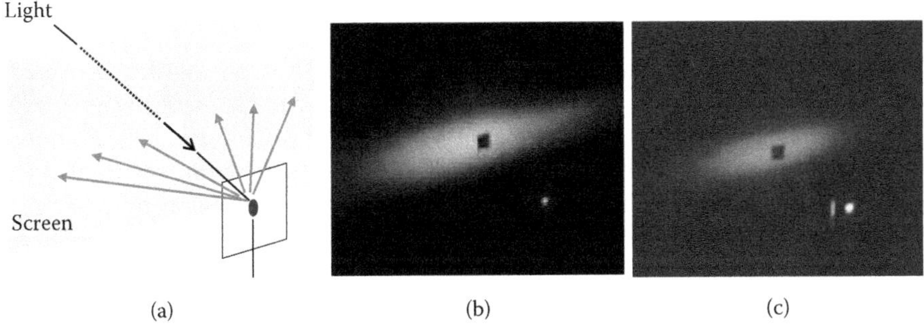

FIGURE 3.38 (See color insert.)
(a) Schematic image of the measurement system. Comparison of patterns of the lights reflected by (b) the original *Morpho* butterfly and (c) the emulated artificial film.

reveal directly the existence of the fabricated discrete multilayer. The step coverage is found to be in a permissible range from the bottom to the top layer, and the steps are successfully formed up to the top layer for both images.

The produced structure showed the basic optical reflecting properties described in Section 3.2 [Saito et al. 2004, Saito 2005]. Concretely, the fabricated film has successfully reproduced the *Morpho* blue with the fundamental characteristics: brilliant blue with high reflectivity, but a wide angular range of the single color, 1D anisotropy of the brilliance, and a smooth hue without a fringe-like sharp flash. Figure 3.38 shows the comparison of patterns of light reflected by the original *Morpho* butterfly, and the emulated artificial film.

The optical properties of the reproduced film were measured quantitatively (Figure 3.39), where an angular dependence of the reflectivity was

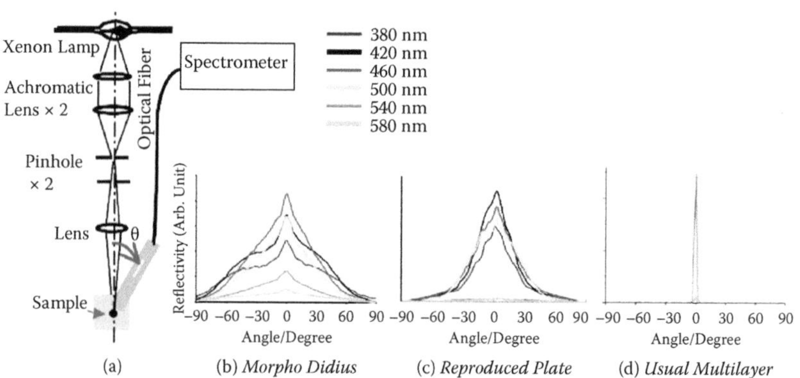

FIGURE 3.39 (See color insert.)
Comparison of measured angular dependence of the reflectivity from normal incident light. The angle was scanned along the lateral axis in Figure 3.38. (a) Schematic image of the measurement system. Angular profiles on (b) *Morpho didius*'s wing, (c) reproduced *Morpho* film, and (d) usual continuous and flat (not discrete) multilayer.

obtained along the lateral axis in Figure 3.33 (b). This axis, which is important because it takes the wide angular dispersion, corresponds to the X-axis in Figure 3.32 (c) and the lateral axis in Figure 3.33 (b). Figure 3.39 (a) shows a schematic image of the measurement setup, where white incident light shines on a sample surface from the normal direction. By scanning an optical fiber along the lateral axis, the angular profile of the reflectivity was measured. Also, a dependence of the reflectivity on various wavelengths was obtained using a spectrometer. Figure 3.39 (b)–(d) shows a comparison of the profiles from the real *Morpho* butterfly's wing, the reproduced *Morpho* film, and a usual (not discrete, but continuous) flat multilayer deposited on a flat substrate, respectively. As foreseen, the usual multilayer shows a sharp peak because of the strong interference effect. On the other hand, the real *Morpho* butterfly's profile (Figure 3.39 (b)) is broad due to the diffraction from the discrete shelves, and smooth (without fringes) due to the randomness. The reproduced *Morpho* film (Figure 3.39 (c)) shows basically common characteristics with the real *Morpho* butterfly's, as mentioned above, in Figure 3.34, far from that of the continuous flat multilayer (Figure 3.39 (d)). The blue is dominant despite the white incident light, the wide angular widths in the blue range, and the smooth angular dependence. In addition, the major wavelength component in the profile of Figure 3.39 (c) is close to the designed value (450 nm).

In order to confirm the optical role of the randomness, the nanopattern in Figure 3.40 (a) was obtained for another multilayer film, which was formed on not a quasi-, but a striped 1D pattern (SEM image is shown in inset). This film lacks randomness along the lines in comparison with a quasi-1D one, shown in Figure 3.40 (b); thus a strong interference is maintained, producing many fringes in the profile differently from the *Morpho*-type films

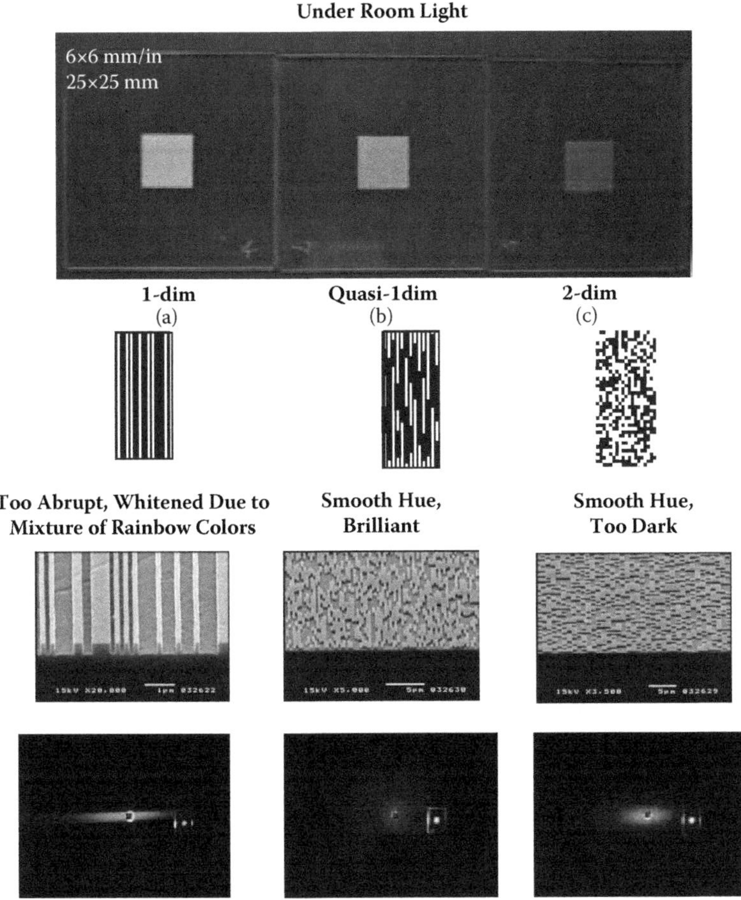

FIGURE 3.40 (See color insert.)
Comparison of the visible aspect (upper), pattern in SEM image (middle), and reflective pattern (lower) among (a) the poorly random pattern having 1D anisotropy, (b) *Morpho*-type pattern having quasi-1D anisotropy, and (c) 2D random pattern.

(Figure 3.40 (b)). Actually, in a visual image (Figure 3.40, upper row), this film looks whitened due to a mixture of reflected rainbow colors. This result indicates an important role of the randomness in a quasi-1D structure to obtain the characteristics of the *Morpho* blue having the smooth angular dependence.

Thus, the model mentioned in Section 3.2 has been experimentally verified by fabricating and testing specific nanostructures, in which several parameters (Figure 3.33 (a) and (b)) were controlled at the 100 nm scale. The basic reflection properties were reproduced, i.e., high reflectivity with a wide angular distribution, anisotropy, and speckle-like brilliance [Saito et al. 2004, Saito 2005].

3.3.5 Mass Production for Applications

The *Morpho* blue, which we have reproduced for scientific purposes to prove the principle, has many potential applications. Colors are produced without pigments, and tones that are qualitatively impossible to produce by pigments (brilliant luster with high reflectivity, and speckle-like aspects) can be realized. Moreover, this color has advantages in lifetime in comparison with conventional pigments, because it is resistant to discoloration due to chemical change even over a period of 100 years, as long as the structure is maintained [Ingram and Parker 2008]. The *Morpho* color thus has an ecological merit, because it does not need chemical pigments and can provide any color by controlling the composition or thickness of only two different materials. Thus, a variety of applications are relevant to the structural color: cosmetics, decorations, textures, paints, holograms (for security), and so on. In addition, the *Morpho*-type color shows both wide reflective angular range and high reflectivity that are usually not compatible with each other. It gives a single color by an anti-interference effect that prevents reflection of multiple colors. These three characters make structural colors ideal candidates for use in posters or displays.

As a next step, we developed a mass production process, because it is important from the industrial viewpoint that a control of the nanoscale structure is achieved on a significantly large substrate size. Thus, a mass production method was developed using nanoimprinting (Figure 3.41) [Saito et al. 2006a, 2006b]. The essence of the progress is to use a nanoimprinting lithography (NIL) technique, which enables easy replication of the nanopattern shown in Figure 3.35 (a) without using expensive EB exposure systems or excimer laser steppers, once the mold is fabricated. Furthermore, by use of

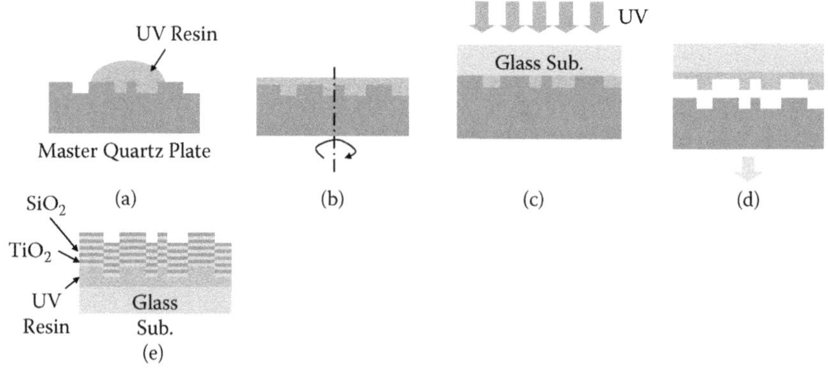

FIGURE 3.41
Proposed process to reproduce the *Morpho* blue structures. The master plate is replicated by nano casting lithography (NCL) using the UV curable resin (a)–(d), and the SiO_2/TiO_2 layers are deposited on the cured resin pattern (e). (a) Drop UV resin. (b) Spin coating. (c) A glass plate is placed and UV is exposed. (d) Release of the master plate. (e) Deposition of multilayered thin films on the replicated resin plate.

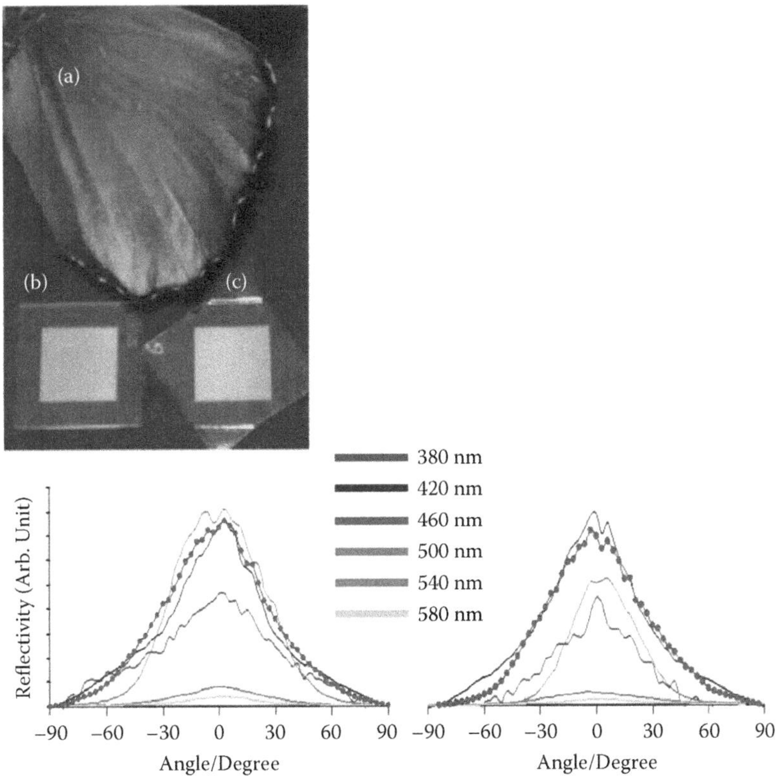

FIGURE 3.42 (See color insert.)
Photographs of (a) *Morpho* butterfly wing, (b) replicated *Morpho* blue plate, (c) synthetic *Morpho* blue plate developed by nanoimprinting for mass production (left, prototype; right, plate fabricated by nanoimprinting method for mass production). Also, corresponding optical property (reflective angular dependence) is shown.

a heat-resistant resin for NIL, we were able to directly deposit multilayers on the nanopatterned resin using EB deposition, which allowed us to leave out the dry etching process, too. As a result, the new process gave us a thousandfold throughput resulting in a low-cost production. The replicated mass-produced *Morpho* blue (Figure 3.42, lower right) was found to have basically the same nanostructures as the prototype, which was verified by AFM and SEM. The optical properties of the low-cost *Morpho* blue were found to be the same as those of the prototype artificial *Morpho* blue, which was estimated by the optical measurement shown in Figure 3.42.

It is worth noting that the area surrounding the centered bright square on this plate is composed of a usual flat multilayer. The difference between the *Morpho* film and the surrounding usual multilayer can be easily distinguished, because the latter reflects the blue in a quite limited angular range.

3.3.6 Control of Optical Properties (Angular Dispersion)

After we reproduced the basic optical properties of the *Morpho* blue and developed the low-cost high-speed replication method, still we have some milestones to pass before real-world applications can be achieved. Particularly, considering a standard of quality available in a general market, one of the needed progresses is to control the optical properties in both meanings of spatial (angular) distribution and wavelength distribution (color phase) quantitatively and precisely. In fact, the artificial film (Figure 3.39 (c)) is still wanting compared to the real *Morpho* blue (Figure 3.39 (b)): one is the angular width in the reflective profile. Although the broad angular dispersion with a single color was artificially achieved, the angular width for the artificial film is still smaller than that for the real *Morpho* blue. Another is a relationship of the angular profiles at different wavelengths. For the artificial film, the relationship of the profiles at different wavelengths is close to each other, whereas it changes gradually for the real *Morpho* blue. Thus, in the next step, we attempted to control the optical properties by changing some structural parameters in order to make the artificial *Morpho* blue closer to the natural one.

For the former problem (the angular dispersion), we have optimized successfully the structural parameters semiempirically to obtain a more natural *Morpho* blue. In this process, conventional analytical optical simulations and microstructural observations were taken in account to consider the parameters that affect the reflective properties. By combining the theoretical analysis with a comparison of a series of films fabricated with different parameters, the relationship between the structural parameters and the optical properties was investigated [Saito et al. 2007].

In brief, the parameter W (unit width and gap of a discrete shelf, shown in Figure 3.35 (b)) dominates directly the angular dispersion by the diffraction effect. This effect was verified by examining the relationship between the measured and simulated angular width by changing the parameter W. Our simulation was based on Fraunhofer diffraction from the averaged width of shelves and the spectrum derived from a designed (continuous) multilayer producing the blue coloration. In spite of its simplicity, this simulation could explain and predict well the angular width of the fabricated film. The best angular width was estimated to be given by a W of 190 nm, considering the value of full width half maximum (FWHM) of the angular profile.

Next, another parameter, D (Figure 3.35 (b)), the depth of the nanopattern, was taken in account. Since the parameter D cannot be contained in the above simulation, the influence of D to the optical properties was examined empirically by comparing films fabricated with different D values. The results on the angular profiles for various films fabricated with different W and D revealed that the aspect ratio D/W affects the quality of the multilayer film. Finally, for the optimized value W at 190 nm, the value D was optimized at 70 ~ 100 nm, by referring to the profile of the real *Morpho* blue. This analytical process could open a way to control efficiently the optical properties by feedback of the optical measurements to the nanostructure information. Not only to make the properties close to the real *Morpho* blue, such optical analysis and feedback will also serve to optimize and design arbitrary reflective properties as we like.

3.3.7 Control of Optical Properties (Spectra)

Since the coloration is determined by the multilayer composition, the spectra might be controlled by designing the multilayer. Consequently, we improved successfully the replicated spectra (Figure 3.39 (c)) closer to the original color (Figure 3.39 (b)) by designing a new multilayer composition, simultaneously maintaining the wide angular dispersion of the *Morpho* blue [Saito et al. 2009].

Hence, since we achieved the improvement of controls in both the angular dispersion and spectra, fundamental tools were obtained, not only to make an optical film close to the real *Morpho* blue, but also to optimize the reflective properties by deposition of TiO_2 and SiO_2 onto the nanopatterns. Actually, by controlling and designing the coloration with conventional optical simulation based on the multilayer interference (Figure 3.43), we have obtained successfully the primary colors (RGB) with the properties of the *Morpho* coloration (Figure 3.44).

However, we observed a slight difference of the optical properties, i.e., the coordinates in the color phase, which can be attributed to the effect of the disorder in the structure that should be taken into account at the stage of the initial optical design of the nanopattern with a single engraving depth. Thus, we took a simulation approach and attempt a theoretical and flexible method using the finite-difference time-domain (FDTD) analysis that enables us to treat directly randomness in the structures (see Chapter 6). Although the optical properties of the *Morpho* blue still contain a few unsolved questions for details, the improvement by the above-mentioned analyses and the structural viewpoint can give a clue to control precisely the color phase and total reflective properties.

3.3.8 Recent Progress for Industrial Applications

For industrial applications, several more realistic optimizations are required on the production cost and time, size, shape, and optical properties (angular dispersion, colors) of the produced nanostructures. For this purpose, we have recently developed a new process where the fabrication process was combined with FDTD simulations [Saito et al. 2009] used for generating the random structure. Actually, the functions of each part in the collective shelf structure have been discussed in references [Gralak et al. 2001, Plattner 2004, Lee and Smith 2009, Kambe et al. 2011]. For example, the role

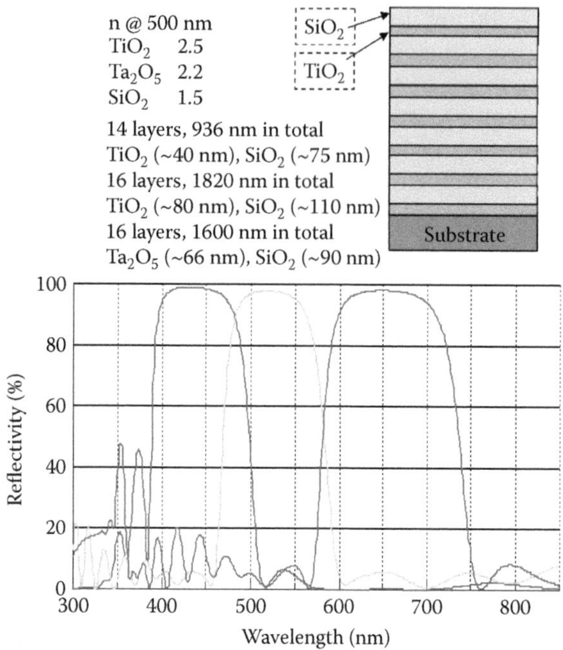

FIGURE 3.43 (See color insert.)
Optical design of the deposited multilayers that determine the artificial coloration and resultant spectra for primary colors (RGB).

FIGURE 3.44 (See color insert.)
Results of the artificial *Morpho* color in RGB primary colors.

of different kinds of disorder in the *Morpho* structure was investigated using a finite-difference time-domain (FDTD) method that enables analysis even for nonanalytical objects. The results showed an essential role of the incoherence in the incident light. Also, the lateral and vertical randomness in the structure and the number of random components were found to have their own roles in realizing the specific *Morpho* color, resulting in the anomalously low angular dependence of the color, without sharp fringes, despite an interference effect. Such simulations allow not only analyzing random optical structures but also designing their optical properties. The described above process illustrates the recent tendency in biophotonics to target not only pure science but also technology.

3.3.9 Summary and Outlook

Significant progress has been achieved recently in biomimetics and its applications. Structural colors have been explored theoretically, analyzed experimentally, and manufacturing techniques have been developed, although many details need further efforts aiming at industrial applications. Structural colors offer several promising advantages, such as the ability to produce tunable color without pigment, with a tone that cannot be achieved using pigments—brilliant luster with high reflectivity, and speckle-like reflection. The absence of pigments makes structural color environmentally friendly and resistant to fading.

The artificially reproduced *Morpho* blue has shown the basic optical properties and features of a typical *Morpho* butterfly. The artificial *Morpho* color was found to have wide potential applications because of the high efficiency of the reflection, single coloration in wide angular range, specific hue without fringes, long lifetime, saving of materials, compactness, and controllability. The reflected light has both wide angular range and high reflectivity, and this rare combination is invaluable for display applications. They can be based on robust inorganic materials that can offer strength, resistance to heat and fire, and can be produced with a wide range of refractive indices. The applications of the structural color include cosmetics, decorations, dressing and furnishings, textures, paints, holograms (security), posters, and displays. However, they are hindered by the development of fabrication processes and theoretical design.

For applications, first a mass production process using the NIL technique was developed, and next the optical properties on the angular dispersion and spectra were successfully optimized by controlling the parameters in the nanostructures. In the analysis, a simple simulation was attempted and found to be available to explain the reflective properties, whereas more a flexible method such as FDTD analysis is under progress to take the randomness in account. These analytical processes can open a way to control efficiently the optical properties by feedback of the optical measurements to the nanostructure information. A further research step is to develop a process to fabricate a substrate-free thin color film with areas larger than 1 m^2, which is in progress (Figure 3.45).

Technology demands high-throughput, inexpensive processes. As shown in this section, conventional processes such as advanced lithography, wet processes, etc., which have long been developed in the semiconductor, electronic, or optical industries, were applied to the generation of structural colors, combined with new technologies such as nanoimprinting. However, dynamic control of the structural color remains problematic, and this hinders many display applications that require rapid refresh rates

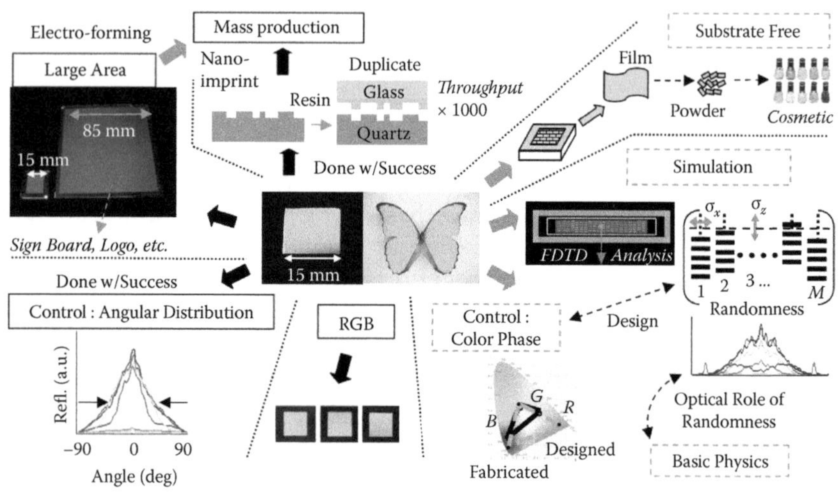

FIGURE 3.45 (See color insert.)
Recent progress (solid frame is accomplished, and dotted frame is in progress).

(more than 50 Hz). This problem may be solved using novel technologies. While the future of display applications will depend on the improvement of the dynamic properties, static structural colors for use in posters, paints, decorations, and cosmetics currently appear as the most promising application area.

Studying nature can provide solutions to the problems in realizing the structural color, and its new types have been found recently [Ingram and Parker 2008] owing to the rising popularity of biomimetics research. New biomimetic approaches are being developed. Although most of these approaches are still at the exploratory stage, they might bring new ideas to existing applications. For example, solar cells can be modified not only with the moth-eye structure, but also with the *Morpho* color, as the solar cell will absorb a wider range of angles and wavelengths of incident light if a *Morpho* colored film is placed at its bottom. Such considerations extend beyond optoelectronic devices; for example, the *Morpho* structure has been applied to gas sensors [Poryrailo et al. 2007] and water-repelling surfaces [Byun et al. 2009], as well as to temperature regulation [Berthier 2005]. Such multifunctions on the "emergence" will provide us with novel ideas, in addition to those presented in this review (Figure 3.46).

The origin of the research on the structural color goes back to the 17th century [Hooke 1665, Newton 1730], but recent developments of the nanotechnology have given new progress to this field. Although this section is focused on the application studies, this field contains many biologically worthy points [Saito 2002]. The topics on the *Morpho* color fit within the field of biomimetics, which also includes ecological importance, and thus are expected to have much importance and development in the near future.

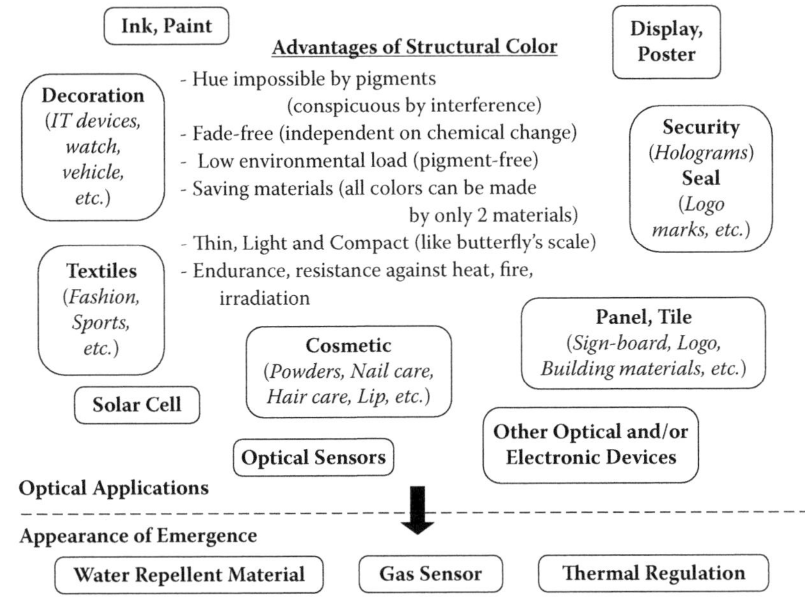

FIGURE 3.46
Advantages of biomimetic structural coloration and potential applications.

References

Anderson, T.F., and Richards Jr., A.G. (1942). An electron microscope study of some structural colors in insects. *J. Appl. Phys.* 13:748–758.

Berthier, S. (2005). Thermoregulation and spectral selectivity of the tropical butterfly *Prepona meander*: a remarkable example of temperature auto-regulation. *Appl. Phys. A* 80:1397–1400.

Berthier, S. (2007). *Iridescence*. Springer Science+Business Media, New York.

Biro, L.P., and Vigneron, J.P. (2011). Photonic nanoarchitectures in butterflies and beetles: valuable sources for bioinspiration. *Laser Photonics Rev.* 5:27–Byun, D., Hong, J., Saputra, Ko, J.H., Lee, Y.J., Park, H.C., Byun, B.K., and Lukes, J.R. (2009). Wetting characteristics of insect wing surfaces. *J. Bionic Eng.* 6:63–70.

Fudouzi, H. (2011). Tunable structural color in organisms and photonic materials for design of bioinspired materials. *Sci. Technol. Adv. Mater.* 12:064704.

Ghiradella, H. (1998). Hairs, bristles, and scales in microscopic anatomy of invertebrates. In Harrison, F.N., and Locke, M. (eds.), *Insecta*, 11A. Wiley-Liss, New York.

Gralak, B., Tayeb, G., and Enoch, S. (2001). Morpho butterflies wings color modeled with lamellar grating theory. *Opt. Express* 9:567–578.

Hooke, R. (1665). *Micrographia*. Martyn and Allestry, London.

Huang, J.Y., Wang, X.D., and Wang, Z.L. (2006). Controlled replication of butterfly wings for achieving tunable photonic properties. *Nano Lett.* 6:2325–2331.

Ingram, A.L., and Parker, A.R. (2008). A review of the diversity and evolution of photonic structures in butterflies, incorporating the work of John Huxley (the Natural History Museum, London from 1961 to 1990). *Phil. Trans. R. Soc. B* 363:2465–2480.

Kambe, M., Zhu, D., and Kinoshita, S. (2011). Origin of retroreflection from a wing of the Morpho butterfly. *J. Phys. Soc. Jpn.* 80:054801.

Kinoshita, S., and Yoshioka, S. (eds.). (2005). *Structural colors in biological systems*. Osaka University Press, Osaka.

Kinoshita, S., Yoshioka, S., and Kawagoe, K. (2002). Mechanisms of structural color in the Morpho butterfly: cooperation of regularity and irregularity in an iridescent scale. *Proc. R. Soc. Lond. B* 269:1417–1422.

Lee, R.T., and Smith, S.G. (2009). Detailed electromagnetic simulation for the structural color of butterfly wings. *Appl. Opt.* 48:4177–4190.

Liu, F., Dong, B.Q., Liu, X.H., Zheng, Y.M., and Zi, J. (2009). Structural color change in longhorn beetles *Tmesisternus isabellae*. *Opt. Express* 17:16183–16191.

Liu, F., Liu, Y.P., Huang, L., Hu, X.H., Dong, B.Q., Shi, W.Z., Xie, Y.Q., and Ye, X. (2011). Replication of homologous optical and hydrophobic features by templating wings of butterflies *Morpho menelaus*. *Opt. Commun.* 284:2376–2381.

Mason, C.W. (1927). Structural colors in insects. *J. Phys. Chem.* 31:321–354.

Michelson, A.A. (1911). On metallic colouring in birds and insects. *Phil. Mag.* 21:554–566.

Newton, I. (1730). *Opticks*. 4th ed., reprinted. Dover, New York.

Parker, A.R. (2000). 515 million years of structural colour. *J. Opt. A* 2:R15–28.

Parker, A.R., McPhedran, R.C., McKenzie, D.R., Botten, L.C., and Nicorovici, N.-A.P. (2001). Aphrodite's iridescence. *Nature* 409:36–37.

Plattner, L. (2004). *J. R. Soc. Interface* 1:49.

Poryrailo, R.A., Ghiradella, H., Vertiatchikh, A., Dovidenko, K., Cournoyer, J.R., and Olson, E. (2007). *Morpho* butterfly wing scales demonstrate highly selective vapour response. *Nat. Photon.* 1:123–128.

Rayleigh, J.W.S. (1918). On the optical character of some brilliant animal colours. *Phil. Mag.* 37:98–111.

Saito, A. (2002). Mimicry in butterflies: microscopic structure. *Forma* 17:1–9.

Saito, A. (2005). Reproduction of Morpho-blue by artificial substrate. In Kinoshita, S., and Yoshioka, S. (eds.), *Structural colors in biological systems*. Osaka University Press, Osaka, pp. 287–295.

Saito, A., Ishikawa, Y., Miyamura, Y., Akai-Kasaya, M., and Kuwahara, Y. (2007). Optimization of reproduced Morpho-blue coloration. *Proc. SPIE* 6767:1–9.

Saito, A., Miyamura, Y., Ishikawa, Y., Murase, J., Akai-Kasaya, M., and Kuwahara, Y. (2009). Reproduction, mass production, and control of the Morpho butterfly's blue. *Proc. SPIE* 7205:1–9.

Saito, A., Miyamura, Y., Nakajima, M., Ishikawa, Y., Sogo, K., Akai-Kasaya, M., Kuwahara, Y., and Hirai, Y. (2006b). Reproduction of the Morpho blue by nano casting lithography. *J. Vac. Sci. Technol. B* 24:3248–3251.

Saito, A., Miyamura, Y., Nakajima, M., Ishikawa, Y., Sogo, Y., and Hirai, Y. (2006a). Morpho-blue reproduced by nanocasting lithography. *Proc. SPIE* 6327:1–9.

Saito, A., Yoshioka, S., and Kinoshita, S. (2004). Reproduction of the Morpho butterfly's blue: arbitration of contradicting factors. *Proc. SPIE* 5526:188–94, and references therein.

Saranathan, V., Osuji, C.O., Mochrieb, S.G.J., Noh, H., Narayanan, S., Sandy, A., Dufresne, E.R., and Prum, R.O. (2010). Structure, function, and self-assembly of single network gyroid (I4132) photonic crystals in butterfly wing scales. *Proc. Natl. Acad. Soc. USA* 107:11676–11681.

Tabata, H. (2005). Structurally colored fibers and applications. In Kinoshita, S., and Yoshioka, S. (eds.), *Structural colors in biological systems*. Osaka University Press, Osaka. pp. 297–308.

Vukusic, P., and Sambles, J.R. (2003). Photonic structures in biology. *Nature* 424: 852–855.

Wang, J., Zhang, Y., Wang, S., Song, Y., and Jiang, L. (2011). Bioinspired colloidal photonic crystals with controllable wettability. *Acc. Chem. Res.* 44:405–415.

Watanabe, K., Hoshino, T., Kanda, K., Haruyama, Y., and Matsui, S. (2005). Brilliant blue observation from a *Morpho*-butterfly-scale quasi-structure. *Jpn. J. Appl. Phys.* 44:L48–50.

4

Photomechanic IR Receptors in Pyrophilous Beetles and Bugs

Herbert Bousack, Helmut Budzier, Gerald Gerlach, and Helmut Schmitz

CONTENTS

4.1 *Melanophila* 117
4.1.1 Structure and Function of the Sensilla 118
4.1.2 Sensitivity of the IR Receptor of the Beetle *Melanophila acuminata* 119
4.2 Principle of a Sensor Based on the IR Receptor 121
4.2.1 Sensor Model 121
4.2.2 Pressure Increase of the Cavity and Deflection of the Membrane 122
4.3 Noise in Golay Cells and in the Photomechanic IR Receptor 128
4.3.1 Noise Sources in Golay Cells 128
4.3.1.1 Temperature Fluctuation Noise 129
4.3.1.2 Mechanical-Thermal Noise 129
4.3.1.3 Readout Noise 130
4.3.1.4 Total Noise 131
4.3.2 Detectivity of Golay Cells 131
4.3.3 Noise Sources in IR Receptors 132
4.4 Measurement of IR Radiation below the Noise Level 132
4.5 Influence of the Fluid in the Cavity 133
4.6 Experimental Results and Devices 135
References 137

4.1 *Melanophila*

Buprestid beetles of the genus *Melanophila* inhabit the palaearctic as well as the nearctic parts of the world. As a special behavioral feature, beetles of both sexes approach forest fires because their brood depends on burnt wood as larval food (Champion 1909, Linsley 1943). Initially, the fresh burnt area serves as a meeting place where the males search for females. After mating, the females deposit the eggs under the bark of the burnt trees. However, the outbreak of a forest fire is highly unpredictable. Because finding a fire is crucial for the survival of all *Melanophila* species, *Melanophila* beetles should be able to detect fires from distances as large as possible. Therefore, it is reasonable to suppose that the sensory organs that are used for fire detection have been subjected to a strong evolutionary pressure especially with regard to sensitivity. For the detection of fires, *Melanophila* beetles are equipped with special infrared (IR) receptors (Evans 1964, 1966, Schmitz and Bleckmann 1997).

4.1.1 Structure and Function of the Sensilla

The IR receptors are situated in two pit organs (Figure 4.1) that are located on the thorax below the wings. Each IR organ houses about 70 IR receptors, which are closely packed together at the bottom of the pit (Schmitz et al. 2007). From the outside, a single receptor (sensillum) can be recognized by a hemispherical dome with a diameter of about 12–15 μm. The dome is built by a thin cuticle that is the outer boundary of a spherical internal cavity. The cavity is almost completely filled out by a tiny cuticular sphere with a diameter of about 10 μm. Thus, the sphere is enclosed in an inner round cavity and from below the innervation is accomplished by a single dendrite of a ciliary mechanosensitive cell. Two important functional components of the sphere have been identified (Figure 4.2): (1) an outer exocuticular shell reinforced by helicoidally arranged layers of chitin fibers, and (2) a microfluidic compartment inside the shell consisting of a spongy mesocuticular layer and a distinct inner pressure chamber that is located in the lower region of the sphere. The microcavities and nanocanals in the intermediate layer communicate with the lumen of the inner pressure chamber and all cavities of the microfluidic compartment are filled with a fluid. The tip of the mechanosensitive dendrite is situated in the pressure chamber. The innermost

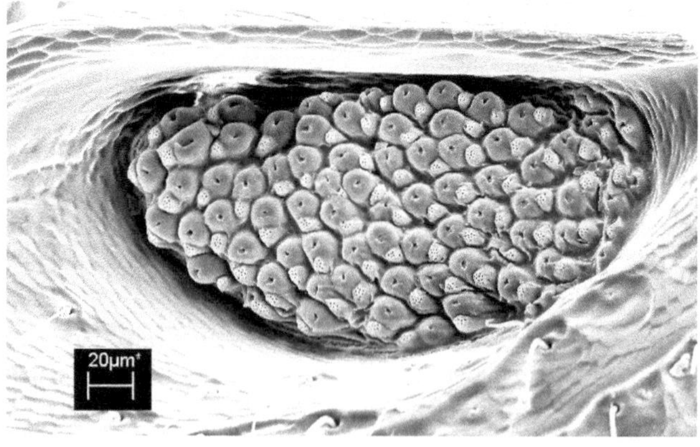

FIGURE 4.1
IR pit organ (size about 0.3 × 0.2 mm) of *Melanophila acuminata* containing about 70 IR receptors.

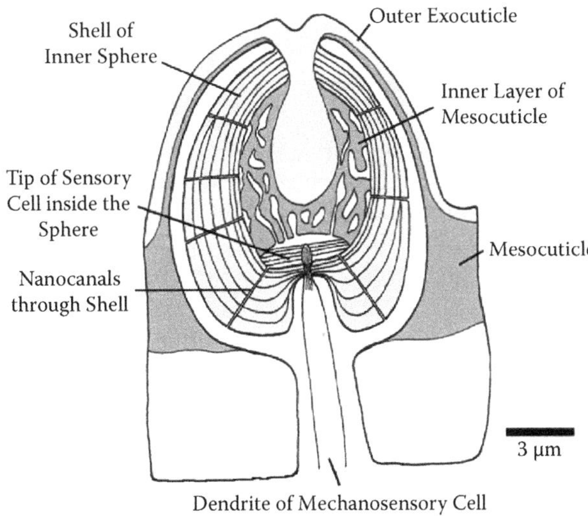

FIGURE 4.2
Reconstruction of an IR receptor. The tip of the mechanosensitive cell is enclosed in the inner pressure chamber inside the internal sphere.

core of the sphere consists of a pear-shaped plug of massive exocuticle. Additionally, about 50 tiny nanocanals radially run through the outer shell. The diameter of these canals is less than 100 nm. It can be hypothesized that these canals may serve for compensation of low-frequency temperature changes.

According to the proposed photomechanic model of IR reception, IR radiation is absorbed by the biopolymers of the cuticle and the fluid (most probably water). Due to heating and thermal expansion, especially of the fluid, an increase in pressure takes place. The fluid transfers the pressure to the tip of the dendrite, resulting in a deformation of the dendritic membrane. Thus, the opening of stretch-activated ion channels is induced. To validate this model, it has to be postulated that the outer shell functions as a pressure chamber and maintains an almost constant volume.

4.1.2 Sensitivity of the IR Receptor of the Beetle *Melanophila acuminata*

Until now the sensitivity of the IR receptor was studied by behavioral experiments (Evans 1964, 1966) or electrophysiological experiments (Hammer et al. 2001, Schmitz and Bleckmann 1998). However, the results often were contradictory and vary between 0.6 W/m^2 (Evans 1966) and 5 W/m^2 (Schmitz and Bleckmann 1998). The calculation of detection distances for forest fires results in distances between 30 km and more (Schmitz et al. 2009) and 5 km (Evans 1966), whereby these results can hardly be compared due to different assumptions with respect to the emissive powers of the forest fire.

In Schmitz and Bousack (2012) an alternative approach was investigated based on an observation of van Dyke (1926), who reported that in August 1925, due to burning oil storage tanks in Coalinga, California, "untold numbers" of *Melanophia consputa* appeared at the fire place. It is of particular interest that within a range of 50 to 100 miles (80 to 160 km) distance from Coalinga no forest suitable for *Melanophia* beetles is situated. This makes it possible to calculate the radiant flux of the fire at these distances and to estimate the sensitivity of the IR receptor.

However, the calculation requires an exact fire model of an oil fire and the determination of the damping of the signal in the atmosphere due to absorption and scattering. It is favorable that quantitative methods are available for the fire safety and risk analysis of buildings or technical equipment as well as for fire scenarios from burning liquid fuels in gas or oil storage tanks in Beyler (2002) and Iqba and Salley (2004). The radiant heat flux could be calculated as a function of the distance using several established fire models. The place of the fire in Coalinga and the place where the beetles must have started their flight toward Coalinga could be identified, resulting in a detection distance of 80 miles (130 km). For the calculation of the damping of the radiant heat flux in the atmosphere, weather data of August 1925 and long-term averages of this month were used. The calculations showed that within the different phases of an oil tank fire the radiant heat flux in a distance of 80 miles varies between to $4 \cdot 10^{-5}$ W/m^2 and $3 \cdot 10^{-4}$ W/m^2. These values are remarkably lower than the sensitivities determined earlier.

These values must be compared with the biological limit of the sensitivity of hair cells as precursors of IR receptors. The deformation of the tip of the dendrite in a hair cell of the cricket *Acheta domestica* of about 0.1 nm produces an action potential. The energy required for this deformation is about 10^{-19} J (Dettner and Peters 1999). In (Shimozawa et al. 2003) even a minimum energy from 10^{-20} to 10^{-21} J is estimated for hair lengths between 100 and 1,000 μm. A simple estimation yields the radiant heat flux $\Phi_{min,bio}$ that is necessary for such an energy increase in the sensillum:

$$\Phi_{min,bio} = \frac{E_{min,bio}}{A_S \cdot t_R} \tag{4.1}$$

with $E_{min,bio}$ the minimal energy to produce an action potential, A_S the cross section of the target (active part of the sensillum), and t_R the response time of the sensillum.

Assuming a cross section A_S of the sensillum of $8 \cdot 10^{-11}$ m^2 (10 μm diameter) and a response time of 3–4 ms until an action potential is produced, Equation (4.1) yields a minimal radiant heat flux of $4 \cdot 10^{-7}$ W/m^2 to $3 \cdot 10^{-9}$ W/m^2. This means that, based on the Coalinga fire, the calculated sensitivities of the IR receptor are several orders of magnitude above the biological limit. The physical limit of the sensitivity due to temperature fluctuation noise is discussed in Section 4.4.

4.2 Principle of a Sensor Based on the IR Receptor

4.2.1 Sensor Model

In order to understand the principle of a technical IR sensor based on the IR sensillum, the system is analyzed mathematically. For this purpose the simple setup of a pneumatic Golay cell detector (Golay 1947) is used (see Figure 4.3(a)). This sensor consists of an internal gas-filled cavity, which is closed on one side by a window and on the other side by a thin membrane. IR radiation enters through the window and heats up the gas by absorption. The deflection of the membrane caused by the expanding gas can be read out by an optical system (Golay 1947), a capacitive detector (Chevrier et al. 1995),

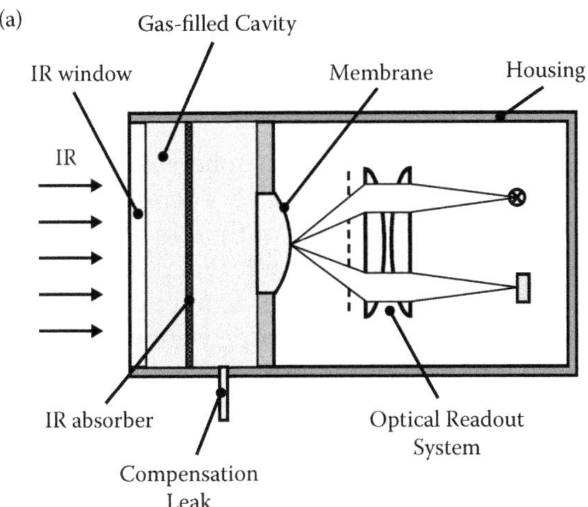

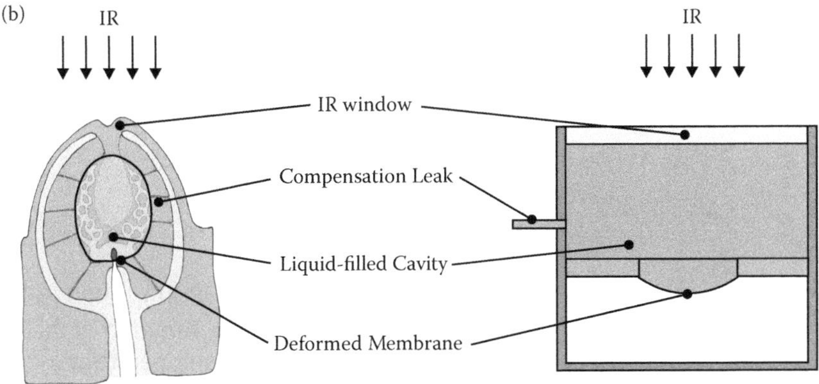

FIGURE 4.3
(a) Principle and (b) model of a gas-filled Golay sensor with optical readout. For explanations see text.

or a tunneling displacement transducer (Kenny et al. 1996). To enhance the IR absorption in the gas the cavity is equipped with an additional absorber, e.g., a thin plastic mesh. Reflecting walls of the cavity are another means to enhance absorption. The temperature changes of the gas caused by the absorbed IR radiation are in the milli-Kelvin range or lower. Because of slow variations of the ambient temperature in the range of a few degrees, it is necessary to integrate a leak, which compensates this influence by an exchange of the gas with a reference volume or a reference pressure.

The model of a sensor based on an IR sensillum is shown in Figure 4.3(b). Similar to the sensillum, the sensor contains an internal water-filled cavity. The different absorber, a liquid instead of a gas, is a major difference of the sensillum compared to the Golay cell detector. The cavity of the technical sensor is etched into a silicon wafer and is closed on one side by a window and on the other side by a thin silicon membrane that acts as a bending plate. The IR radiation being absorbed produces a change in pressure or volume, respectively, of the fluid. The deflection of the membrane caused by this pressure increase can be read out by, e.g., a capacitive detector or a tunneling displacement transducer.

In the *Melanophila* IR receptor the inner sphere is enclosed by a thin layer of fluid, which is always at ambient pressure. Therefore, the nanocanals in the shell of the sphere allow the exchange of fluid in and out of the microfluidic compartment of the sphere. Thus, any internal pressure change that may be caused by the slowly changing ambient temperature can be compensated. Golay cell detectors also use such compensation leaks for compensating changes of ambient temperatures (Golay 1947).

4.2.2 Pressure Increase of the Cavity and Deflection of the Membrane

For calculating the change of the liquid pressure in the cavity based on the temperature profile, the corresponding equation of state must be solved. Because the pressure in the cavity depends upon the two independent variables temperature and volume, its total change is given by

$$\Delta P = \beta \cdot \Delta T_{C,mean} - \frac{1}{\kappa \cdot V} \cdot \Delta V \qquad \Delta T_{C,mean} = \frac{\int_0^{H_C} T_C(z) \cdot dz}{H_C} - T_0 \tag{4.2}$$

where ΔP is the pressure increase in the cavity, $\beta = (\partial P/\partial T)_V$, the isochoric tension coefficient, $T_C(z)$ is the temperature increase as a function of the axial coordinate z of the cavity and time, $\Delta T_{C,mean}$ is the mean temperature increase averaged over the cavity, T_0 is the initial, ambient temperature at $t = 0$, H_C is the height of the cylindrical cavity, $\kappa = -(1/V) \cdot (\partial V/\partial P)_T$ is the isothermal compression coefficient, V is the volume of the cavity, and ΔV is the volume increase. For water (25°C, 1 bar) $\beta = 5.68 \cdot 10^5$ Pa/K and $\kappa = 4.5 \cdot 10^{-10}$ 1/Pa (Weast 1988).

The increase ΔV of the volume of the cavity results in a tiny deflection of the membrane. The local deflection of this membrane caused by a pressure difference can be calculated as a function of the radial distance using the shell theory (Szabo 2000, Timoshenko and Woinowsky-Krieger 1959), which results in a linear relation between deflection and pressure increase ΔP in the cavity (Klocke et al. 2011):

$$\Delta P = \frac{\beta \cdot \Delta T_{C,mean}}{1+\Omega} \qquad \Omega = \frac{R^4 \cdot (1-v^2)}{16 \cdot E \cdot t_P^3 \cdot H_C \cdot \kappa} \tag{4.3}$$

with R the radius of the membrane, E the Young's modulus of elasticity, t_P the thickness of the membrane, and ν the Poisson's ratio.

The factor Ω characterizes the change of state of the liquid inside the cavity due to a temperature increase: for $\Omega \to 0$, which corresponds to an extremely hard membrane, the change of state is isochoric with a maximal pressure increase. For $\Omega \to \infty$, as for an extremely soft membrane, the change of state is isobaric with maximal volume increase. The transition between these two cases takes place at $\Omega \approx 1$. The maximal deflection y_{max} of the circular membrane can be calculated as a function of Ω (Schmitz et al. 2011):

$$y_{max} = \frac{3 \cdot \alpha \cdot \Omega \cdot \Delta T_{C,mean} \cdot H_C}{1+\Omega} \tag{4.4}$$

with $\alpha = \beta \cdot \kappa$ the isobaric thermal expansion coefficient.

Figure 4.4 shows the maximal deflection as a function of Ω according to Equation (4.4) for an IR power density of 10 W/m² (IR window without

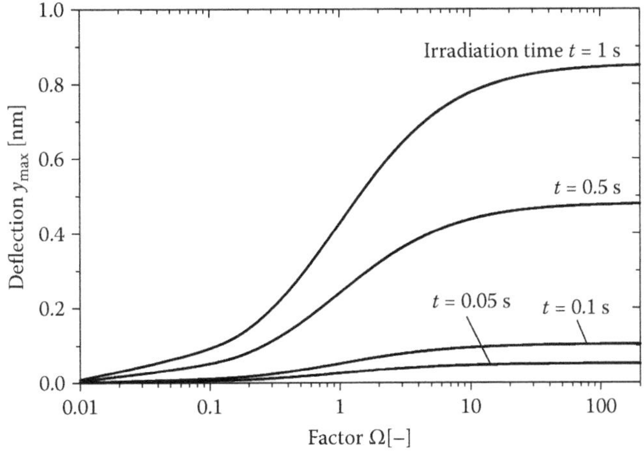

FIGURE 4.4
Maximum central deflection y_{max} of a circular membrane (silicon, thickness 1 µm, diameter 0.5 mm) as function of factor Ω and irradiation time t for a water-filled adiabatic cavity without compensation leak. IR radiation flux 10 W/m², diameter and height of the cavity 0.5 mm each.

absorption loss assumed) for an adiabatic cavity (height and diameter of 0.5 mm each) filled with water. Obviously, the maximal deflection is achieved for the isobaric case with a very soft membrane. Due to the very low deflection of less than 1 nm the readout capacitor has to be extremely sensitive. As an alternative an optical readout or the use of a tunneling displacement transducer may be useful.

The pressure increase in the cavity due to an ambient temperature increase will be considerably larger than the pressure increase due to the absorbed IR radiation. The compensation of ambient temperature changes with an artificial leak or canal acts like a high-pass filter and ensures that the membrane works within its linear range, which means that the deflection depends linearly on the applied pressure. For simulating this effect, the model in Figure 4.5 was analyzed mathematically.

A mass and energy balance for the subsystem's cavity, canal, and reservoir yields two differential equations for the pressure in the cavity and the reservoir for the nonadiabatic cavity (Schmitz et al. 2011). The heat loss is assumed as heat conduction through the cavity walls.

The solutions for the pressure P_C in the cavity and P_R in the reservoir, respectively, are:

$$P_C(t) = P_{C0} + \frac{1}{1+A} \cdot \left\{ \frac{\gamma \cdot \Theta}{1+\Omega} \cdot \left(1 - A \cdot \frac{\tau}{\Theta - \tau}\right) \cdot \left(1 - e^{\frac{t}{\Theta}}\right) - A \cdot (\Delta P_0 - B \cdot \tau) \cdot \left(1 - e^{\frac{t}{\tau}}\right) \right\}$$

$$P_R(t) = P_{R0} + \frac{1}{1+A} \cdot \left\{ (\Delta P_0 - B \cdot \tau) \cdot \left(1 - e^{\frac{t}{\tau}}\right) - B \cdot \Theta \cdot \left(1 - e^{\frac{t}{\Theta}}\right) \right\} \tag{4.5a}$$

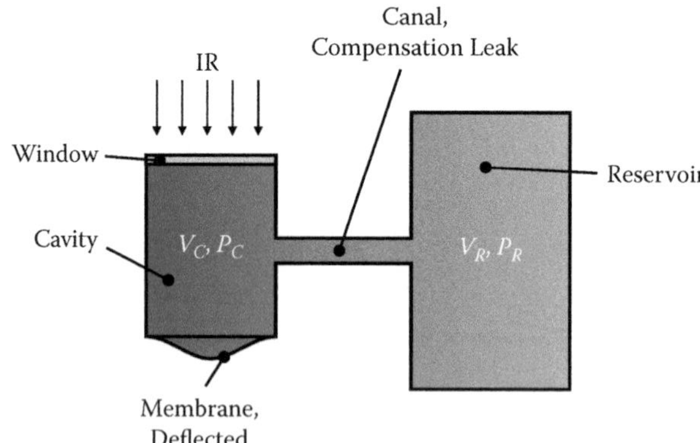

FIGURE 4.5
Model of the IR sensor with cavity (pressure P_C, volume V_C) and reservoir (pressure P_R, volume V_R). R_L and L are the radius and length of the canal of the compensation leak, respectively.

In Equation (4.5a) the following abbreviations have been used:

$$\tau_C = \frac{8 \cdot \eta \cdot L \cdot V_C \cdot \kappa}{\pi \cdot R_L^4}, \qquad \tau_R = \frac{8 \cdot \eta \cdot L \cdot V_R \cdot \kappa}{\pi \cdot R_L^4}$$

$$\Theta = \frac{\rho \cdot cp \cdot R^2}{6 \cdot \lambda}, \qquad \gamma = \frac{\beta \cdot I_0}{\rho \cdot cp \cdot H_C} \tag{4.5b}$$

$$A = \frac{\tau_R}{\tau_C \cdot (1+\Omega)}, \qquad B = \frac{\gamma \cdot \Theta}{(\Theta - \tau) \cdot (1+\Omega)}$$

$$\Delta P = P_{C0} - P_{R0}$$

Here t is time; $P_{C0} = P_C(t = 0)$ and $P_{R0} = P_R(t = 0)$ are the initial pressures in the cavity and the reservoir at $t = 0$; ρ, cp, η, and λ are the material values of the fluid in the cavity (density, heat capacity, dynamic viscosity, and heat conductivity); V_C and V_R are the volumes of the cavity and reservoir; τ_C and τ_R are the time constants of the cavity and reservoir; R_L is the radius of the canal; L is the length of the canal; and I_0 is the IR power density.

The time constant yields

$$\tau = \frac{\tau_C \cdot (1+\Omega)}{1 + \frac{\tau_C}{\tau_R} \cdot (1+\Omega)} \tag{4.5c}$$

Figure 4.6(a) shows the maximal deflection of the membrane as a function of Ω for a nonadiabatic cavity connected to a reservoir volume of $V_R = 100 \cdot V_C$ by a compensation leak. With increasing heat loss (decreasing time constant Θ) the maximum deflection is reduced obviously. This result proves that a good thermal insulation of the cavity and a soft membrane would be beneficial to yield a high deflection.

The design of the compensation, that is, the flow resistance of the compensation canal or the size of the reservoir, has a significant influence on the deflection depending on Ω, as can be seen Figure 4.6(b). It is obvious that for $\Omega < 1{,}000$ a large reservoir volume considerably reduces the maximal deflection. Only for high values of Ω (a very soft membrane) can this influence be neglected. Nevertheless, the compensation is necessary when the operation temperature of the sensor differs from the filling temperature of the sensor. For the water-filled adiabatic cavity of the example from Figure 4.6(a) and (b), without compensation, an increase of the ambient temperature of 10 K will result in a pressure increase in the cavity of about 900 Pa. This pressure difference will cause a maximal deflection y_{max} that is two to four times larger than the thickness t_P of the membrane. However, $y_{max}/t_P > 1$ results in a nonlinear relation between the membrane deflection and the pressure difference because the membrane stress cannot be neglected. In Timoshenko and

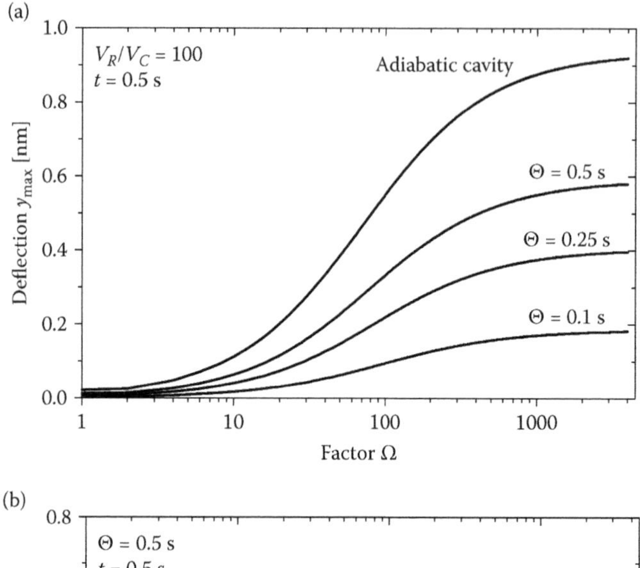

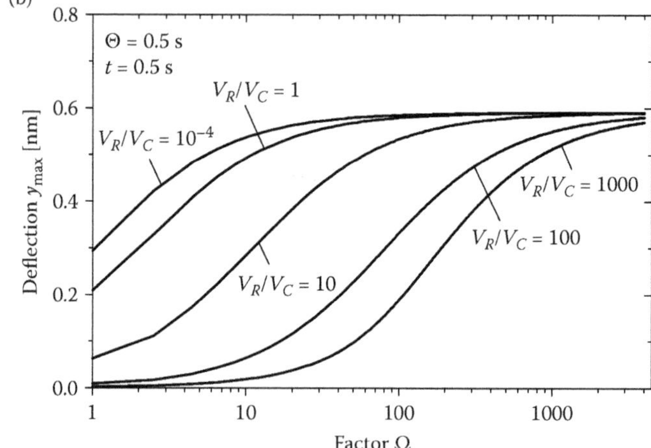

FIGURE 4.6
Maximum central deflection y_{max} of a circular bending plate (silicon, thickness 1 μm, diameter 0.5 mm) after 0.5 s irradiation time (IR radiation flux 10 W/m^2, diameter and height of the cavity 0.5 mm each) as a function of factor Ω. (a) Influence of heat loss time constant Θ for a water-filled nonadiabatic cavity with compensation leak (length 10 mm, diameter 10 μm). Decreasing numbers of Θ mean increasing heat loss. (b) Influence of the relation V_R/V_C of the reservoir volume to cavity volume for a water-filled nonadiabatic cavity with compensation leak (length 10 mm, diameter 10 μm) and heat loss time constant Θ = 0.5 s.

Woinowsky-Krieger (1959) an implicit equation for the nonlinear case is given, applicable for $y_{max}/t_P > 1$:

$$\frac{y_{\max}}{t_P} = \frac{\Delta P}{64 \cdot D \cdot t_P} \cdot \frac{R^4}{1 + 0.488 \cdot \left(\frac{y_{\max}}{t_P}\right)^2} \tag{4.6}$$

with

$$D = \frac{E \cdot t_P^3}{12 \cdot (1 - v^2)}$$

the flexural stiffness of the membrane.

For $y_{max}/t_P \rightarrow 0$ Equation (4.6) changes over to the linear case. Figure 4.7 shows the relative deflection y_{max}/t_P the nonlinear and the linear case. A high-pressure bias due to an ambient temperature change will reduce the sensor characteristic $\Delta y/\Delta p$ at the new operating point compared to an operating point without ambient temperature change ($\Delta p = 0$ in Figure 4.7). This leads to a lower deflection as result of the absorbed IR radiation at the new operating point and makes the sensor operation dependent on the ambient temperature. The nonlinearity is also influenced by the fluid in the cavity because the pressure difference ΔP in Equation (4.3) depends on the mean temperature increase of the fluid, which changes regarding the material values of the fluid (see Section 4.5).

It can be concluded that the compensation leak avoids the nonlinearity, acting as a high-pass filter with the cutoff frequency $1/\tau$ in the case of an

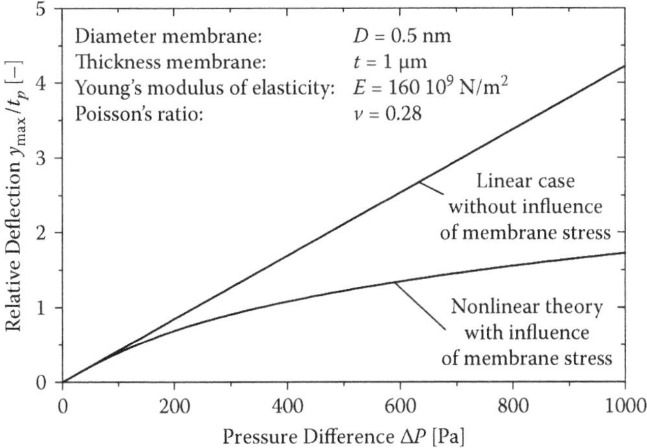

FIGURE 4.7
Relative plate deflection at $r = 0$ as a function of applied pressure difference comparing the influence of a membrane stress for a water-filled cavity.

ambient temperature change. However, the layout of the compensation is critical: a fast compensation, e.g., for high values of V_R/V_C, will also reduce the deflection (see Figure 4.6(b)), and a reduced outflow through the compensation leak due to a higher flow resistance (larger time constant τ_R of the reservoir in Equation (4.5(b))) leads to slower compensation. Furthermore, a nonadiabatic cavity also reduces the deflection of the membrane (see Figure 4.6(a)).

Overall, the compensation leak is an important issue depending on the fluid used and experiments are needed to find the optimum trade-off between sensitivity and linearity.

4.3 Noise in Golay Cells and in the Photomechanic IR Receptor

The performance of an IR sensor is limited by the signal-to-noise ratio (SNR) or the corresponding so-called detectivity *D*. If the sensor signal is very weak, as it is for radiation flux far away from a thermal source, noise has to be reduced as much as possible. The lowest possible noise is given by the physical limit of radiation noise.

4.3.1 Noise Sources in Golay Cells

There are several noise sources inside the Golay cell detector (see Figure 4.8). To calculate the noise we use the root mean square (rms) values of the noise values.

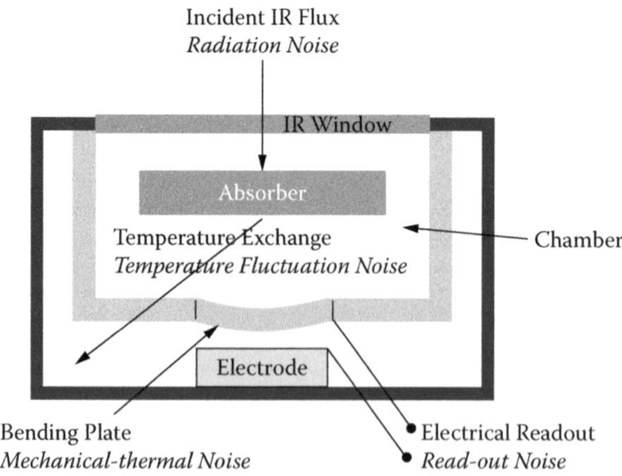

FIGURE 4.8
Noise sources in a Golay cell.

4.3.1.1 Temperature Fluctuation Noise

Temperature fluctuation noise results from the fluctuation of temperature around a mean value (absolute temperature T), even in thermal equilibrium. Temperature fluctuation noise is innate to all thermal sensors. The noise flux is (Budzier and Gerlach 2010)

$$\bar{\Phi}_R^2 = 4k_B G_{th} T^2 \Delta f \tag{4.7}$$

with the thermal conductance G_{th} and the bandwidth Δf. The thermal conductance describes the thermal coupling of the sensor to the environment. In our case that is the thermal conduction $G_{th,G}$ to the surrounding fluids and the radiation coupling $G_{th,R}$ with the environment:

$$G_{th} = G_{th,R} + G_{th,G} \tag{4.8}$$

The part of the temperature fluctuation noise due to radiation coupling is also called radiation noise. The thermal conductance by heat radiation applies

$$G_{th,R} = 4\varepsilon\sigma T^3 A_S \tag{4.9}$$

The radiation noise cannot be avoided and gives the fundamental limit. With Equations (4.7) and (4.9) we get the noise flux by radiation:

$$\bar{\Phi}_{RR}^2 = 16\varepsilon k_B \sigma T^5 A_S \Delta f \tag{4.10}$$

The noise voltage, resulting from temperature fluctuation noise, depends on voltage responsivity R_V:

$$\overline{V_{RR}^2} = R_V \overline{\Delta\Phi_R^2} \tag{4.11}$$

4.3.1.2 Mechanical-Thermal Noise

In microstructures like bending plates mechanical-thermal noise is caused by molecular agitation. Therefore, mechanical-thermal noise is often also called Brownian motion noise or thermoelastic noise. It results in a fluctuation force F related to any mechanical resistance R_M (Gabrielson 1993):

$$\overline{F^2} = 4k_B T R_M \Delta f \tag{4.12}$$

The mechanical resistance depends on the design of the bending plate and the damping of the surrounding gas or liquid. For a bending plate the fluctuation force leads to a noise displacement (Gabrielson 1993):

$$\overline{z^2} = \frac{4k_B T}{\omega_0 k Q} \Delta f \tag{4.13}$$

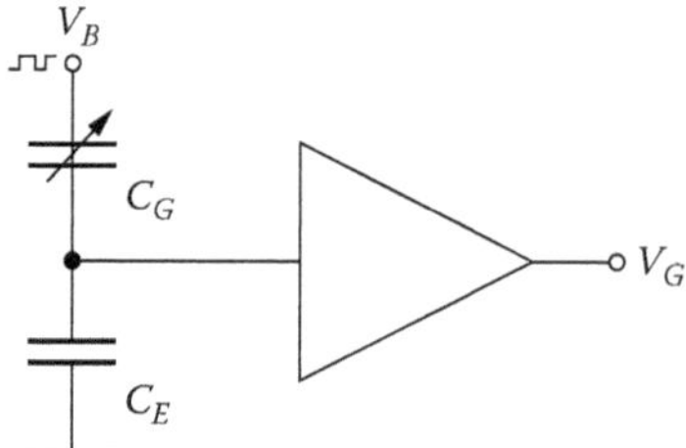

FIGURE 4.9
Simple circuit for capacitive readout of a capacitive Golay cell.

where ω_0 is the resonant frequency, Q the quality factor, and k the spring constant. For circular plates with radius R and thickness h the resonant frequency is calculated as

$$\omega_0 = 2.96\frac{h}{R^2}\sqrt{\frac{E}{\rho(1-v^2)}} \tag{4.14}$$

where E is the Young's modulus, ν the Poisson's ratio, and ρ the density.

Figure 4.9 shows a simple readout circuit for a capacitive readout to calculate the resulting noise. The output voltage V_G depends on the capacity of the Golay cell C_G, a reference capacity C_E, and a pulsed bias voltage V_B:

$$V_G = V_B\frac{C_G}{C_E} \tag{4.15}$$

Using Equations (4.8) and (4.9), the noise voltage change becomes (Budzier and Gerlach 2010)

$$\overline{V_{RM}^2} = \left(V_B\frac{C_G}{C_E}\right)^2\frac{\overline{z^2}}{z_0^2} \tag{4.16}$$

with mean displacement z_0.

Normally, mechanical-thermal noise is very small and hence negligible.

4.3.1.3 Readout Noise

Also, the conversion of the mechanical displacement into an electrical signal generates noise. In the case of a capacitive readout the output shows a thermal noise of the capacitor, the so-called kTC noise. Without bandwidth limitation it applies that (Budzier and Gerlach 2010)

$$\overline{V_{RC}^2} = \frac{k_BT}{C_E + C_G} \tag{4.17}$$

4.3.1.4 Total Noise

Since the individual noise sources are independent of each other, the corresponding noise powers can be added, leading to

$$\overline{V_R^2} = \overline{V_{RT}^2} + \overline{V_{RM}^2} + \overline{V_{RC}^2} \tag{4.18}$$

All noise voltages are "white," which means that they are independent of frequency.

4.3.2 Detectivity of Golay Cells

Detectivity D is a parameter describing the ratio between the sensor signal ΔV_S due to an incident radiation flux $\Delta \Phi_S$ and the rms noise voltage V_R:

$$D = \frac{\Delta V_S}{\Delta \Phi_S} \Big/ V_{Rn} = \frac{R_V}{V_{Rn}} \tag{4.19}$$

D is also known as the inverse of the so-called noise-equivalent power NEP:

$$D = 1/NEP \tag{4.20}$$

Because usually the sensor signal scales with the square root of the sensor size $\sqrt{A_S}$, often the so-called specific detectivity D^* (D star) is used to describe the performance of a sensor. D^* is used favorably to compare detectors. We can calculate the specific detectivity from basic detector parameters:

$$D^* = \frac{\sqrt{A_S}}{\overline{V_{Rn}}} R_V \tag{4.21}$$

with the voltage responsivity R_V, the detector area A_S, and the normalized noise voltage. For white noise, the noise voltage scales with bandwidth Δf:

$$\overline{V_{Rn}} = \frac{\overline{V_R}}{\Delta f} \tag{4.22}$$

It is of major interest to determine the maximum achievable specific detectivity. That is given by the minimal possible noise due to radiation noise:

$$D^*_{max} = \frac{1}{\sqrt{16 \varepsilon k_B \sigma T^5}} \tag{4.23}$$

So the maximally attainable detectivity at $T = 300$ K amounts to $1.8 \cdot 10^{10}$ cm $Hz^{1/2}/W$.

4.3.3 Noise Sources in IR Receptors

In principle, the IR receptor is comparable with a Golay cell detector, and hence also limited by temperature radiation noise, as can be seen in Equation (4.10). It can be assumed that the beetles use flame pulsation frequencies to distinguish a fire from a radiating hot spot on the ground. In addition, smaller bandwidths reduce the thermal noise. For oil pool fires (Bejan 1991) like the Coalinga fire and also for forest fires (Finney and McAllister 2011) the pulsation frequency f depends on the fire diameter d with $f \sim d^{1/2}$. For oil tank fires with a diameter of 30–50 m, the pulsation frequency amounts to 0.2–0.3 Hz results (Schmitz and Bousack 2012). Using Equation (4.10) with $\varepsilon = 1$ (black-body radiation), the minimal detectable radiant flux Φ_S due to temperature noise yields

$$\Phi_S = \frac{\sqrt{\overline{\Phi_{RR}^2}}}{A_S} \tag{4.24}$$

with a cross section A_S of the sensillum of $8 \cdot 10^{-11}$ m^2 (10 µm diameter), a temperature T of the sensillum of 300 K, and a bandwidth $\Delta f = 10$ Hz, which is sufficient to detect the pulsation frequencies. The resulting minimal detectable radiant flux Φ_S due to temperature noise is $2 \cdot 10^{-3}$ W/m^2.

This value is one or two orders of magnitude higher than the radiant heat flux of $4 \cdot 10^{-5}$ to $3 \cdot 10^{-4}$ W/m^2 most probably detected by the beetles at the Coalinga fire. Nevertheless, the sensitivity of hair mechanoreceptors as evolutionary precursors of the IR sensillum was thoroughly investigated by other authors (e.g., Bialek 1987). Here, signals that are three orders of magnitude lower than the broadband ciliary displacement noise could be reliably detected from single hair cells.

4.4 Measurement of IR Radiation below the Noise Level

The detection of a weak, periodic, or pseudoperiodic signal hidden under a noisy background is a task that often occurs. Many methods were developed to solve the problem depending on the characteristics of the signal and the nature of the detector. In our case, the IR sensillum is characterized by (1) a weak periodic signal due to the pulsating fire, (2) a noise level that is larger than the periodic signal, and (3) a sufficiently nonlinear system, e.g., due to a level or threshold that the signal has to exceed before it will be detected by the sensor. The last requirement is fulfilled because the dendrite inside the sensillum produces only an action potential when a certain stress level of the dentritic membrane is exceeded. In addition we must note that the looked for data are a real-life signal. That means that the signal is non-Gaussian, noncircular, nonstationary, and nonlinear.

The detection of such signals is possible with stochastic resonance (Hänggi 2002, Wiesenfeld and Moss 1995) in connection with adaptive signal processing (Adali and Haykin 2010, Jafari and Chambers 2003) or phase space projection (Johnson and Povinelli 2005). Since the beetle has several tens of IR sensilla present, the signal processing will also use neuronal networks. Stochastic resonance is based on increasing and matching of a weak periodic input signal and the wideband noise of a detector in nonlinear threshold measurement systems like natural cells. The descriptive signal processing methods try to recognize certain samples based on the physical characteristics of the expected signal. In addition, the receiving signal is transformed into the complex domain or projected into the phase space. The looked for information, i.e., the interrelationship of the real and imaginary parts or the magnitude and phase, can be exploited and interpreted then.

The enumerated methods are simple to implement into technical systems. Unfortunately, no experimental results are published about the signal processing in the beetle. In fact, stochastic resonance was observed experimentally for different types of hair cells. For hair cells in the inner ear, the addition of noise leads to an improvement in the output signal-to-noise ratio (Bennett et al. 2004, Lindner et al. 2005, Zhao et al. 2008). Experiments with hydrodynamically sensitive mechanoreceptor hair cells located in the tailfans of crayfish *Procambarus clarkii* showed very clearly that weak signals can be enhanced by an optimal level of external noise in single sensory neurons (Douglass et al. 1993). In experiments on the cercal system of the cricket *Acheta domestica* it was demonstrated clearly that a significant degree of encoding enhancement through stochastic resonance occurred (Levin and Miller 1996).

4.5 Influence of the Fluid in the Cavity

The cavity of the sensillum is filled by a spongy intermediate layer from soft mesocuticle and a liquid, most probably water (Schmitz et al. 2007). However, when the principle of the sensillum should be transferred to a technical sensor no biological material can be used and the influence of the filling material of the cavity has to be investigated. For example, the fluid in the cavity requires a sufficient absorption coefficient for IR radiation, appropriate thermodynamic properties resulting in a large deflection of the membrane, safe and simple handling during the filling of the cavity, and compatibility with the other materials used.

The absorption coefficient of different fluids differs considerably, depending also on the considered wavelength window (see Table 4.1).

In the case of water as filling of the cavity, this means that the radiation is absorbed in a very thin layer directly behind the window, which results in

TABLE 4.1

Absorption Coefficient of Selected Media (3–5 μm)

Medium	Absorption Coefficient (1/cm)
Water	10^3–10^4
Hydrocarbons (methanol, n-pentane, toluene)	1–10
Air, depending on humidity	10^{-2}–10^{-3}

a significant loss of the absorbed energy due to heat conduction through the window. Gas as filling needs an additional absorber in the cavity due to its very low absorbance. However, hydrocarbons allow use of the whole cavity depth of 0.1 to 1 mm for an absorption of the radiation, whereas the energy loss due to the window is reduced.

The influence of the thermodynamic properties can be estimated when the deflection of the membrane is maximized. Based on Equation (4.4), both a high thermal expansion coefficient and a high mean temperature increase in the cavity yield a large deflection. The mean temperature increase in the cavity can be calculated for an adiabatic cavity with

$$\Delta T_{C,mean}(t) = \frac{\Phi_{IR} \cdot t}{H_C \cdot \rho \cdot cp} \tag{4.25}$$

where Φ_{IR} is the IR radiation absorbed in the cavity, and ρ and cp the density and heat capacity of the fluid, respectively.

Introducing Equation (4.25) in Equation (4.4) yields

$$\frac{y_{\max}}{\Phi_{IR}}(t) = \frac{\Omega}{1+\Omega} \cdot \frac{3 \cdot \alpha}{\rho \cdot cp} \cdot t = \delta \cdot t \tag{4.26}$$

with Δ the relative deflection, which represents a key figure for the maximal deflection upon the absorption of 1 J/m^2 in the cavity. Figure 4.10 shows a comparison of the relative deflection for different liquids.

Obviously, hydrocarbons are the best choice to yield a large deflection that is one order of magnitude larger than for water. However, the most appropriate liquids, like n-pentane or diethyl ether, have some serious drawbacks, like low evaporation temperature, incompatibility with other materials, or toxicity. Therefore, methanol seems to be a good compromise regarding the thermodynamic properties.

The relative deflection for gas yields a surprisingly high value of about 10^3 nm/J/m^2 due to the high thermal expansion coefficient and a low density compared to liquids. But this is only true for the assumption of an adiabatic cavity. However, the need of an additional absorber leads to a considerably increasing heat loss of the cavity, depending on the arrangement of the absorber and an increasing product of density and heat capacity due

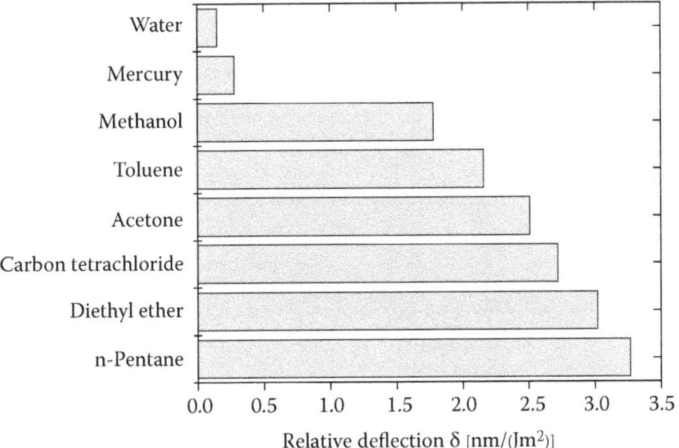

FIGURE 4.10
Comparison of the relative deflection δ of the bending plate for different liquids.

to the absorber material. If, for instance, the absorber is deposited as a film on the inside of the window, then only less than 1% of the expected temperature increase was achieved due to an adiabatic cavity (Klocke et al. 2011). It must be concluded that the evaluation of a suitable fluid as filling medium for the cavity has to take into account the design of the cavity, especially with regard to the heat loss.

The filling procedure of the cavity with a liquid is a very critical issue because any remaining gas bubbles in the edges of the cavity must be avoided. Here, liquids with low surface tension like hydrocarbons are advantageous.

4.6 Experimental Results and Devices

A model of a gas-filled sensillum, a miniaturized Golay cell prototype was developed and studied (Schossig et al. 2011). Basically, it consists of a sealed cavity made of lithium tantalate (LT) and a capacitive displacement transducer (Figure 4.11). IR radiation is absorbed in the gas chamber with a volume of approximately $2.5 \times 2.5 \times 0.03$ mm^3. That results in a warming of the gas and, consequently, an increase of gas pressure resulting in a deflection of a thin membrane. The membrane forms one electrode of a parallel plate capacitor with an active area of 2×2 mm. The capacitance of the manufactured prototypes amounts to about 3.5 – 7 pF. To measure the capacitance, a commercial 24-bit capacitance-to-digital converter (CDC) is used. The miniaturized Golay cell detector and the CDC are assembled on a printed circuit board (PCB) in a small transistor housing (TO-8 package), as can be seen in Figure 4.12.

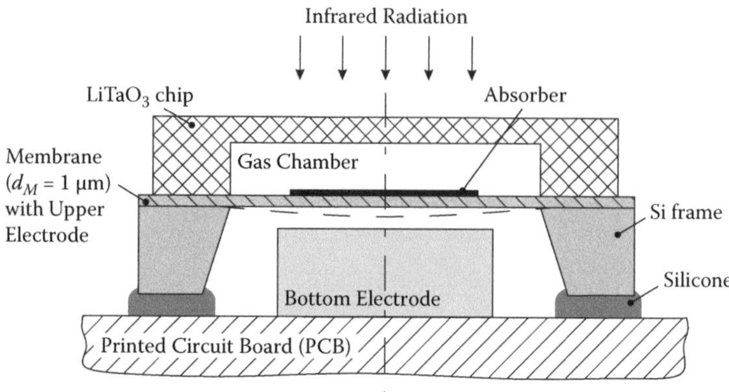

FIGURE 4.11
Structure of a miniaturized Golay cell detector (From Schossig, M. et al., A pneumatic infrared detector with capacitve read-out circuit. Paper presented at SENSOR+TEST Conference 2011, Nürnberg).

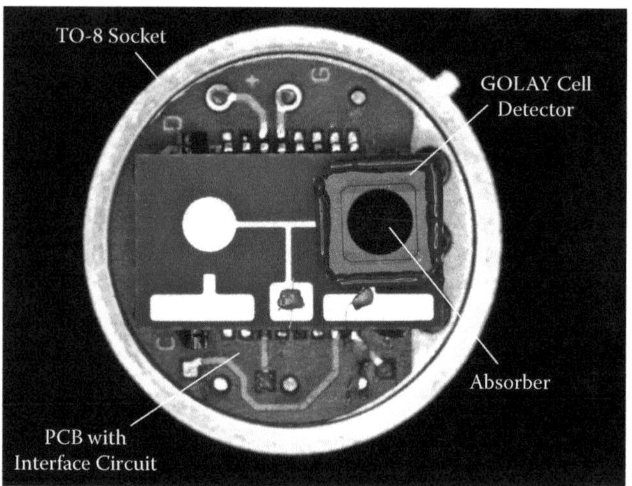

FIGURE 4.12
Prototype assembly (Schossig, M. et al., A pneumatic infrared detector with capacitve read-out circuit. Paper presented at SENSOR+TEST Conference 2011, Nürnberg).

Schossig et al. (2011) estimated a responsivity of about 5 aF/μW based on a simple thermal model. Hence, the calculated minimum noise-equivalent power (*NEP*) is 0.9 μW. That yields a specific detectivity D^* of $4.3 \cdot 10^5$ cm Hz$^{1/2}$/W. The quite low detectivity value results from rather low thermal isolation of the gas chamber.

Preliminary measurements of the prototype showed a *NEP* of 14 μW and a specific detectivity of $4 \cdot 10^4$ cm Hz$^{1/2}$/W. The detectivity is one order of magnitude lower than the calculated value, what could be caused by the estimated values of thermal resistance and unknown values of materials properties. However, other authors demonstrated Golay cells with similar results. For example, Yamashita et al. (1998) reported a detectivity of $1.3 \cdot 10^3$ cm $Hz^{1/2}/W$.

References

Adali, T., and Haykin, S.S. (2010). *Adaptive signal processing: next generation solutions*. Wiley-IEEE, Hoboken, NJ.

Bejan, A. (1991). Predicting the pool fire vortex shedding frequency. *J. Heat Transfer Trans. ASME*, 113:261–263.

Bennett, M., Wiesenfeld, K., and Jaramillo, F. (2004). Stochastic resonance in hair cell mechanoelectrical transduction. *Fluctuation Noise Lett.*, 4:L1–10.

Beyler, C.L. (2002). Fire hazard calculations for large, open hydrocarbon fires. In Andrau, R. (ed.), *SFPE handbook of fire protection engineering*. National Fire Protection Association, Quincy, MA. pp. 268–314.

Bialek, W. (1987). Physical limits to sensation and perception. *Annu. Rev.Biophys. Biophys. Chem.*, 16:455–478.

Budzier, H., and Gerlach, G. (2010). *Thermal infrared sensors: Theory, optimization, and practice*. Wiley, Chichester, UK.

Champion, G.C. (1909). A buprestid and other coleoptera on pine injured by "heath fires" in Surrey. *W. Entomol. Mon. Mag.*, 45:247–250.

Chevrier, J.B., Baert, K., and Slater, T. (1995). An infrared pneumatic detector made by micromachining technology. *J. Micromechanics Microeng.*, 5:193–195.

Dettner, K., and Peters, W. (1999). *Lehrbuch der Entomologie*. Stuttgart, Gustav Fischer-Verlag.

Douglass, J.K., Wilkens, L., Pantazelou, E., and Moss, F. (1993). Noise enhancement of information-transfer in crayfish mechanoreceptors by stochastic resonance. *Nature*, 365:337–340.

Evans, W.G. (1964). Infrared receptors in *Melanophila acuminata* De Geer. *Nature*, 202:211.

Evans, W.G. (1966). Perception of infrared radiation from forest fires by *Melanophila acuminata* De Geer (Buprestidae, Coleoptera). *Ecology*, 47:1061–1065.

Finney, M.A., and McAllister, S.S. (2011). A review of fire interactions and mass fires. *J. Combustion*, 2011:548328.

Gabrielson, T.B. (1993). Mechanical-thermal noise in micromachined acoustic and vibration sensors. *IEEE Trans. Electron Devices*, 40:903–909.

Golay, M.J.E. (1947). A pneumatic infra-red detector. *Rev. Sci. Instrum.*, 18:357–362.

Hammer, D.X., Schmitz, H., Schmitz, A., Rylander III, H.G., and Welch, A.J. (2001). Sensitivity threshold and response characteristics of infrared perception in the beetle *Melanophila acuminata* (Coleoptera: Buprestidae), *Comp. Biochem. Physiol. A*, 128:805–819.

Hänggi, P. (2002). Stochastic resonance in biology—how noise can enhance detection of weak signals and help improve biological information processing. *ChemPhysChem*, 3:285–290.

Iqba, N.L., and Salley, M.H. (2004). Fire dynamics tools: quantitative fire hazard analysis methods for the U.S. Nuclear Regulatory Commission Fire Protection Inspection Program. U.S. Nuclear Regulatory Commission. Available from http://www.nrc.gov/reading-rm/doc-collections/nuregs/staff/sr1805/final-report/ (accessed January 24, 2012).

Jafari, M.G., and Chambers, J.A. (2003). Adaptrive noise cancellation and blind source separation. In *4th International Symposium on Independent Component Analysis and Blind Signal Separation (ICA2003)*, Nara, Japan, pp. 627–632.

Johnson, M.T., and Povinelli, R.J. (2005). Generalized phase space projection for non-linear noise reduction. *Physica D Nonlinear Phenomena*, 201:306–317.

Kenny, T.W., Reynolds, J.K., Podosek, J.A., Vote, E.C., Miller, L.M., Rockstad, H.K., and Kaiser, W.J. (1996). Micromachined infrared sensors using tunneling displacement transducers. *Rev. Sci. Instrum.*, 67:112–128.

Klocke, D., Schmitz, A., Soltner, H., Bousack, H., and Schmitz, H. (2011). Infrared receptors in pyrophilous ("fire loving") insects as model for new un-cooled infrared sensors. *Beilstein J. Nanotechnol.*, 2:186–197.

Levin, J.E., and Miller, J.P. (1996). Broadband neural encoding in the cricket cercal sensory system enhanced by stochastic resonance. *Nature*, 380:165–168.

Lindner, J.F., Bennett, M., and Wiesenfeld, K. (2005). Stochastic resonance in the mechanoelectrical transduction of hair cells. *Phys. Rev. E*, 2005:72.

Linsley, E.G. (1943). Attraction of *Melanophila* beetles by fire and smoke. *J. Econ. Entomol.*, 36:341–342.

Schmitz, A., Sehrbrock, A., and Schmitz, H. (2007). The analysis of the mechanosensory origin of the infrared sensilla in *Melanophila acuminata* (Coleoptera; Buprestidae) adduces new insight into the transduction mechanism. *Arthropod Struct. Dev.*, 36:291–303.

Schmitz, H., and Bleckmann, H. (1997). Fine structure and physiology of the infrared receptor of beetles of the genus *Melanophila* (Coleoptera: Buprestidae). *Int. J. Insect Morphol. Embryol.*, 26:205–215.

Schmitz, H., and Bleckmann, H. (1998). The photomechanic infrared receptor for the detection of forest fires in the buprestid beetle *Melanophila acuminata*. *J. Comp. Physiol. A*, 182:647–657.

Schmitz, H., and Bousack, H. (2012). Modelling of a historic oil-tank fire allows an estimation of the sensitivity of the infrared receptors in pyrophilous *Melanophila* beetles. *PLoS One*, 7:e37627.

Schmitz, H., Kahl, T., Soltner, H., and Bousack, H. (2011). Biomimetic infrared sensors based on the infrared receptors of pyrophilous insects. Paper presented at Smart Structures/NDE 2011, San Diego.

Schmitz, H., Norkus, V., Hess, N., and Bousack, H. (2009). The infrared sensilla in the beetle *Melanophia acuminata* as model for new infrared sensors. In Rodríguez-Vázquez, Á.B., ed., *Microtechnologies for the new millennium*. Bioengineered and Bioinspired Systems IV, Dresden, Germany.

Schossig, M., Norkus, V., and Gerlach, G. (2011). A pneumatic infrared detector with capacitive read-out circuit. Paper presented at SENSOR+TEST Conference 2011, Nürnberg.

Shimozawa, T., Murakami, J., and Kumagai, T. (2003). Cricket wind receptors: thermal noise for the highest sensitivity known. In Barth, F.G., Humphrey, J.A.C., and Secomb, T.W. (eds.), *Sensors and sensing in biology and engineering*. Springer, New York. pp. 145–157.

Szabo, I. (2000). *Höhere technische Mechanik: Nach Vorlesungen*. Berlin, Springer.

Timoshenko, S.P., and Woinowsky-Krieger, S. (1959). *Theory of plates and shells*. McGraw-Hill, Singapore.

Van Dyke, E.C. (1926). Buprestid swarming. *Pan Pacific Entomologist*, 1926:3.

Weast, R.C. (1988). *CRC handbook of chemistry and physics*. CRC Press, Boca Raton, FL.

Wiesenfeld, K., and Moss, F. (1995). Stochastic resonance and the benefits of noise—from ice ages to crayfish and squids. *Nature*, 373:33–36.

Yamashita, K., Murata, A., and Okuyama, M. (1998). Miniaturized infrared sensor using silicon diaphragm based on Golay cell. *Sensors Actuators A*, 66:29–32.

Zhao, X.H., Long, Z.C., Zhang, B., and Yang, N. (2008). Suprathreshold stochastic resonance in mechanoelectrical transduction of hair cells. *Chin. Phys. Lett.*, 25:1490–1493.

5

Toward Industrial Production of Biomimetic Photonic Structures

Hiroshi Fudouzi, Tsutomu Sawada, and Yoshihiro Uozu

CONTENTS

5.1 Tunable Structural Color in Colloidal Photonic Crystal 141
5.1.1 1D Photonic Crystal Materials with Tunable Structure Color 144
5.1.2 Structural Color of 3D Colloidal Crystals 144
5.1.3 3D Photonic Crystal Materials with Tunable Structure Color 146
5.1.4 3D Photonic Crystal Materials with Tunable Structure Color by Swelling 147
5.1.5 Colloidal Crystal Gels with Tunable Structural Color by Applying Mechanical Stress 149
5.1.6 Summary and Outlook 152
References 152
5.2 Moth-Eye Antireflection Surface Using Anodic Porous Alumina 155
5.2.1 Review of Films for Diminishing Reflection 156
5.2.2 Production Process of the Moth-Eye Antireflection Film 157
5.2.3 Characteristics of the Moth-Eye Antireflection Film 160
5.2.4 Summary 162
References 163

5.1 Tunable Structural Color in Colloidal Photonic Crystal

Hiroshi Fudouzi and Tsutomu Sawada

In the early 1970s, Michael F. Land concluded in a review paper that multilayer interference plays an important role for structural colors in animals [Land 1972]. The multilayer consists of nanoplates of guanine or chitin that are high refractive index materials. Furthermore, it is known that the skin of some fish and insects can change their structural color [Berthier 2007, Oshima 2005]. There are different mechanisms of tuning structural color

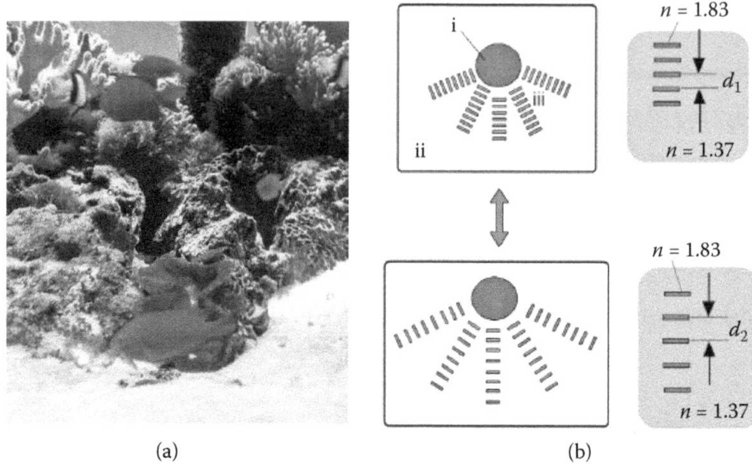

FIGURE 5.1
Tunable structural color of organism. (a) Photography of tropical blue damselfish. The structural color is due to multilayers of high refractive index guanine plates. (b) Schematic of the mechanism of active tunable color in iridophores via the reversible change of the interspace of the guanine plates on the nanoscale.

in nature. Three key factors, refractive index, periodic nanostructure, and incident light angle, affect the change of structural color. Figure 5.1(a) shows a tropical damselfish (Cobalt Blue, Pomacentridae). This fish can reversibly change its skin color from cobalt blue to green in response to changing surrounding conditions. Kasukawa et al. [1987] studied the mechanism of color change in damselfish and proposed the mechanism shown in Figure 5.1(b). In the iridophore, nanosized single-crystalline reflector plates of guanine are arrayed regularly and show tuned interspacing. The refractive index of the reflector is 1.83, and the one of the surrounding cytoplasm is 1.37 [Oshima 2005]. The structural color change is induced by the motility of the reflector plates. The skin color of the damselfish reversibly changed from blue at d_1 to green at d_2.

Other mechanisms are shown in Figure 5.2 in which three key factors are incident light angle, periodicity of the nanostructure, and refractive index. Neon tetra (*Paracheirodon innesi*) changes structural color from green in daytime to violet-blue at night. This color change is caused by a change in the tilting angle of the nanoplatelets shown in Figure 5.2(a) [Nagaishi et al. 1990, Yoshioka et al. 2011]. The motile iridophore of neon tetra works as a Venetian blind. Cephalopoda, such as octopus and squid, can change their color by changing the thickness of nanoplates shown in Figure 5.2(b) [Tao et al. 2010]. In the iridophore, cell membrane-enclosed nanoplates of guanine form a multilayer structure. The reversibly tunable structural color is caused by the thickness change of the space between d_1 and d_2. The iridophore is chemically tuned by exposure to acetylcholine.

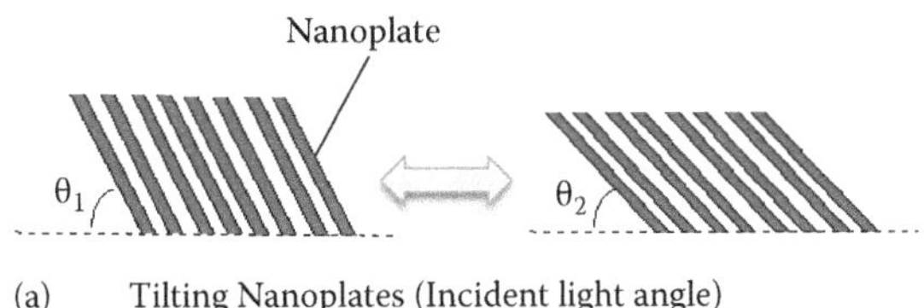

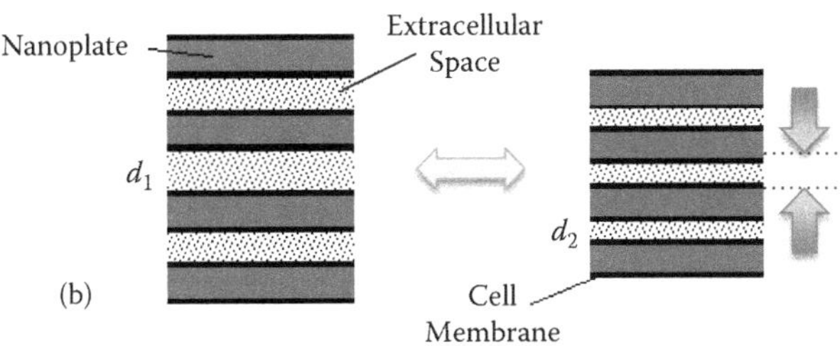

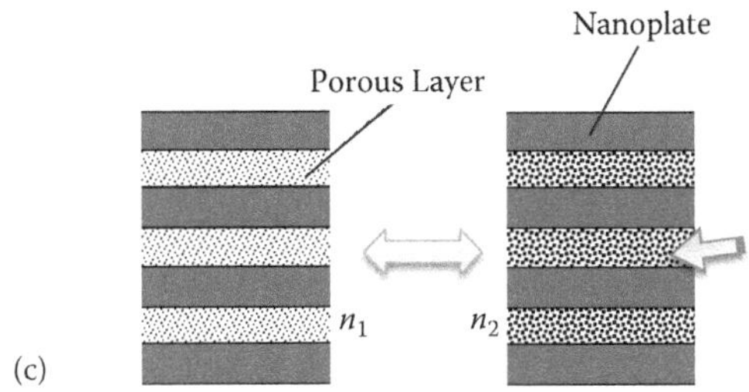

FIGURE 5.2
Schematic of other types of tuning structural colors. (a) Neon tetra can change its skin color by changing the tilt angle of guanine plates. (b) Compressing the space of the guanine plates in cephalopods. (c) Absorption of water in the cuticula of coleoptera causes color change due to a change in refractive index.

The interspace of nanoplates influences the structural color in a wide visible spectrum in such seawater life-forms. In contrast, some beetles can change the structural color of their cuticle skin by varying its refractive index. Berthier [2007] described the tunable color mechanism of beetle as shown in Figure 5.2(c). At rest, the cuticle skin of a porous laminate layer structure is dry, while under stress water will penetrate into the porous layer and the refractive index will increase from n_1 to n_2. The tortoise beetle can reversibly change its skin color from metallic yellow to a more diffusive red as a result of an external disturbance [Vigneron et al. 2007]. When the nanoporous layer of cuticle absorbs water, the backscattering caused by this structure disappears.

5.1.1 1D Photonic Crystal Materials with Tunable Structure Color

Mechanisms of tunable structural color in nature give us inspiration to design artificial photonic materials. Thus bioinspired materials or biomimetic materials are emerging and have prospective wide applications [Sanchez et al. 2005, Parker 2009, Biro and Vigneron 2010, Sato et al. 2009, Walsh et al. 2009, Graham-Rowe 2009, Kolle 2011]. There are mainly two types of nanostructures, 1D photonic crystals made from multilayers of sheets and 3D photonic crystals assembled from colloidal spherical particles. To begin with 1D photonic crystals, the multilayer interference is a typical structural color in animals [Land 1972]. The more or less flexible composite films are 1D photonic crystals with a tunable photonic stop band. Multilayer interference structures with tunable structural color are mainly fabricated via two approaches. One is the cyclic coating of two materials with different refractive indices by repetitive spin coating, dip coating, or layer-by-layer deposition. Calvo et al. [2009] have reported multilayer films of high and low refractive index soft materials that exhibit a tunable structural color. The mechanism of color tuning is the expansion of the lattice distance. On the other hand, the refractive index is also one parameter to change structural color. Inorganic mesoporous films can be applied as color-tunable sensors. These multilayers were also fabricated by repetitive spin coating [Choi et al. 2006], dip coating [Fuertes et al. 2007], and layer-by-layer assembly [Wu et al. 2007] of precursor solutions. A multilayer film can even be fabricated by just casting a single solution that contains blends of block-copolymers, followed by annealing in chloroform vapor. By choosing the appropriate volume fractions of the block-copolymers, a thermodynamically stable lamellar structure is formed by microphase separation [Edrington et al. 2001, Bockstaller et al. 2005]. Kang et al. [2007] demonstrated a chemically tunable block-copolymer multilayer film with a broad wavelength range of the reflected light from 364 nm to 1,627 nm. The ordered lamellar structure exhibits a reversibly tunable reflection peak by swelling and de-swelling. An electrochemically tunable block-copolymer changes its structural color from red to green [Walsh et al. 2009]. Haque et al. [2010] demonstrated a different type of self-assembly soft material with unidirectional aligned lamellar bilayers in a polyacrylamide hydrogel. The membrane-like lamellar bilayer causes Bragg diffraction of visible light. The hydrogel color can be reversibly tuned by swelling, anisotropic modulus, and stress/strain.

5.1.2 Structural Color of 3D Colloidal Crystals

First we will describe 3D photonic crystals assembled with colloidal particles. Recently Parker et al. [2003] has reported the opal analogue nanostructure in a weevil (Curculionoidea). This finding was a breakthrough for research about structural colors generated by 3D photonic crystals in living organisms. The most well-known 3D photonic structure in nature is opal. Opals show iridescent rainbow colors due to Bragg diffraction at arrays of colorless silica colloidal particles with random orientation of their crystal planes [Sanders 1968]. From thereon the 3D crystalline diffraction of opals has been studied for potential applications in photonic crystal engineering [Ozin and Arsenault 2008, Galisteo-Lõpez et al. 2011, Xia et al. 2000, Vlasov et al. 2001].

Colloidal crystals are usually categorized as either non-closely packed or closely packed. Non-closely packed colloidal crystals have been used in colloid science and solid-state physics as models of phase transition for the last 40 years. Therefore this type can be regarded as a typical, traditional colloidal crystal. Colloidal particles with a small polydispersity that are suspended in deionized water can spontaneously form 3D periodic arrays [Ise et al. 1983, Matsuoka et al. 1994]. However, these self-assembled colloidal crystals are generally polycrystalline with a random orientation of crystal grains of various sizes that can slowly grow by Ostwald ripening [Yoshida et al. 1991]. Recently, Kanai et al. [2005] developed a technique for quick fabrication of uniform and high-quality single colloidal crystal sheets. In their method, a suspension of charged silica or polystyrene (PS) colloids was forced to flow in a flat capillary cell using an air-pulse drive system. Above a critical air-pulse pressure, almost the entire capillary space was filled with a single-domain crystal with high spectral quality. Figure 5.3(a) shows the crystallized monodispersed PS colloids inside the cell. Figure 5.3(b) shows a confocal optical microscope image of a hexagonal array of 300 nm PS colloids with a spacing of 470 nm. This lattice spacing results in the photonic bandgap effect in the visible range, causing a structural color.

In contrast, a closely packed colloidal crystal, i.e., opal, film is shown in Figure 5.3(c). The opal film is made of PS, colloids of 200 nm diameter, and shows a uniform green structural color over the whole area of the silicon wafer substrate. This opal film was created by an oil covering method [Fudouzi 2004]. Figure 5.3(d) shows a scanning electron microscope (SEM) image of the opal film in which PS colloids are hexagonally closely packed. From the cross section of the state of colloids arrangement, the colloidal crystal has a 3D array lattice structure of cubic close packing (ccp). Furthermore, the ccp (111) planes are parallel to the substrate, and vertical stacking causes monochromatic structural color of the opal thin film. The reflected wavelength λ is expressed by the following Bragg equation combined with Snell's law.

$$\lambda = 2d_{111}\sqrt{\left(n_{eff}^2 - \sin^2\theta\right)}$$

Here λ is the wavelength of the Bragg diffraction peak, d_{111} indicates the interspace of ccp (111) planes, n_{eff} is the average refractive index, and θ is the incident angle. It is clear from the equation that the structural color depends

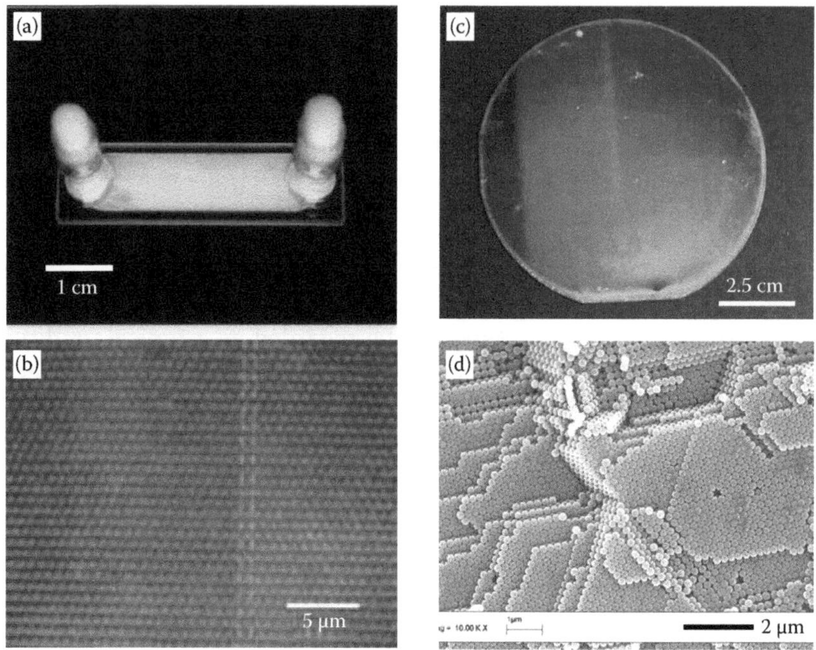

FIGURE 5.3 (See color insert.)
Two types of colloidal crystals generated by original fabrication techniques in the author's group. (a) A red structural color of non-closely packed colloidal crystal inside a flat cell. (b) Confocal laser micrograph indicates hexagonal array of 300 nm PS colloidal particles. (c) Green structural color of colloidal crystal film coated on a silicon wafer. (d) SEM image shows a 3D opal film assembled from 200 nm PS colloidal particles.

not only on the interspace of ccp (111) planes, but also on their tilting angle toward the incident light beam.

5.1.3 3D Photonic Crystal Materials with Tunable Structure Color

This research field became one of the hot topics in material chemistry recently. Here we will focus on two types of colloidal crystals with tunable structural colors: elastic opal photonic crystals and colloidal crystal gels.

Figure 5.4 shows the tunable structural color in colloidal crystals due to Bragg diffraction from multilayers of ccp (111) planes. Here, the elastomer or hydrogel plays an important role in the reversible tuning of the lattice. The structural color can be changed by controlling three parameters, d_{111}, n_{eff}, and θ. Especially the tunable d_{111} is a key factor to change the structural color, as shown in Figure 5.4(a). Non-closely packed and closely packed colloidal crystals are embedded in an elastomer or hydrogel. Figure 5.4(b) shows the SEM surface image of an opal film made of 200 nm PS colloids that is filled with a polydimetylsilicone (PDMS) elastomer [Fudouzi 2009]. Figure 5.4(c) illustrates a cross-sectional image of such an opal film with variable ccp (111)

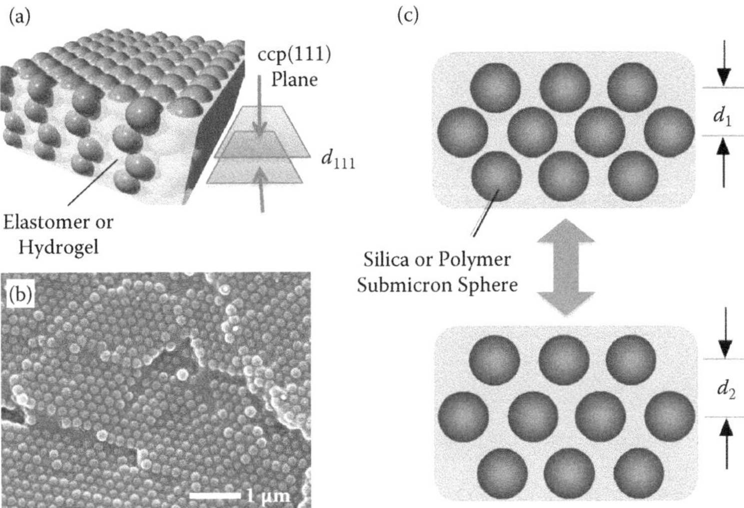

FIGURE 5.4
Bioinspired material design for colloidal crystals with tunable lattice spacing. (a) 3D array colloidal particles embedded in elastomer or hydrogel. Cubic closely packed (111) planes are stacked vertically to a supporting substrate. (b) SEM surface image shows an opal type colloidal crystal consisting of 200 nm PS particles in a PDMS elastomer. (c) A cross section concept image of colloidal crystal hybrid material that is reversibly tunable due to the changing interspace of d_{111}.

distances. The bioinspired idea from Figure 5.1 is to control the interspace of colloidal crystal planes reversibly to tune between d_1 and d_2.

5.1.4 3D Photonic Crystal Materials with Tunable Structure Color by Swelling

Figure 5.5 shows structural color change and its reversibility by swelling [Fudouzi and Sawada 2006]. Figure 5.5(a) and (b) shows photographs of an opal photonic crystal before and after swelling with isopropanol. The elastic opal crystal film is made of PS colloidal particles with an average diameter of 200 nm on a 4-inch silicon wafer. When isopropanol was sprayed onto the surface of the dry film, its structural color changed from green to red due to the swelling of the PDMS network, as illustrated in Figure 5.5(c). The surface of this elastic opal film was sufficiently uniform and smooth that the reflected light also clearly displayed an image of the camera used to shoot the photographs. After the isopropanol had completely evaporated, the film went back to its original green color (as the PDMS matrix shrank to its original state). Figure 5.5(c) shows the reflectance spectra taken from before and after swelling. The Bragg diffraction peak was initially located at 545 nm (green color) and shifted to 604 nm (red color) after swelling with isopropanol. As isopropanol was evaporating, the diffraction peak gradually moved back to its

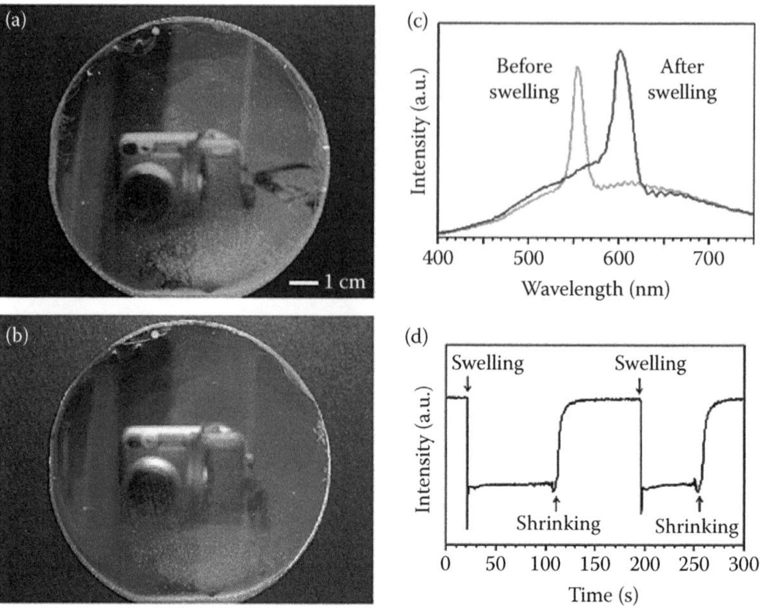

FIGURE 5.5 (See color insert.)
Tuning structural color in a closely packed colloidal crystal, i.e., opal photonic crystal, film by swelling and shrinking. (a, b) Structural colors before and after wetting the surface with isopropanol. (c) Reflectance spectra taken from the surface of the film before and after swelling. (d) A plot showing the variation of the reflectance intensity measured at 550 nm with time. (From Fudouzi, H., and Xia, Y., *Langmuir*, 19:9653–9660, 2003. Reproduced with permission. Copyright © 2003, American Chemical Society.)

original position at 545 nm. Figure 5.5 shows the intensity of reflected light measured at 545 nm as a function of time. The intensity dropped quickly during the swelling of the elastic opal film, essentially within a few seconds. This time is mainly determined by the diffusion of the solvent molecules into the PDMS elastomer and the expansion dynamics of the PDMS network. On the other hand, the time required for recovering was largely controlled by the amount of the liquid deposited on the surface of the film and the evaporation rate of the liquid under ambient conditions. For isopropanol, the intensity at 545 nm could be completely recovered within a period of 10s after the excess amount of solvent had evaporated. This reversible change between colors could be repeated for more than 50 times without observing any deterioration in the quality of the displayed color.

Figure 5.6 shows that the swelling volume of the hydrogel is dependent on the solvent type and its composition [Toyotama et al. 2005]. The colloidal crystal gel can tune the lattice constant by selecting the solvent composition. The gelled colloidal crystals shrank with increasing ethanol (EtOH) concentration, whereas the iridescent color remained almost uniform. The blue shift of the reflection color indicates a reduction in the lattice constant.

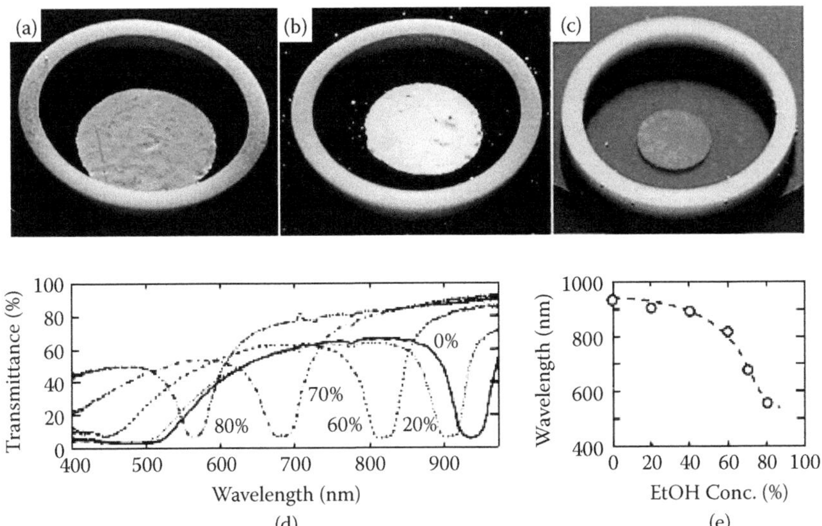

FIGURE 5.6 (See color insert.)
Tuning structural color by swelling of a hydrogel colloidal crystal. Change in the Bragg reflection color of the gelled colloidal crystals for different ethanol (EtOH) concentrations: (a) 20%, (b) 65%, and (c) 80%. The diameter of the Teflon ring is 12.3 mm. (d) Transmittance spectra for various EtOH concentrations. The gelled colloidal crystals shrink with an increase in EtOH concentration, and a blue shift is observed in its spectrum. (e) The dip wavelength in the transmittance spectra shown in (a) as a function of the EtOH concentration. (From Toyotama, A., *Langmuir*, 21:10268–10270, 2005. Reproduced with permission. Copyright © 2005, American Chemical Society.)

Solvent exchange can be accomplished by merely soaking the gel in a solvent with a different composition for a few minutes. Figure 5.6(a)–(c) shows photographs of the same hydrogel change in various EtOH-water concentration liquids: (a) 20%, (b) 65%, and (c) 80%. Figure 5.6(d) shows the transmittance spectra for various EtOH concentrations. The spectral profiles simply shift to a shorter wavelength. Figure 5.6(e) shows the dip wavelength due to Bragg's diffraction as a function of the EtOH concentration. The dip wavelength can be reduced to almost half its initial value. This shrinking-swelling process resulting from the solvent substitution was reversible and repeatable.

5.1.5 Colloidal Crystal Gels with Tunable Structural Color by Applying Mechanical Stress

Figure 5.7 shows a structural color change and its reversibility by mechanical forces [Fudouzi and Sawada 2006]. Figure 5.7(a) shows the color change of the elastic opal film coated on a supporting carbon containing a silicone rubber sheet by elongation. When the sheet was stretched in the horizontal direction, the lattice of colloidal crystal, d_{111}, was reduced in the vertical

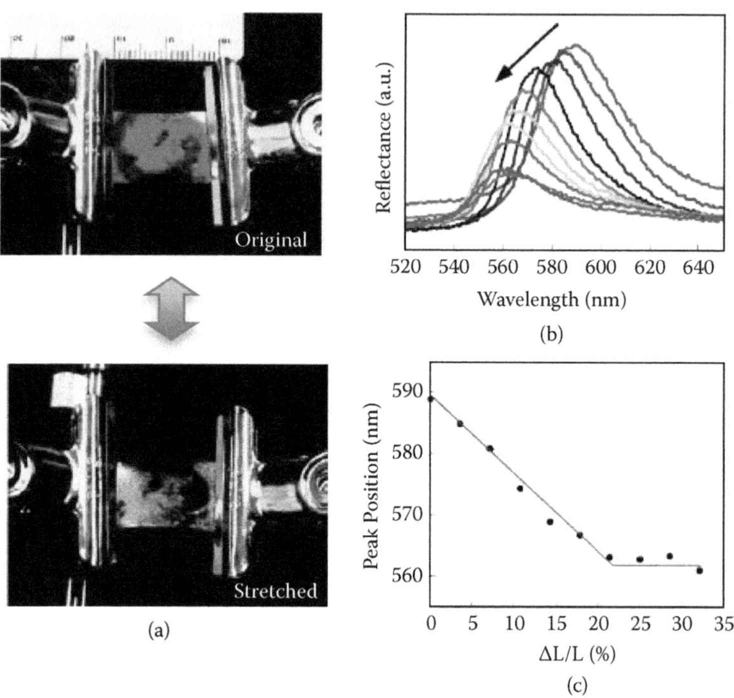

FIGURE 5.7 (See color insert.)
Tuning structural color by mechanical deformation of an elastic opal film coated on a silicone rubber sheet. (a) The reversible color change in the initial and stretched condition. The structural color of the elastic opal film reversibly changes from red to green. (b) Reflectance spectra change under mechanical strain on the rubber sheet. Bragg diffraction peak shifts to shorter wavelengths. (c) Relationship between the peak position and elongation ratio $\Delta L/L_0$. (From Fudouzi, H., and Sawada, T., *Langmuir*, 22:1365–1368, 2006. Reproduced with permission. Copyright © 2006, American Chemical Society.)

direction. The colloidal crystal film layer is less than 20 μm thickness formed on a PDMS elastomer sheet of 5 mm in thickness. After releasing the mechanical strain on the rubber sheet, the sheet quickly shrank, and its color reverted from green to its original color, red. The tuning was reversible and repeatable during the cycle of application and release of mechanical strain. Furthermore, the mechanical strain on the rubber sheet can be estimated by the position of the Bragg diffraction peak. Figure 5.7 shows that the peak position shifts to shorter wavelengths by increasing elongation. The peak of reflection can be changed from 590 nm to 560 nm when increasing the sheet elongation, while the reflectance intensity decreased gradually. The relationship between peak position as a function and elongation ratio is shown in Figure 5.7(c). The peak position decreased linearly with deformation for elongations of less than 20%. Tuning the color of the PDMS sheet is a reversible and repeatable process. The novel PDMS sheet has the potential to be applied to mechanical strain sensing [Fudouzi and Sawada 2006, Pursiainen et al. 2005].

Other applications using structural colors of colloidal crystals are, for example, photonic paper [Fudouzi and Xia 2003] and P-ink [Arsenault et al. 2003]. Furthermore, using opal as a template, an inverse opal structure shows a porous microstructure with tunable color. Many review papers describe the tunable structural color on inverse opal photonic crystals [Aguirre et al. 2010, Zhao et al. 2010, Harun-ur-rashie et al. 2009, Marlow et al. 2009, Furumi et al. 2010, Ge and Yin 2011, Kim et al. 2011]. Inverse opal hydrogels have potential application as chemical sensors [Lee and Braun 2003, Takeoka and Watanabe 2003]. Elastomer inverse opal film can be used for mechanical finger printing [Arsenault et al. 2006].

Figure 5.8 shows tuning of structural color by applying mechanical stress to a colloidal crystal hydrogel made of 110 nm silica particles embedded in a polyacrylamide matrix [Iwayama et al. 2003]. The colloidal crystal gel compressed between two parallel quartz plates was observed with an optical microscope, and measured by reflection microspectroscopy. The four images correspond to the single-crystal grain of colloidal crystal gel at different compression ratios, t/t_0 (1.0, 0.85, 0.8, and 0.75). Figure 5.8(b) shows the reflection spectrum of the single grain. The continuous blue shift of the

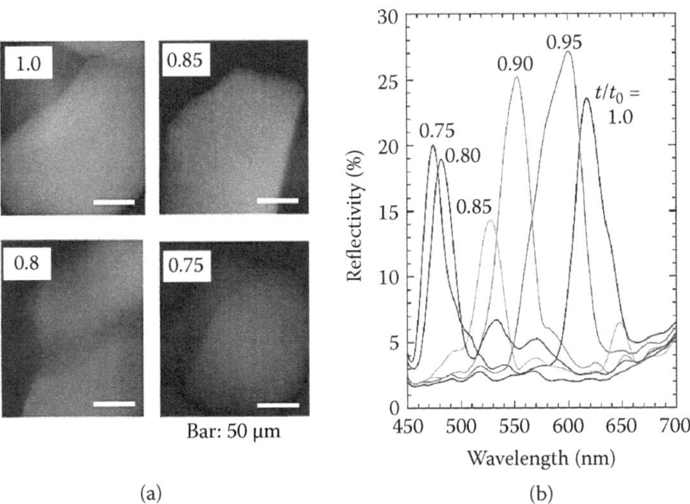

FIGURE 5.8 (See color figure at http://www.crcpress.com/product/isbn/9781439877463)
Tuning structural colors by mechanical deformation on a non-closely packed colloidal crystal. (a) Reflection optical microscope images of single grains under compressed conditions. Colloidal photonic crystal incorporated in a disk-shaped polyacrylamide hydrogel matrix, showing a Bragg diffraction of visible light. The compression ratio of the thickness of the hydrogel is changed: 1.0, 0.85, 0.80, and 0.75 as t/t_0. (b) Reflection spectrum from the single grains under mechanical compression. (From Iwayama, Y. et al. *Langmuir*, 19:977–980, 2003. Reproduced with permission. Copyright © 2003, American Chemical Society.)

Bragg peak was observed under the compression, and the Bragg diffraction peak shifted from 620 nm to 475 nm, i.e., covering nearly all wavelengths in the visible light spectrum.

In colloidal crystal gels, Asher demonstrated that hydrogel colloidal crystals are able to detect variations of pH, and ion concentration [Holtz and Asher 1997, Hu et al. 2001]. Colloidal crystals embedded in hydrogels can serve as mechanical sensors to measure strains due to 1D stretching or compression [Xia et al. 2005]. Foulger reported stimulatory response hydrogel colloidal crystals for electric tunable structural color [Lawrence et al. 2006], tunable laser emission [Jiang et al. 2005], and structural color printing [Kim et al. 2009]. The recent topics are color-tunable hydrogel materials by magnetic fields, called M-ink [Ge et al. 2007], and a tunable visible wavelength laser device by compression with mechanical stress [Furumi et al. 2011].

5.1.6 Summary and Outlook

In nature, more complex nanostructures cause smart and high functional photonic properties. Recently, unique and useful nanostructures, such as opals [Sanders 1968] and gyroids [Michielsen and Stavenga 2008, Saranathan et al. 2010], were discovered in organisms. Furthermore, Bartl demonstrated diamond-structured 3D photonic crystals by replica of a beetle surface using a sol-gel technique [Galusha et al. 2010]. In addition, even more useful but unknown mechanisms of tuning structure colors may be discovered in the near future. As described in this chapter, the bioinspired or biomimetic approach is useful for photonic material design.

References

Aguirre, C.I., Reguera, E., and Stein, A. (2010). Tunable colors in opals and inverse opal photonic crystals. *Adv. Funct. Mater.*, 20:2565–2568.

Arsenault, A.C., Clark, T.J., Von Freymann, G., Cademartiri, L., Sapienza, R., Bertolotti, J., Vekris, E., Wong, S., Kitaev, V., Manners, I., Wang, R.Z., John, S., Wiersma, D., and Ozin, G.A. (2006). Elastic photonic crystals: from color fingerprinting to control of photoluminescence. *Nature Mater.*, 5:179–184.

Arsenault, A.C., Miguez, H., Kitaev, V., Ozin, G.A., and Manners, I. (2003). A polychromic, fast response metallopolymer gel photonic crystal with solvent and redox tunability: a step towards photonic ink (P-ink). *Adv. Mater.*, 15:503–507.

Berthier, S. (2007). *Iridescences—the physical colors of insects*. Springer International, Dordrecht.

Biro, L.P., and Vigneron, J.P. (2010). Photonic nanoarchitectures in butterflies and beetles: valuable sources for bioinspiration. *Laser Photonics Rev.*, 5:27–51.

Bockstaller, M.R., Mickiewicz, R.A., and Thomas, E.L. (2005). Block copolymer nanocomposites: perspectives for tailored functional materials. *Adv. Mater.*, 17:1331–1349.

Calvo, M.E., Sánchez, S.O., Lozano, G., and Míguez, H. (2009). Molding with nanoparticle-based one-dimensional photonic crystals: a route to flexible and transferable Bragg mirrors of high dielectric contrast. *J. Mater. Chem.*, 19:3144–3148.

Choi, S.Y., Mamak, M., Von Freymann, G., Chopra, N., and Ozin, G.A. (2006). Mesoporous Bragg stack color tunable sensors. *Nano Lett.*, 6:2456–2461.

Edrington, A.C., Urbas, A.M., Derege, P., Chen, C.X., Swager, T.M., Hadjichristidis, N., Xenidou, M., Fetters, L.J., Joannopoulos, J.D., Fink, Y., and Thomas, E.L. (2001). Polymer-based photonic crystals. *Adv. Mater.*, 13:421–425.

Fudouzi, H. (2004). Fabricating high-quality opal films with uniform structure over a large area. *J. Colloid Inter. Sci.*, 275:277–283.

Fudouzi, H. (2009). Optical properties caused by periodical array structure with colloidal particles and their applications. *Adv. Powder Technol.*, 20:502–508.

Fudouzi, H., and Sawada T. (2006). Photonic rubber sheets with tunable color by elastic deformation. *Langmuir*, 22:1365–1368.

Fudouzi, H., and Xia, Y. (2003). Colloidal crystals with tunable colors and their use as photonic papers. *Langmuir*, 19:9653–9660.

Fuertes, M.C., López-Alcaraz, F.J., Marchi, M.C., Troiani, H.E., Luca, V., Míguez, H., and Soler-Illia, G.J.D.A. (2007). Photonic crystals from ordered mesoporous thin-film functional building blocks. *Adv. Funct. Mater.*, 17:1247–1254.

Furumi, S., Fudouzi, H., and Sawada, T. (2010). Self-organized colloidal crystals for photonics and laser applications. *Laser Photonics Rev.*, 4:205–220.

Furumi, S., Kanai, T., and Sawada, T. (2011). Widely tunable lasing in a colloidal crystal gel film permanently stabilized by an ionic liquid. *Adv. Mater.*, 23:3815–3820.

Galisteo-Lõpez, J.F., Ibisate, M., Sapienza, R., Froufe-Pérez, L.S., Blanco, Ú., and Lõpez, C. (2011). Self-assembled photonic structures. *Adv. Mater.*, 23:30–69.

Galusha, J.W., Jorgensen, M.R., and Bartl, M.H. (2010). Diamond-structured titania photonic band gap crystals from biological templates. *Adv. Mater.* 22:107–110.

Ge, J., Hu, Y., and Yin, Y. (2007). Highly tunable superparamagnetic colloidal photonic crystals. *Angew. Chem. Int. Ed.*, 46:7428–7431.

Ge, J., and Yin, Y. (2011). Responsive photonic crystals. *Angew. Chem. Int. Ed.*, 50:1492–1522.

Graham-Rowe, D. (2009). Tunable structural colour. *Nature Photonics*, 3, 551–553.

Haque, M.A., Kamita, G., Kurokawa, T., Tsujii, K., and Gong, J.P. (2010). Unidirectional alignment of lamellar bilayer in hydrogel: one-dimensional swelling, anisotropic modulus, and stress/strain tunable structural color. *Adv. Mater.*, 22:5110–5114.

Harun-ur-rashie, M., Seki, T., and Takeoka, Y. (2009). Structural colored gels for tunable soft photonic crystals. *Chem. Rec.*, 9:87–105.

Holtz, J.H., and Asher, S.A. (1997). Polymerized colloidal crystal hydrogel films as intelligent chemical sensing materials. *Nature*, 389:829–832.

Hu, Z., Lu, X., and Gao, J. (2001). Hydrogel opals. *Adv. Mater.*, 13:1708–1712.

Ise, N., Okubo, N., Sugimura, M., Ito, K., and Nolte, H.J. (1983). Ordered structure in dilute suspensions of highly charged polymer latices as studied by microscopy. I. Interparticle distance as a function of latex concentration. *J. Chem. Phys.* 78:536–540.

Iwayama, Y., Yamanaka, J., Takiguchi, Y., Takasaka, M., Ito, K., Shinohara, T., Sawada, T., and Yonese, M. (2003). Reentrant behavior in the order-disorder phase transition of a charged monodisperse latex. *Langmuir*, 19:977–980.

Jiang, P., Smith Jr., D.W., Ballato, J.M., and Foulger, S.H. (2005). Multicolor pattern generation in photonic bandgap composites. *Adv. Mater.*, 17:179–184.

Kanai, T., Sawada, T., Toyotama, A., and Kitamura, K. (2005). Air-pulse-drive fabrication of photonic crystal films of colloids with high spectral quality. *Adv. Funct. Mater.*, 15:25–29.

Kang, Y., Walsh, J.J., Gorishnyy, T., and Thomas, E.L. (2007). Broad-wavelength-range chemically tunable block-copolymer photonic gels. *Nature Mater.*, 6:957–960.

Kasukawa, H., Oshima, N., and Fujii, R. (1987). Mechanism of light reflection in blue damselfish motile iridophore. *Zool. Sci.*, 4:243–257.

Kim, H., Ge, J., Kim, J., Choi, S., Lee, H., Lee, H., Park, W., Yin, Y., and Kwon, S. (2009). Structural colour printing using a magnetically tunable and lithographically fixable *photonic* crystal. *Nature Photonics*, 3:534–540.

Kim, S.H., Lee, S.Y., Yang, S.M., and Yi, G.R. (2011). Self-assembled colloidal structures for photonics. *NPG Asia Mater.*, 3:25–33.

Kolle, M. (2011). *Photonic structures inspired by nature*. Springer International, Dordrecht.

Land, M.F. (1972). The physics and biology of animal reflectors. *Progr. Biophys. Mol. Biol.*, 24:75–106.

Lawrence, J.R., Ying, Y., Jiang, P., and Foulger, S.H. (2006). Dynamic tuning of organic lasers with colloidal crystals. *Adv. Mater.*, 18:300–303.

Lee, Y.J., and Braun, P.V. (2003). Tunable inverse opal hydrogel. *Adv. Mater.*, 15:563–566.

Marlow, F., Muldarisnur, S.P., Brinkmann, R., and Mendive, C. (2009). Opals: status and prospects. *Angew. Chem. Int. Ed.*, 48:6212–6233.

Matsuoka, H., Harada, T., and Yamaoka, H. (1994). An exact evaluation of salt concentration dependence of interparticle distance in colloidal crystals by ultra-small-angle x-ray scattering. *Langmuir* 10:4423–4425.

Michielsen, K., and Stavenga, D.G. (2008). Gyroid cuticular structures in butterfly wing scales: biological photonic crystals. *J. R. Soc. Interface*, 5:85–94.

Nagaishi, H., Oshima, N., and Fujii, R. (1990). Light-reflecting properties of the iridophores of the neon tera, *Paracheirodon innesi*. *Comp. Biochem. Physiol. A Physiol.*, 95:337–341.

Oshima, N. (2005). Light reflection in motile iridophores of fish. In Kinoshita, S., and Yoshioka, S. (eds.), *Structural colors in biological systems*. Osaka University Press, Osaka. pp. 211–229.

Ozin, G.A., and Arsenault, A.C. (2008). *Nanochemistry*. 2nd ed. RCS Publishing, Cambridge.

Parker, A.R. (2009). Natural photonics for industrial inspiration. *Phil. Trans. R. Soc. A*, 367:1759–1782.

Parker, A.R., Welch, V.L., Driver, D., and Martini, N. (2003). Structural colour: opal analogue discovered in a weevil. *Nature*, 426:786–787.

Pursiainen, O.L.J., Baumberg, J.J., Ryan, K., Bauer, J., Winkler, H., Viel, B., and Ruhl, T. (2005). Compact strain-sensitive flexible photonic crystals for sensors. *Appl. Phys. Lett.*, 87:101902.

Sanchez, C., Arribart, H., and Guille, M.M.G. (2005). Biomimetism and bioinspiration as tools for the design of innovative materials and systems. *Nature Mater.*, 4:277–288.

Sanders, J.V. (1968). Diffraction of light by opals. *Acta Cryst.*, A24:427–434.

Saranathan, V., Osujib, C.O., Mochrieb, S.G.J., Nohb, H., Narayananf, S., Sandyf, A., Dufresneb, E.R., and Pruma, R.O. (2010). Structure, function, and self-assembly of single network gyroid (I4132) photonic crystals in butterfly wing scales. *Proc. Natl. Acad. Soc. USA*, 107:11676–11681.

Sato, O., Kubo, S., and Gu, Z.Z. (2009). Structural color films with lotus effects, superhydrophilicity, and tunable stop-bands. *Acc. Chem. Res.*, 42:1–10.

Takeoka, Y., and Watanabe, M. (2003). Template synthesis and optical properties of chameleonic poly(N-isopropylacrylamide)gels using closest-packed self-assembled colloidal silica crystals. *Adv. Mater.*, 15:199–201.

Tao, A.R., DeMartini, D.G., Izumi, M., Sweeney, A.M., Holt, A.L., and Morse D.E. (2010). The role of protein assembly in dynamically tunable bio-optical tissues. *Biomaterials*, 31:793–801.

Toyotama, A., Kanai, T., Sawada, T., Yamanaka, J., Ito, K., and Kitamura, K. (2005). Gelation of colloidal crystals without degradation in their transmission quality and chemical tuning. *Langmuir*, 21:10268–10270.

Vigneron, J.P., Pasteels, J.M., Windsor, D.M., Vértesy, Z., Rassart, M., Seldrum, T., Dumont, J., Deparis, O., Lousse, V., Biró, L.P., Ertz, D., and Welch, V. (2007). Switchable reflector in the Panamanian tortoise beetle *Charidotella egregia* (Chrysomelidae: Cassidinae). *Phys. Rev. E*, 76:031907.

Vlasov, Y.A., Bo, X.Z., Sturm, J.C., and Norris, D.J. (2001). On-chip natural assembly of silicon photonic bandgap crystals. *Nature*, 414:289–293.

Walsh, J.J., Kang, Y., Mickiewicz, R.A., and Thomas, E.L. (2009). Bioinspired electrochemically tunable block copolymer full color pixels. *Adv. Mater.*, 21:3078–3081.

Wu, Z., Lee, D., Rubner, M.F., and Cohen, R.E. (2007). structural color in porous, superhydrophilic and self-cleaning SiO2/TiO2 Bragg stacks. *Small*, 3:1445–1454.

Xia, Y., Gates, B., Yin, Y., and Lu, Y. (2000). Monodispersed colloidal spheres: old materials with new applications. *Adv. Mater.*, 12:693–713.

Xia, J., Ying, Y., and Foulger, S.H. (2005). Electric-field-induced rejection-wavelength tuning of photonic-bandgap composites. *Adv. Mater.* 17:2463–2467.

Yoshida, H., Ito, K., and Ise, N. (1991). Colloidal crystal growth. *J. Chem. Soc. Faraday Trans.*, 87:371–378.

Yoshioka, S., Matsuhana, B., Tanaka, S., Inouye, Y., Oshima, N., and Kinoshita, S. (2011). Mechanism of variable structural colour in the neon tetra: quantitative evaluation of the Venetian blind model. *J. R. Soc. Interface*, 8:56–66.

Zhao, Y., Zhao, X., and Gu, Z. (2010). Photonic crystals in bioassays. *Adv. Funct. Mater.*, 20:2970–2988.

5.2 Moth-Eye Antireflection Surface Using Anodic Porous Alumina

Yoshihiro Uozu

Figure 5.9 is a photograph of a natural moth-eye structure. Moth eyes have an uneven surface that shows low reflection of light. This structure serves as a survival mechanism by making the animal less vulnerable to predators [Clapham and Hultley 1973]. Recently, research and development for the industrial production of the moth-eye structures for antireflection coatings has been advanced. The continuous photo-nanoimprinting of moth-eye structures has been developed with a seamless anodic porous alumina mold.

FIGURE 5.9
Natural moth-eye structure.

5.2.1 Review of Films for Diminishing Reflection

Generally, there are two types of commercial films that aim at diminishing reflection, antireflection (AR) and antiglare (AG) films. An AR film reduces reflection itself to improve the image quality of plasma displays (PDs). An AG film scatters the reflected light to reduce glare of liquid crystal displays (LCDs). Flat panel display (FPD) screens have become increasingly large, and the improvement in image quality is becoming more important. For an LCD with an AG film, improvement of clearness of a picture without the whitish appearance of the scattered light is required. For PD with an AR film, diminishing reflection is required. Extremely low reflectance films fulfill both requirements. Most currently marketed AR films are of the multilayer type. The multilayer type AR film can improve its characteristics by increasing layer numbers [Dobrowolski et al. 2002], but because low cost is crucial for mass-produced FPDs, simple two-layer type films are commonly used.

Figure 5.10 is the image of an artificial moth-eye surface produced in our group. The moth-eye surface has a minute, uneven structure on the nanometer scale. There is a continuous decrease in refractive index from 1.5 at the bottom, near the plastic substrate, to 1.0 at the upper surface, which is the value of air, giving a very high difference of the refractive index of 0.5. Individual protrusions in the film have a cone-like shape with a diameter of less than 250 nm and an aspect ratio of larger than unity [Ibn-Elhaj and Schadt 2001].

When the pitch between adjacent protrusions is lower than the value of wavelength of incident light divided by refractive index of the substrate, reflection becomes negligible (Figure 5.10). In the case of a wavelength of

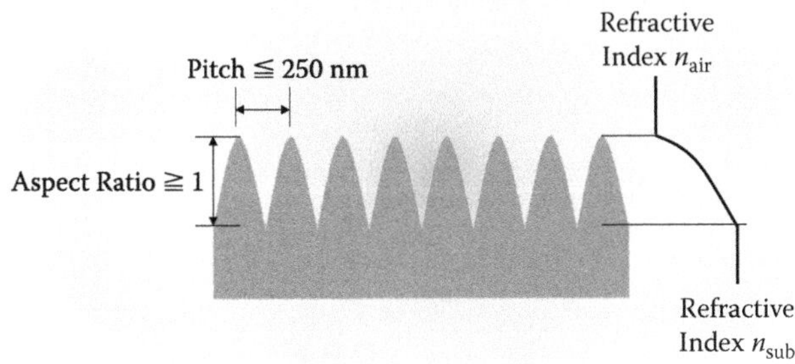

Pitch Realizing Non-reflection

$$\Lambda \leqq \frac{\lambda}{n}$$

Λ Pitch (nm)
λ Wavelength (nm)
n Refractive Index

FIGURE 5.10
Characteristics of our artificial moth-eye surface.

400 nm and the refractive index of 1.5, the pitch is required to be lower than 267 nm.

5.2.2 Production Process of the Moth-Eye Antireflection Film

Nanoimprinting, which generates fine patterns of polymers using molds, is a promising candidate for a high-throughput patterning process of large-area films. When the pattern for imprinting is required to be fine, the manufacture of the mold is one of the most important assignments of the technology. The mold used for nanoimprinting is usually prepared by electron beam (EB) lithography. However, this technique has the disadvantage of low throughput for preparing molds of large size.

The anodization of aluminum in acidic solution forms anodic porous alumina by self-organization, as illustrated in Figure 5.11. The porous alumina has cells of a uniform size with hexagonal packing of pores. Every cell has a similar size of 50–500 nm controlled by the anodization voltage and has a pore of the size of one-third of the cell distance. Generally, carbon is used as the cathode and several acids, such as sulfuric acid, oxalic acid, etc., can be chosen, depending on the applied voltage [Masuda and Fukuda 1995]. The nanoimprinting processes based on such highly ordered anodic porous alumina had been researched in detail [Masuda et al. 2001, Yanagishita et al. 2006, 2007a, b].

Table 5.1 compares the characteristics of two moth-eye producing processes. Lithography enables the production of moth-eye structures on a small area but is generally too expensive. Also, lithography cannot produce moth-eye structures on large areas and on curved surfaces. On the other hand, porous alumina can be formed on a large area and on curved

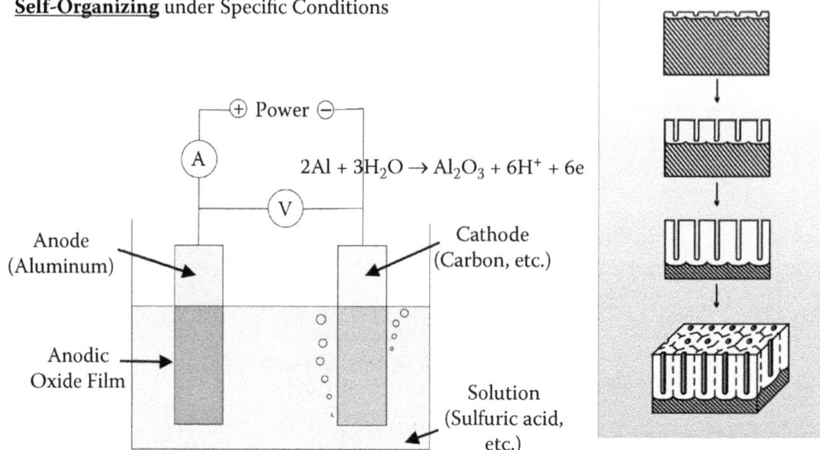

FIGURE 5.11
Electrochemical formation of porous alumina.

TABLE 5.1

Characteristics of Moth-Eye Film Production Processes

	Anodic Porous Alumina	Lithography (electron beam drawing, interference exposure)
Structure formation	Possible	Possible
Wide area	Possible	Difficult
Roll-type mold	Possible	Difficult
Production cost	Low cost	Expensive

surfaces. Therefore, porous alumina can be processed on a large size roll mold required for a roll-to-roll process with high productivity.

Figure 5.12 illustrates the moth-eye alumina mold production process. An oxalic acid aqueous solution is used as the electrolyte. High-purity aluminum is anode-oxidized under constant voltage to form the porous alumina. Next, this porous alumina is etched with a phosphoric acid aqueous solution to enlarge pore diameters. By repeating these processes several times, a porous alumina mold with a taper shape can be fabricated.

Figure 5.13 shows a scheme of a photo-imprinting process with the moth-eye mold [Yanagishita et al. 2007a, 2008]. At first, the photopolymer is filled on the mold and the mold with the photopolymer is covered with the transparent substrate, for example, poly(ethylene terephthalate) (PET) film, poly(methyl methacrylate) (PMMA) film, or glass sheet. Next, it is irradiated with UV light over the covered transparent substrate, and last, the moth-eye

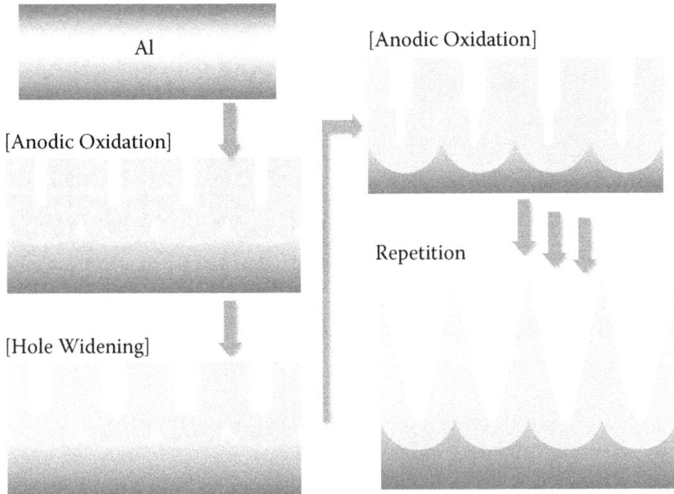

FIGURE 5.12
Production process of the moth-eye alumina mold.

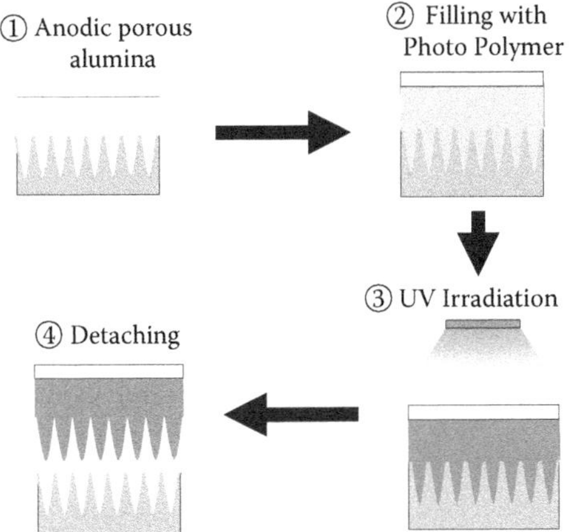

FIGURE 5.13
Photo-imprinting process.

sheet is detached from the mold. Figure 5.14(a) is the cross section image and Figure 5.14(b) is the surface image of a typical taper porous alumina mold. Figure 5.14(c) shows the field emission (FE)-SEM image of a moth-eye surface on PET film. Thus we could confirm that the shape of the taper porous alumina mold can be properly transferred to the moth-eye surface.

Continuous production processes of functional films with roll molds have also been developed. Figure 5.15 illustrates an image of a continuous roll photo-imprinting. The photopolymer is spread on the PET film. Then,

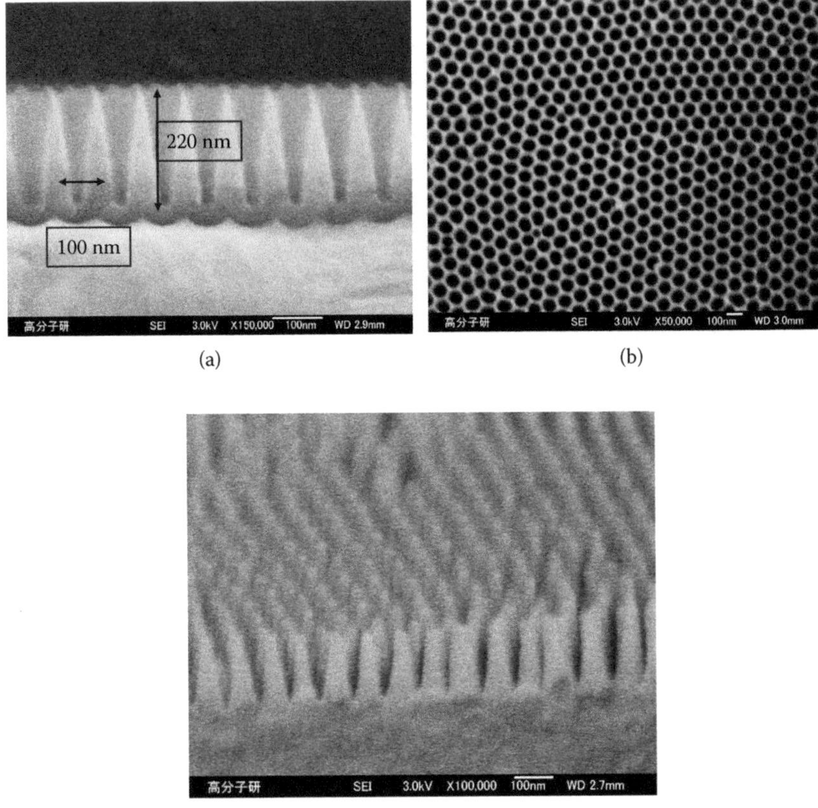

FIGURE 5.14
(a) Cross section TEM image of the tapered alumina mold. (b) Surface TEM image of the tapered porous alumina mold. (c) Cross section SEM image of the moth-eye surface.

the film is pressed on the roll mold and irradiated with UV light. Figure 5.16(a) shows the seamless roll mold for a moth-eye surface. The diameter is 200 mm and the width is about 300 mm. Figure 5.16(b) is the photograph of the transferred moth-eye film. The moth-eye film is highly transparent (the blue color comes from the protection film).

5.2.3 Characteristics of the Moth-Eye Antireflection Film

Figure 5.17 illustrates reflectance of a multilayer type AR film and two moth-eye AR films at an angle of incidence of 5°. Generally, multilayer type AR films prevent reflection mainly in the green, yellow, and red part of the spectrum, for which humans have the highest visual sensitivity. Both moth-eye AR films have lower reflection than multilayer type and have

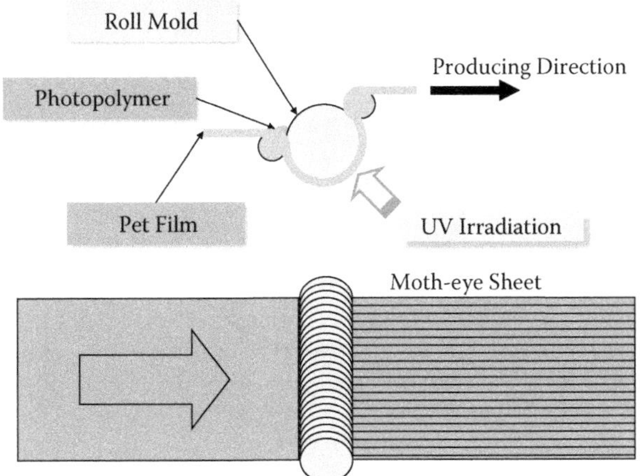

FIGURE 5.15
Image of continuous roll photo-imprinting.

FIGURE 5.16
(a) Seamless roll mold for a moth-eye surface. (b) Transferred moth-eye film.

low reflection throughout the whole visible spectrum. Figure 5.18 shows incident angle dependence on reflectance. In the case of the multilayer type AR film, reflectance becomes abruptly larger at 50°, reaching nearly 4% at 60°. For the moth-eye AR film, the tendency of increasing reflection with increasing incident angle is the same, but reflectance is lower than 1% even at an angle of 60°.

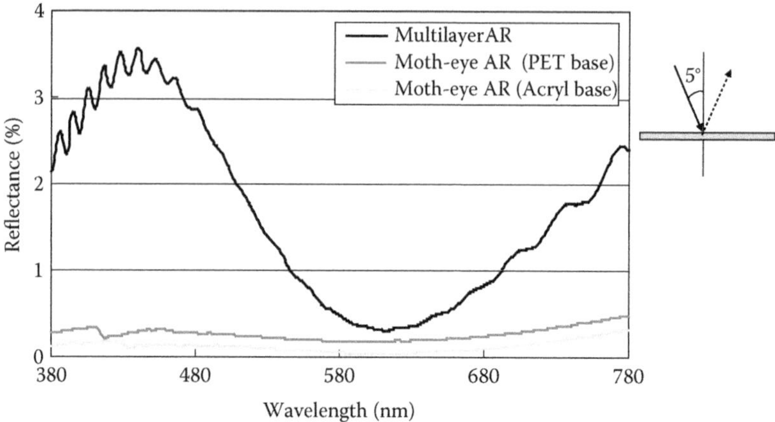

FIGURE 5.17
Reflectance of AR films for an angle of incidence of 5°.

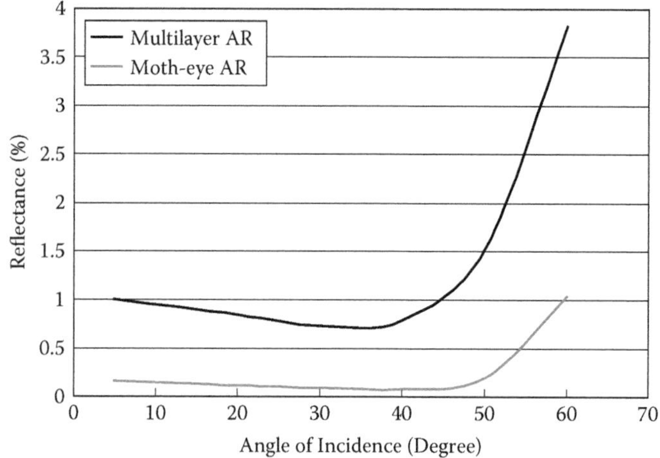

FIGURE 5.18
Incident angle dependence on reflectance at 550 nm light.

Figure 5.19 shows a picture of the background reflections of fluorescent lamps on the AR films. On the right is the multilayer type AR film and on the left is the moth-eye AR film. The right image shows a strong reflection of the lamps, which cannot be seen on the left image.

5.2.4 Summary

Self-organization is the bottom-up production technology completely opposite of the conventional top-down technology. Self-organization is widely

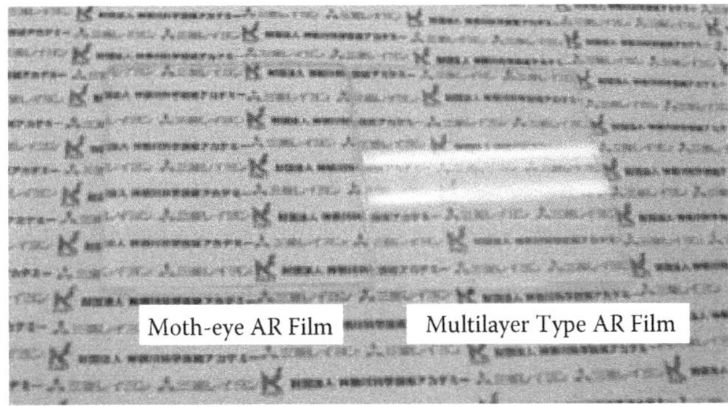

FIGURE 5.19
Background reflections on AR films.

noticed as the indispensably required processes in the bottom-up construction of nanostructured materials and devices. The fabricating process of the porous alumina mold is a representative applied example of self-organization.

The study of biomimetics in imitation of creatures is becoming popular, to express new functions in artificial media and to reduce the production cost of nanostructures [Shimomura 2010]. Our moth-eye antireflection film shows a higher antireflection property than conventional multilayer type antireflection films while being lower in production energy and cost.

References

Clapham, P.B., and Hultley, M.C. (1973). Reduction of lens refluxion by the "moth eye" principle. *Nature*, 244:281–282.

Dobrowolski, J.A., Poitras, M.D., Vakil, P.H., and Acree, M. (2002). Toward perfect antireflection coatings: numerical investigation. *Appl. Opt.*, 41:3075–3083.

Ibn-Elhaj, M., and Schadt, M. (2001). Optical polymer thin films with isotropic and anisotropic nano-corrugated surface topologies. *Nature*, 410:796–799.

Masuda, H., and Fukuda, K. (1995). Ordered metal nanohole arrays made by a two-step replication of honeycomb structures of anodic alumina. *Science*, 268:1466–1468.

Masuda, H., Yotsuya, M., Asano, M., Nishio, K., Nakao, M., Yokoo, A., and Tamamura, T. (2001). Self-repair of ordered pattern of nanometer dimensions based on self-compensation properties of anodic porous alumina. *Appl. Phys. Lett.*, 78:826–828.

Shimomura, M. (2010). The new trends in next generation biomimetics material technology: learning from biodiversity. *Nistep Q. Rev.*, 37:53–75.

Yanagishita, T., Nishio, K., and Masuda, H. (2006). Nanoimprinting using Ni molds prepared from highly ordered anodic alumina. *Jpn. J. Appl. Phys.*, 45:L804–806.

Yanagishita, T., Nishio, K., and Masuda, H. (2007a). Polymer through-hole membrane fabricated by nanoimprinting using metal molds with high aspect ratios. *J. Vac. Sci. B*, 25, L35–38.

Yanagishita, T., Nishio, K., and Masuda, H. (2008). Optimization of antireflection structures of polymer based on nanoimprinting using anodic porous alumina. *J. Vac. Sci. B*, 26:1856–1859.

Yanagishita, T., Yasui, K., Kondo, T., Kawamoto, K., Nishio, K., and Masuda, H. (2007b). Antireflection polymer surface using anodic porous alumina molds with tapered holes. *Chem. Lett.*, 36:530–531.

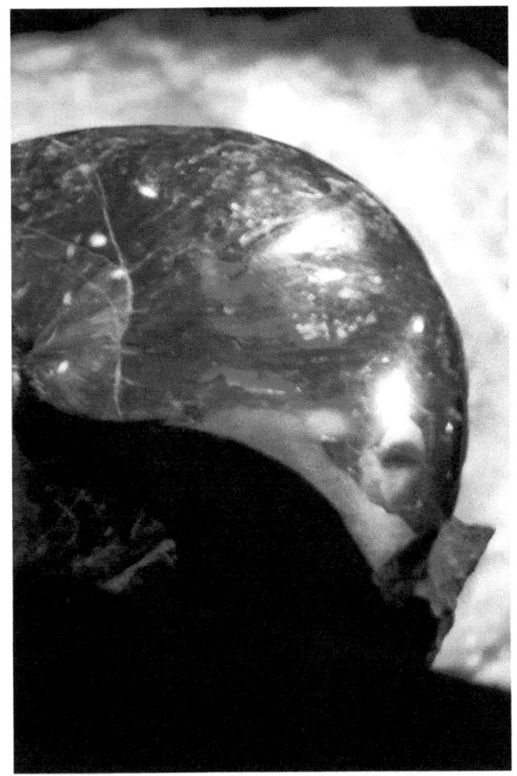

FIGURE 2.1
A fossil ammonite (Museum Zeche Zollverein, Essen, Germany). Such gemstones with color effects are called ammolites.

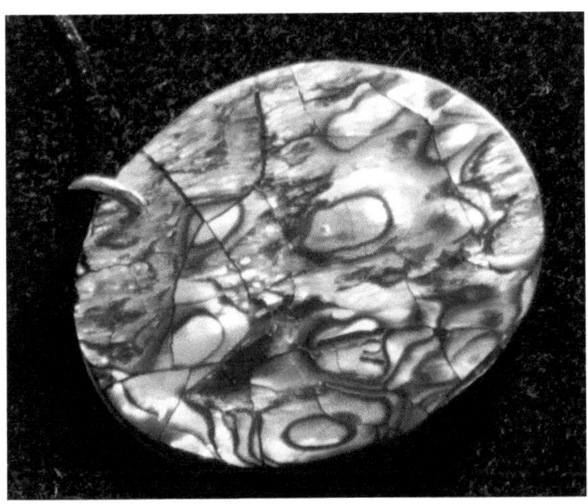

FIGURE 2.5
Jewelry made from the nacreous parts of abalone shells.

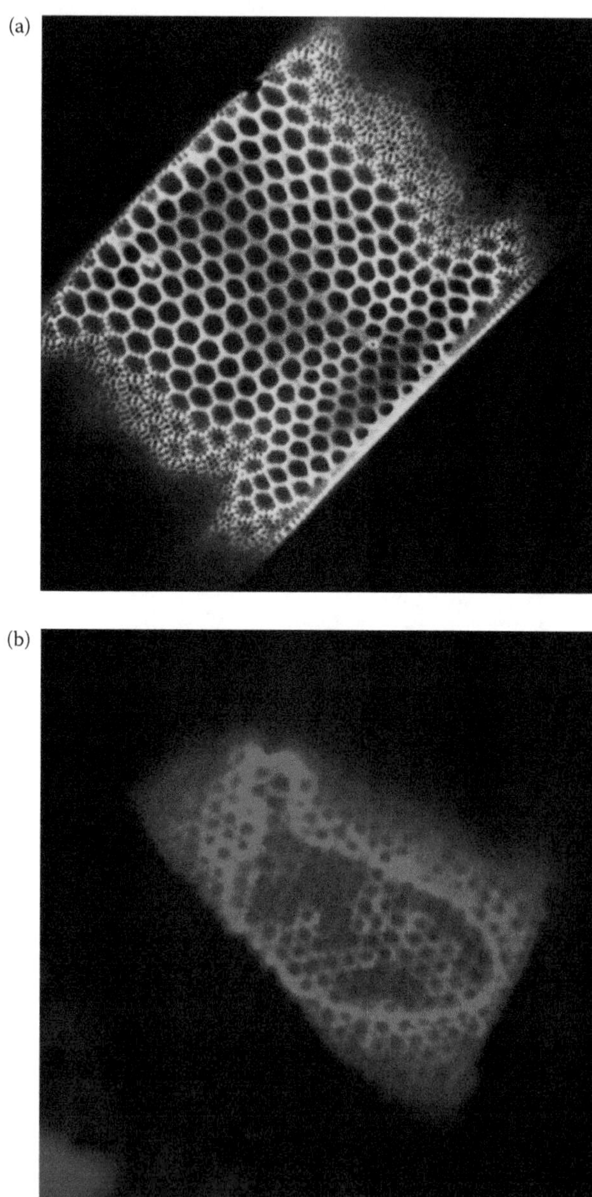

FIGURE 2.16
Confocal laser scanning microscope images of diatom shells (*Coscinodiscus granii*) with (a) an incorporated rhodamine dye after in vivo fluorochromation [Kucki 2009] (b) with a rhodamine dye (Rhodamine B) covalently attached to the silica surface.

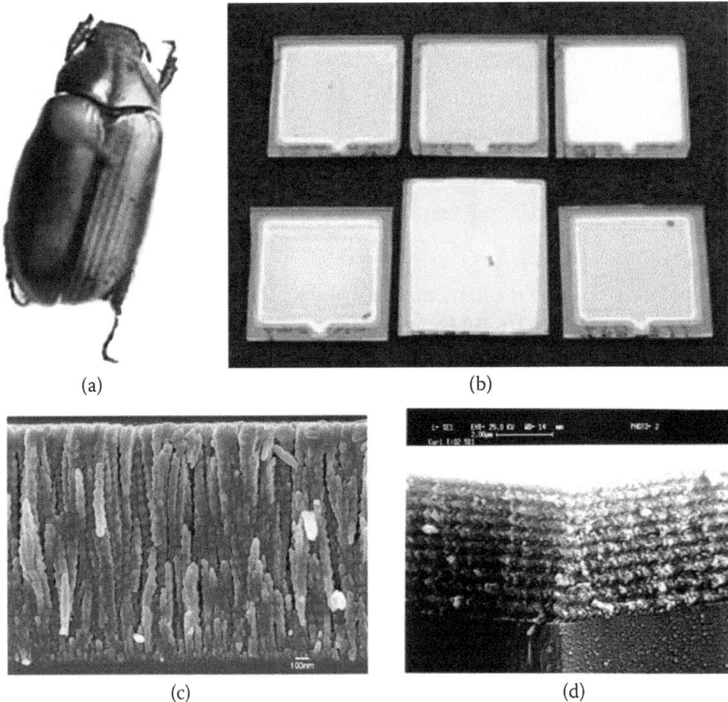

FIGURE 3.3
(a) A Manuka (scarab) beetle with (b) titania mimetic films of slightly different pitches. (c) Scanning electron micrograph of the chiral reflector in the beetle's cuticle. (d) Scanning electron micrograph of the titania mimetic film. Images by L. DeSilva and I. Hodgkinson, reproduced with permission.

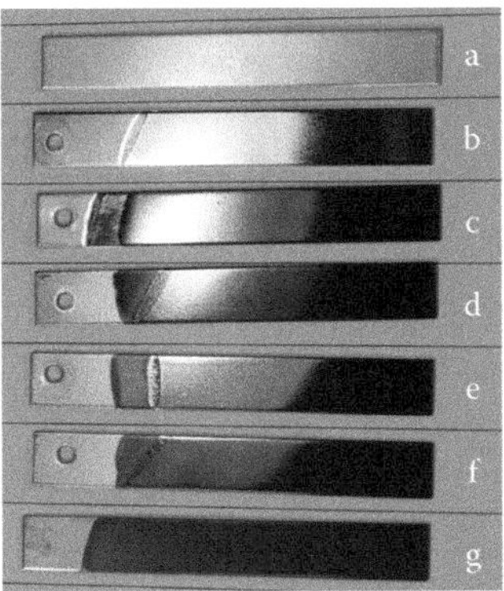

FIGURE 3.13
Stainless steel samples electropolished and coated with a single transparent layer resulting in interference colors. (a) Raw 304 stainless steel, (b) electropolished, (c)–(g) electropolished and coated with increased layer thickness. (Photograph courtesy of Per Møller and Torben Lenau, Technical University of Denmark.)

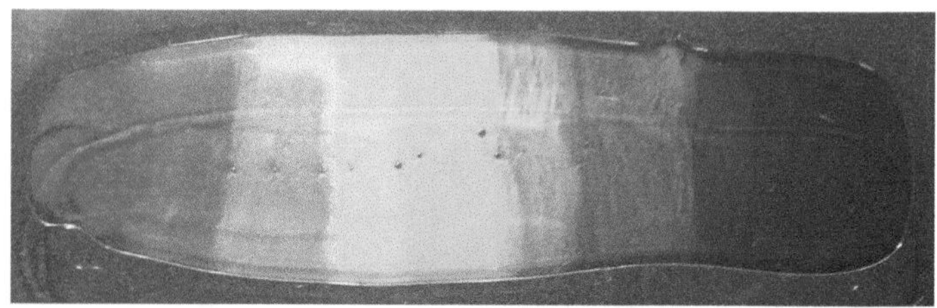

FIGURE 3.19
Range of colors that can be made by mixing two block copolymers in varying proportions. (With kind permission from Dr. Andrew Parnell, Sheffield University.)

FIGURE 3.20
The golden beetle *Anoplagnathus parvulus*. (With kind permission from Robert Perger from the Coleop-Terra Organization.)

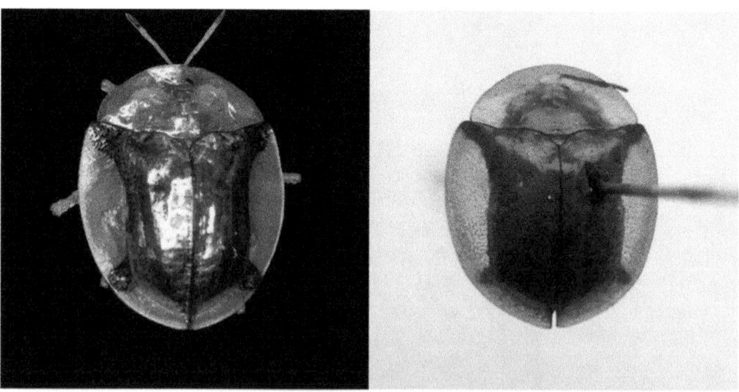

FIGURE 3.21
The *Aspidomorpha tecta* beetle at left is alive and has a golden appearance, while the museum specimen at the right has lost the golden color. The beetles are about 10 mm long. (Photograph courtesy of the Aquazoo/Löbbecke-Museum in Düsseldorf and the Zoological Museum at Copenhagen University.)

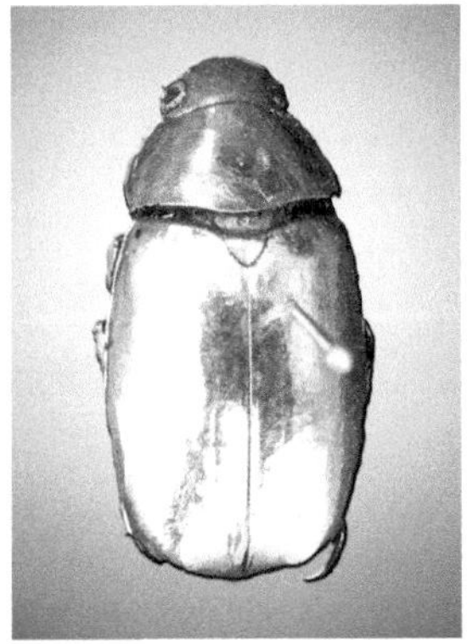

FIGURE 3.23
The golden beetle *Plusiotis resplendens*. (The beetle is kindly provided from the collections of the Zoological Museum, Copenhagen University.)

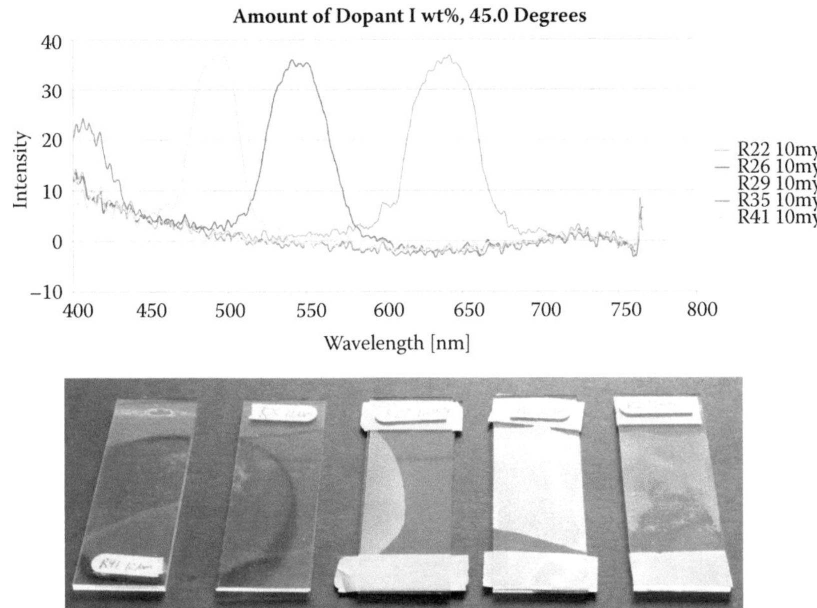

FIGURE 3.29
Reflectance spectra at 45° incidence and observation angle for Merck MLC 6608 doped with (from right to left) 22% (red curve), 26% (yellow curve with black spots), 29% (green curve), and 35% (blue curve) Merck MLC 6248. Measurements are normalized regarding light source and glass reflections. The picture shows slightly "colder" colors due to a different angle of incidence.

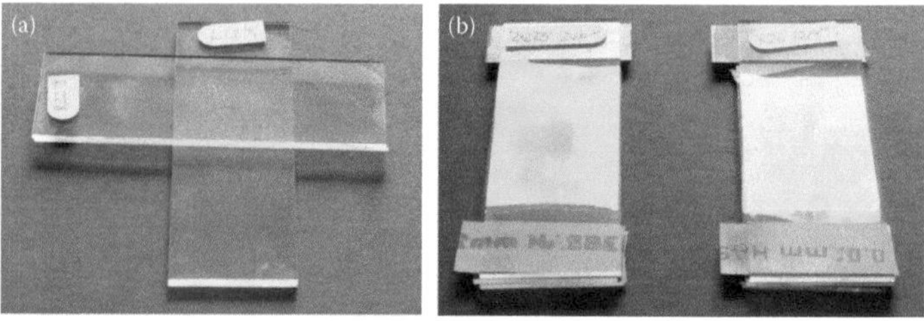

FIGURE 3.30
(a) A left-handed and a right-handed CLC placed between glass plates. Apparently the reflection looks larger where the two samples are superimposed. Samples have 22% dopant. (b) The samples are placed between three glasses. The left one has 2 × 25% left-handed dopant and the one to the right, which looks brighter, has 25% left-handed and 26% right-handed dopant.

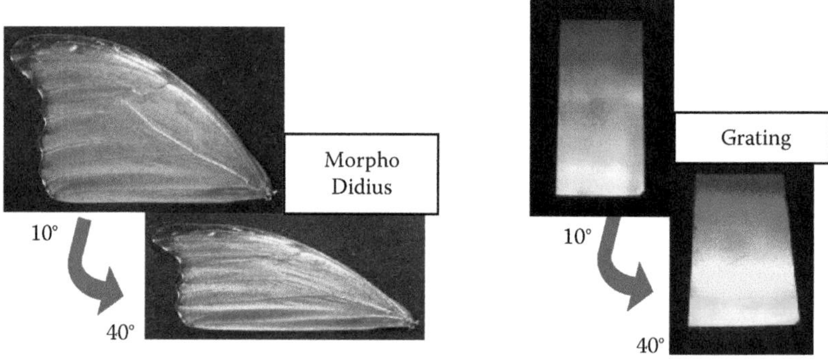

FIGURE 3.31
Angular dependences of the coloration on the viewing angle: (left) *Morpho* butterfly and (right) grating.

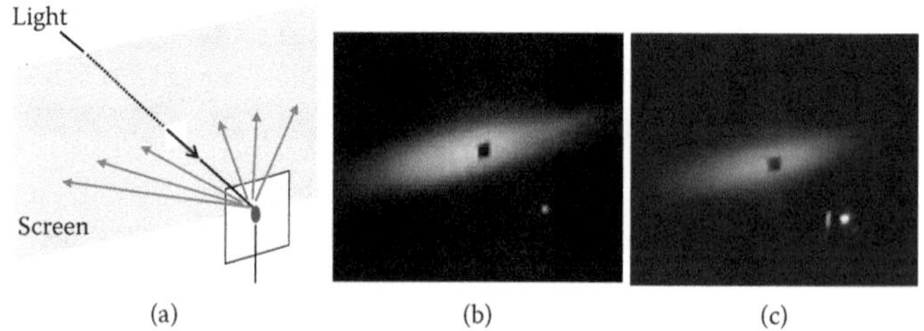

FIGURE 3.38
(a) Schematic image of the measurement system. Comparison of patterns of the lights reflected by (b) the original *Morpho* butterfly and (c) the emulated artificial film.

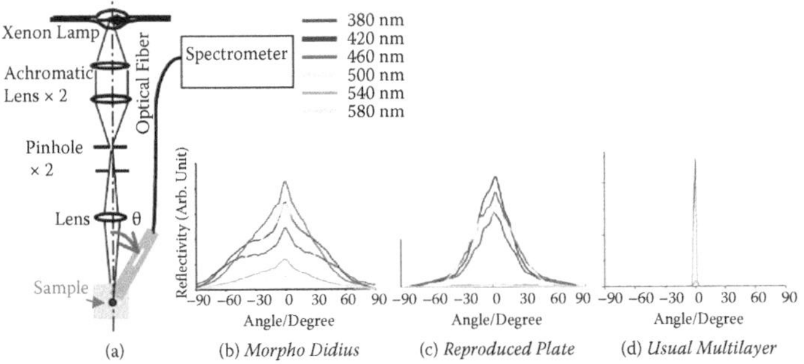

FIGURE 3.39
Comparison of measured angular dependence of the reflectivity from normal incident light. The angle was scanned along the lateral axis in Figure 3.38. (a) Schematic image of the measurement system. Angular profiles on (b) *Morpho didius*'s wing, (c) reproduced *Morpho* film, and (d) usual continuous and flat (not discrete) multilayer.

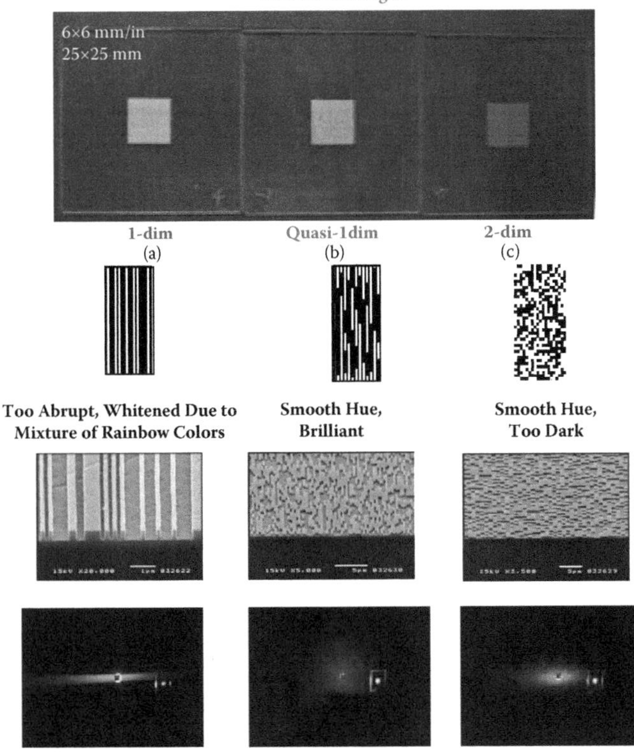

FIGURE 3.40
Comparison of the visible aspect (upper), pattern in SEM image (middle), and reflective pattern (lower) among (a) the poorly random pattern having 1D anisotropy, (b) *Morpho*-type pattern having quasi-1D anisotropy, and (c) 2D random pattern.

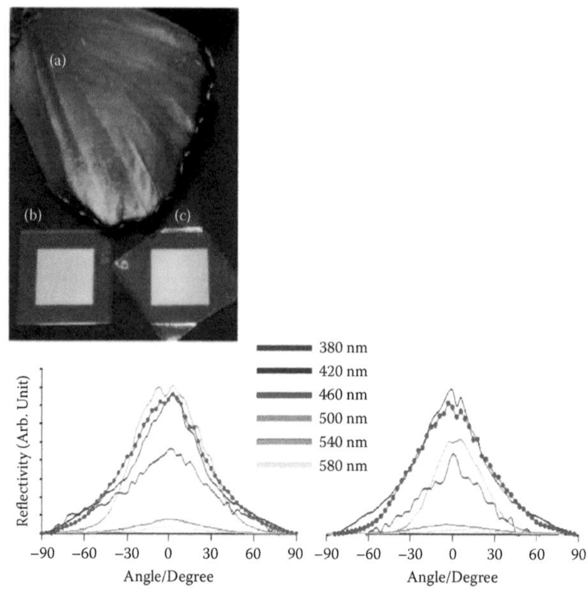

FIGURE 3.42
Photographs of (a) *Morpho* butterfly wing, (b) replicated *Morpho* blue plate, (c) synthetic *Morpho* blue plate developed by nanoimprinting for mass production (left, prototype; right, plate fabricated by nanoimprinting method for mass production). Also, corresponding optical property (reflective angular dependence) is shown.

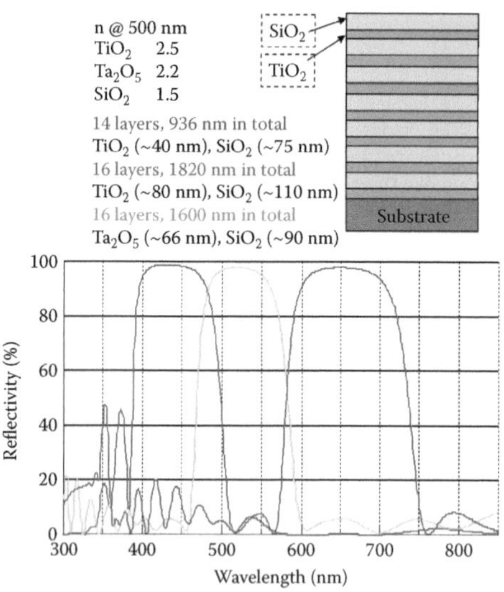

FIGURE 3.43
Optical design of the deposited multilayers that determine the artificial coloration and resultant spectra for primary colors (RGB).

FIGURE 3.44
Results of the artificial *Morpho* color in RGB primary colors.

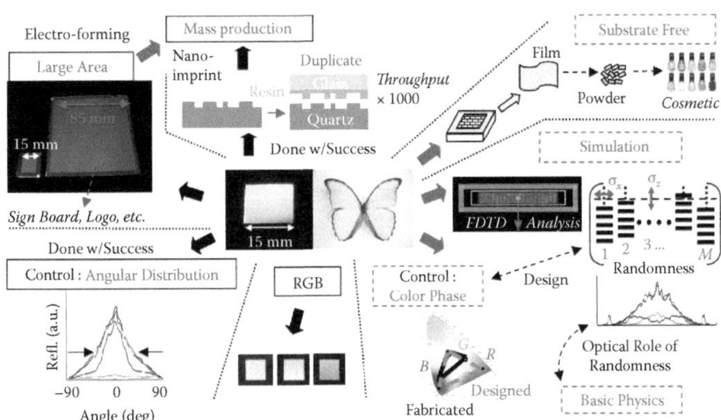

FIGURE 3.45
Recent progress (solid frame is accomplished, and dotted frame is in progress).

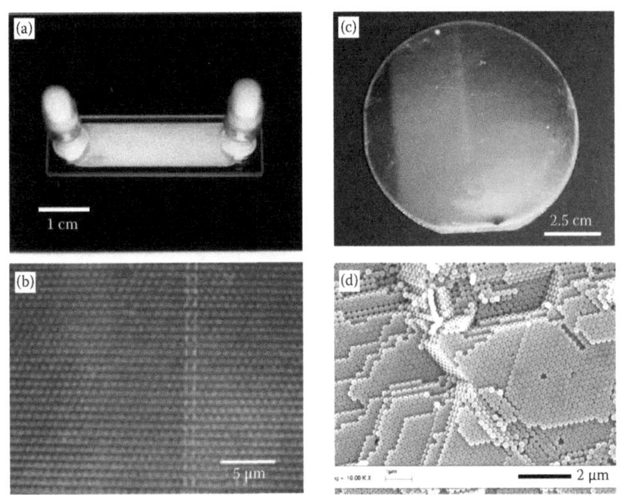

FIGURE 5.3
Two types of colloidal crystals generated by original fabrication techniques in the author's group. (a) A red structural color of non-closely packed colloidal crystal inside a flat cell. (b) Confocal laser micrograph indicates hexagonal array of 300 nm PS colloidal particles. (c) Green structural color of colloidal crystal film coated on a silicon wafer. (d) SEM image shows a 3D opal film assembled from 200 nm PS colloidal particles.

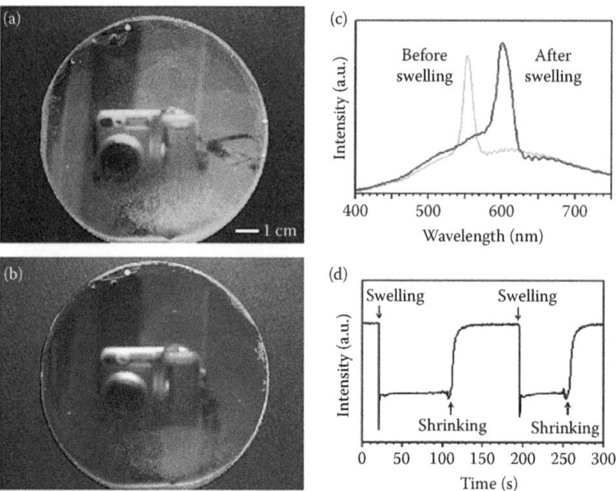

FIGURE 5.5
Tuning structural color in a closely packed colloidal crystal, i.e., opal photonic crystal, film by swelling and shrinking. (a, b) Structural colors before and after wetting the surface with isopropanol. (c) Reflectance spectra taken from the surface of the film before and after swelling. (d) A plot showing the variation of the reflectance intensity measured at 550 nm with time. (From Fudouzi, H., and Xia, Y., *Langmuir*, 19:9653–9660, 2003. Reproduced with permission. Copyright © 2003, American Chemical Society.)

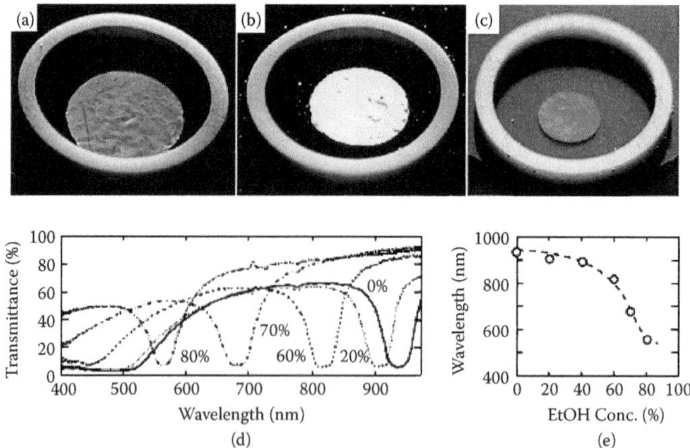

FIGURE 5.6
Tuning structural color by swelling of a hydrogel colloidal crystal. Change in the Bragg reflection color of the gelled colloidal crystals for different ethanol (EtOH) concentrations: (a) 20%, (b) 65%, and (c) 80%. The diameter of the Teflon ring is 12.3 mm. (d) Transmittance spectra for various EtOH concentrations. The gelled colloidal crystals shrink with an increase in EtOH concentration, and a blue shift is observed in its spectrum. (e) The dip wavelength in the transmittance spectra shown in (a) as a function of the EtOH concentration. (From Toyotama, A., *Langmuir*, 21:10268–10270, 2005. Reproduced with permission. Copyright © 2005, American Chemical Society.)

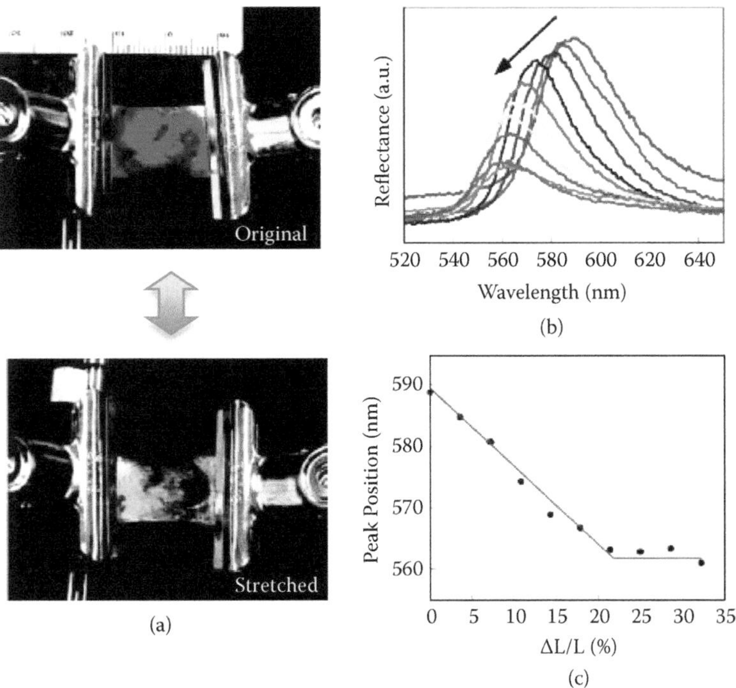

FIGURE 5.7
Tuning structural color by mechanical deformation of an elastic opal film coated on a silicone rubber sheet. (a) The reversible color change in the initial and stretched condition. The structural color of the elastic opal film reversibly changes from red to green. (b) Reflectance spectra change under mechanical strain on the rubber sheet. Bragg diffraction peak shifts to shorter wavelengths. (c) Relationship between the peak position and elongation ratio $\Delta L/L_0$. (From Fudouzi, H., and Sawada, T., *Langmuir*, 22:1365–1368, 2006. Reproduced with permission. Copyright © 2006, American Chemical Society.)

FIGURE 6.5
One frame of a color sequence obtained by a Sony DCR-HC23 handheld consumer camera. The resolution of this clip is 360 × 288. (a) An image from the original dark sequence. (b) The contrast-enhanced image, utilizing contrast-limited histogram equalization. (c) An image taken from the sequence after the adaptive smoothing approach has been applied.

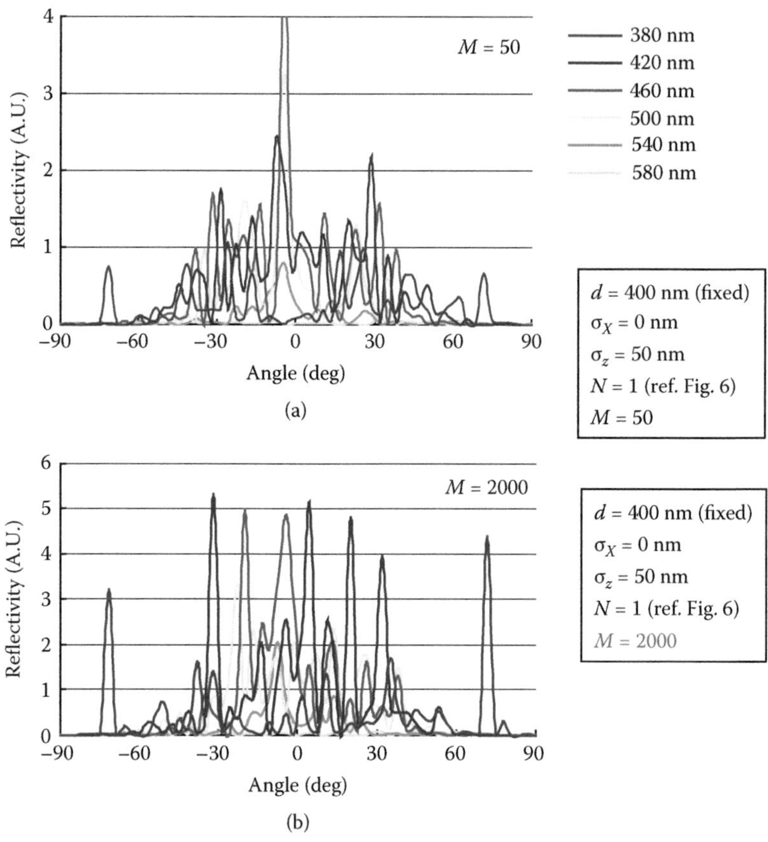

FIGURE 7.19
Comparison of the optical effects of the structural randomness. Simulated angular dependence of reflectivity for the model shown in Figure 7.18(b) for M = 50 and M = 2,000.

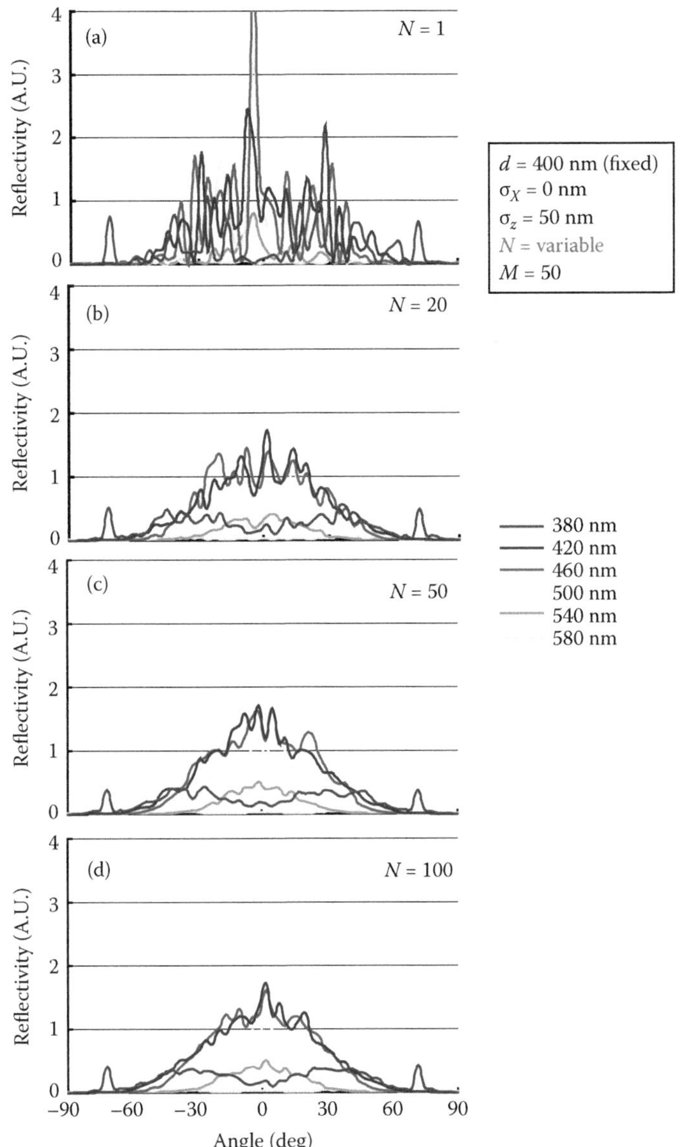

FIGURE 7.20
Optical effect of the incoherence (division) parameter N. Dependence of simulated angular profile of reflectivity on N. (a) N = 20, (b) N = 50, (c) N = 100. The results on the independent N blocks (one block is modeled in Figure 7.18(b), where M = 50, σ_z = 50 nm, d = 400 nm) were averaged.

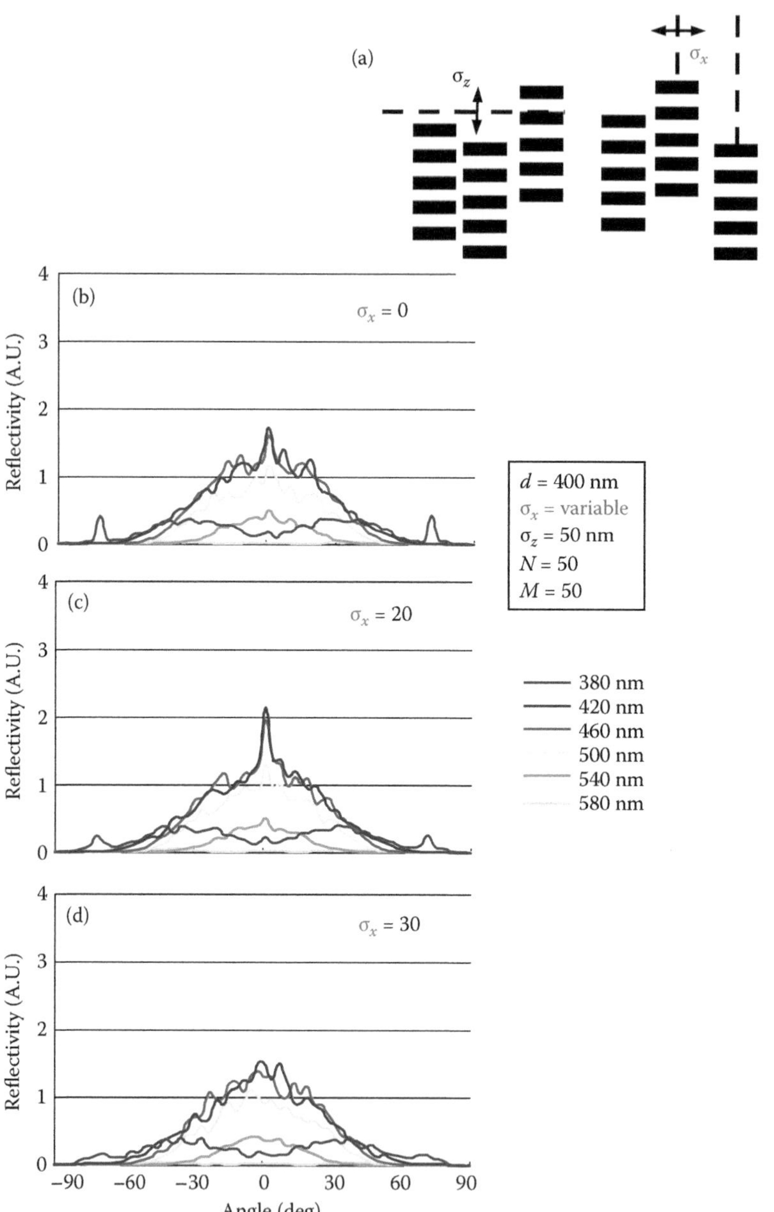

FIGURE 7.22
(a) Concept of the lateral randomness σ_x (b)–(d) Optical effect of the lateral randomness σ_x. Simulated angular profile of reflectivity for the models that have common averaged lateral distance d = 400 nm and the different lateral randomness σ_x: (b) σ_x = 0 nm, (c) σ_x = 20 nm, (d) σ_x = 30 nm. N = 50, M = 50, and σ_z = 50 nm were used.

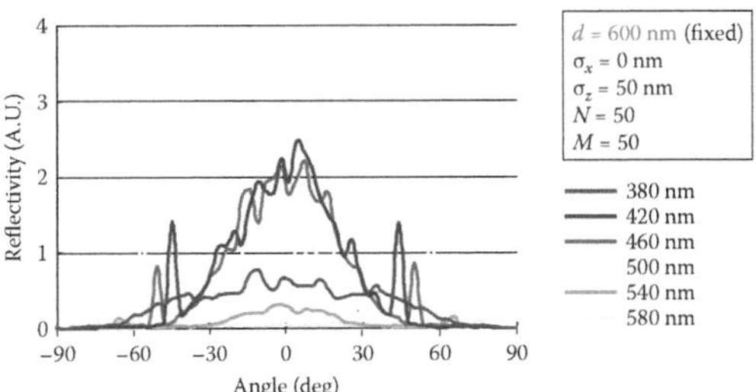

FIGURE 7.23
Example of the grating effect under the condition that $\sigma_x = 0$ nm, where side peaks appear dependent on wavelength. Simulated angular profile of reflectivity for the model in which the averaged lateral distance d = 600 nm, M = 50, N = 50, and $\sigma_z = 50$ nm.

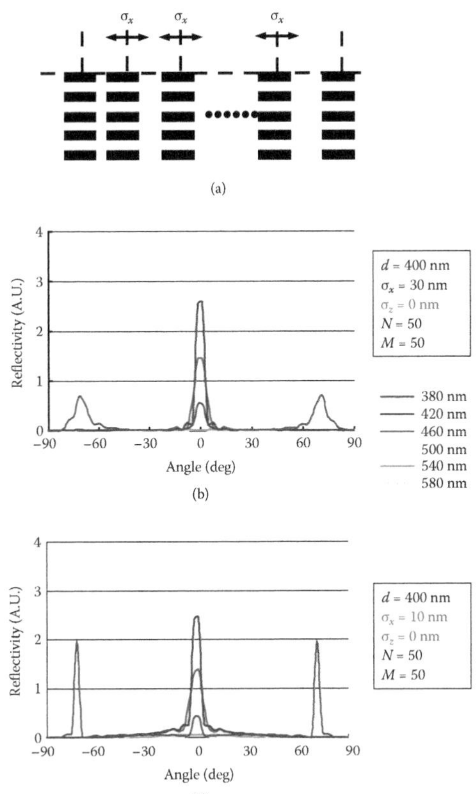

FIGURE 7.24
Optical effect of the vertical randomness σ_z. (a) Concept of the model that considers only the lateral randomness σ_x (average lateral distance d = 400 nm), whereas vertical randomness $\sigma_z = 0$ nm. (b, c) Simulated angular profiles of reflectivity for the model shown in (a), with the values of (b) $\sigma_x = 30$ nm and (c) $\sigma_x = 10$ nm. M = 50, N = 50 for both.

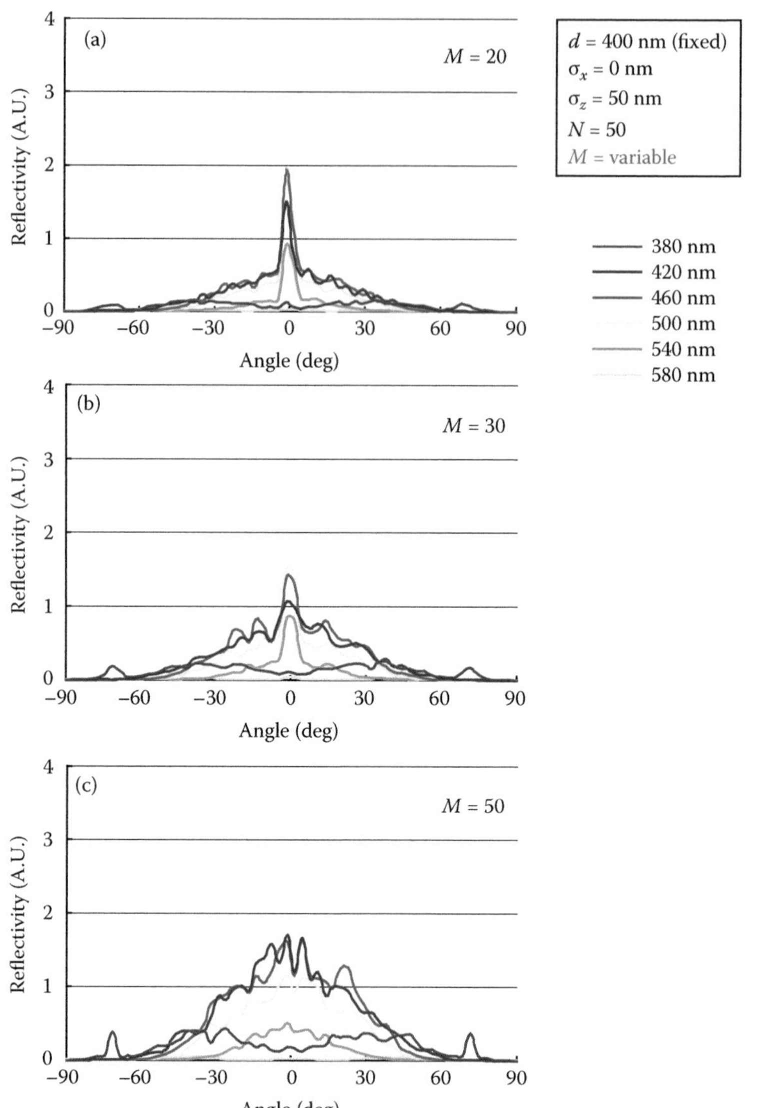

FIGURE 7.25
Optical effect of the number M of the random components (Figure 7.18(b)). Dependence of the simulated angular profile of reflectivity on M. (a) M = 20, (b) M = 30, (c) M = 50. N = 50, $\sigma_x = 0$ nm, $\sigma_z = 50$ nm, and d = 400 nm, for all.

6

A Night Vision Algorithm Inspired by the Visual System of a Nocturnal Bee

Eric Warrant, Magnus Oskarsson, and Henrik Malm

CONTENTS

6.1 Vision and Visual Processing in the Nocturnal Bee *Megalopta genalis* ... 166
6.1.1 The Nocturnal Bee *Megalopta genalis* ... 166
6.1.2 Vision Is Unreliable in Dim Light ... 167
6.2 Neural Strategies for Reliable Nocturnal Vision in *Megalopta* ... 169
6.2.1 Photoreceptor Strategies ... 169
6.2.2 Higher-Order Strategies ... 171
6.3 A Biologically Inspired Night Vision Algorithm ... 172
6.3.1 Overview of Algorithm ... 173
6.3.2 Contrast Enhancement ... 174
6.3.3 Noise Reduction ... 176
6.4 Consideration for Color ... 180
6.5 Implementation on a Graphics Processing Unit ... 181
6.6 Results and Comparison to Other Methods ... 182
6.7 Summary ... 185
References ... 185

The diversity of different life-forms, their conquest of virtually all available habitats, and their seemingly limitless ability to adapt to new ecological challenges show that structural and physiological specializations of a few basic building plans uniquely match plants and animals to the specific lives they lead.

From a human perspective, these specializations often appear remarkable—not only do they allow organisms to solve difficult physical problems, such as to withstand extreme heat or cold, or to detect the weak mating calls of a distant conspecific, but they are also highly adaptable and robust, allowing animals to cope with large deviations in normal environmental conditions in an automatic and self-organizing fashion. Such qualities represent the "holy grail" in engineering design, where sensors and actuators not only solve specific tasks well, but also are robust against unexpected and unpredictable changes in the system. Such qualities are particularly desirable in robotics where engineers have for decades strived to create fully autonomous machines that can perform faultlessly in the absence of human intervention.

In this chapter we describe a new computer algorithm that has taken its inspiration from biological vision, and in particular the vision of animals that are active in extremely dim light. Many nocturnal and deep-sea animals see well at light levels that are several orders of magnitude too dim for us. As day-active (diurnal) beings, our visual system is adapted for vision in bright daylight. At night, our visual capacities are severely reduced, with a complete loss in our ability to see color and a severe loss in our ability to see fine spatial and temporal details. This is not the case for many nocturnal animals, notably insects. Recent work—particularly on fast-flying and highly aerodynamic moths and bees and on ball-rolling dung beetles—has shown that nocturnal animals are able to distinguish colors, to detect faint movements, to learn visual landmarks, to orient to the faint polarization pattern produced by the moon, and to navigate using the constellations of stars in the sky (reviewed in Warrant 2008a). These impressive visual abilities are the result of exquisitely adapted eyes and visual systems. Nocturnal animals typically have highly sensitive eye designs and visual neural circuitry that is optimized for extracting reliable information from dim and very noisy signals. Thus, even at light levels in which we are essentially unable to see, nocturnal animals enjoy the benefits of reliable vision, which allows them to orient, to gather food, and to pursue suitable mates. Moreover, this vision is robust, being able to automatically adjust to changing light levels caused by the passage of clouds or by movement between open and forested habitats.

The new algorithm is inspired by the neural principles that are present in nocturnal animals to produce reliable vision in dim light. Before discussing the details of the algorithm, we will begin by explaining the biological principles on which it is based. These principles have been derived from more than 20 years of basic research on the vision of nocturnal insects, particularly the tropical sweat bee *Megalopta genalis*, a nocturnal bee that is native to the rain forests of Central America.

6.1 Vision and Visual Processing in the Nocturnal Bee *Megalopta genalis*

6.1.1 The Nocturnal Bee *Megalopta genalis*

The nocturnal bee *Megalopta genalis* (Figure 6.1) is a semisocial bee of the family Halictidae (the sweat bees) that lives in groups of up to 11 females in hollowed-out sticks suspended in the rain forest understory (Wcislo et al. 2004, Wcislo and Tierney 2009). A nocturnal lifestyle is rare in bees, and has occurred

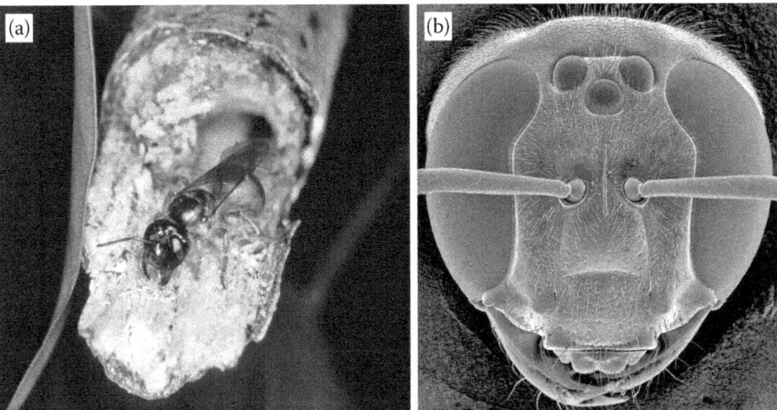

FIGURE 6.1 (See color figure at http://www.crcpress.com/product/isbn/9781439877463) The central American sweat bee *Megalopta genalis* (Halictidae), whose sensitive apposition eyes allow them to forage at night by visually learning landmarks along the foraging route and around the nest entrance. (a) A female at the entrance to the nest (a hollowed-out stick). (Reproduced with the kind permission of the photographer, Dr. Michael Pfaff.) (b) A colored scanning electron microscope image of the head of *Megalopta*, showing the prominent compound eyes and the three ocelli.

during evolution within only a few lineages (Warrant 2008b, Tierney et al. 2011), and mostly in the tropical regions of the world. Reduced competition for floral nectar resources and the reduced risk of predation are thought to be the main reasons why nocturnality in bees has arisen (Wcislo et al. 2004). *Megalopta* leaves its nest and forages on flowers under the thick rain forest canopy during two short time windows after dusk and before dawn, when light levels are comparable to starlight levels above the forest canopy (Warrant et al. 2004, Kelber et al. 2006). Vision plays a central role in this behavior, both in the control of flight and in the learning of spatial landmarks used during homing (Warrant et al. 2004, Baird et al. 2011).

6.1.2 Vision Is Unreliable in Dim Light

To see well in dim light, a visual system needs to extract reliable information from what may be an unreliable visual signal, that is, to extract information from a visual signal that is contaminated by visual "noise." Part of this noise arises from the stochastic nature of photon arrival and absorption: each sample of absorbed photons (or signal) has a certain degree of uncertainty (or noise) associated with it. The relative magnitude of this uncertainty is greater at lower rates of photon absorption, and these quantum fluctuations set an upper limit to the visual signal-to-noise ratio (Rose 1942, de Vries 1943, Land 1981). As light levels fall, the fewer the number of photons that are absorbed, the greater the noise relative to the signal and the less that can be seen. This noise limitation on detection reliability is equally problematic for artificial imaging systems, such as cameras, as it is for eyes.

There are also two other sources of noise that further degrade visual discrimination by photoreceptors in dim light. The first of these, referred to as transducer noise, arises because photoreceptors are incapable of producing an identical electrical response, of fixed amplitude, latency, and duration, to each (identical) photon of absorbed light. This source of noise, originating in the biochemical processes leading to signal amplification, degrades the reliability of vision (Lillywhite 1977, Lillywhite and Laughlin 1979, Laughlin and Lillywhite 1982). The second source of noise, referred to as dark noise, arises because the biochemical pathways responsible for transduction are occasionally activated—even in perfect darkness (Barlow 1956). These "dark events," electrical responses that are indistinguishable from those produced by real photons, are more frequent at higher retinal temperatures. At very low light levels this dark noise can significantly contaminate visual signals. In insects and crustaceans dark events are rare, only around 10 per hour at 25°C (Lillywhite and Laughlin 1979, Dubs et al. 1981, Doujak 1985). But in nocturnal toad rods the rate is much higher—360 per hour at 20°C (Baylor et al. 1980)—and this sets the ultimate limit to visual sensitivity (Aho et al. 1988, 1993). Thermal noise is also a severe limitation to signal reliability in digital imaging systems.

Visual signal reliability in dim light can be improved with an eye design that maximizes light capture. In nocturnal insects, including most moths and many beetles, a refracting superposition compound eye allows single photoreceptors in the retina to receive focused light from hundreds (and in some extreme cases, thousands) of corneal facet lenses (Figure 6.2(b)). This design represents a vast improvement in sensitivity over the apposition compound eye (Figure 6.2(a)), a design in which single photoreceptors receive light only from the corneal facet lens residing in the same ommatidium. Not surprisingly, apposition eyes are typical of diurnal insects active in bright sunlight, and this includes all diurnal bees. Strangely, apposition eyes are also found in the nocturnal bees such as *Megalopta*. Even stranger, despite the poor sensitivity afforded by apposition eyes, nocturnal bees see extremely well, with documented abilities to distinguish color (Somanathan et al. 2008) and to learn visual landmarks during foraging and homing (Warrant et al. 2004).

Nonetheless, the apposition compound eyes of nocturnal bees have optical adaptations that make them more sensitive to light than the apposition eyes of their diurnal relatives. For instance, *Megalopta* has larger eyes and larger corneal facet lenses (with diameters up to 36 μm) than the strictly day-active European honeybee *Apis mellifera* (with lens diameters up to 20 μm). Moreover, in *Apis* the light-sensitive rhabdom in each ommatidium (which collects, absorbs, and transduces incoming light) has a width of only 2 μm, whereas in *Megalopta* they reach an extraordinary 8 μm. These wide rhabdoms collect light from regions of space that are seven times wider than in

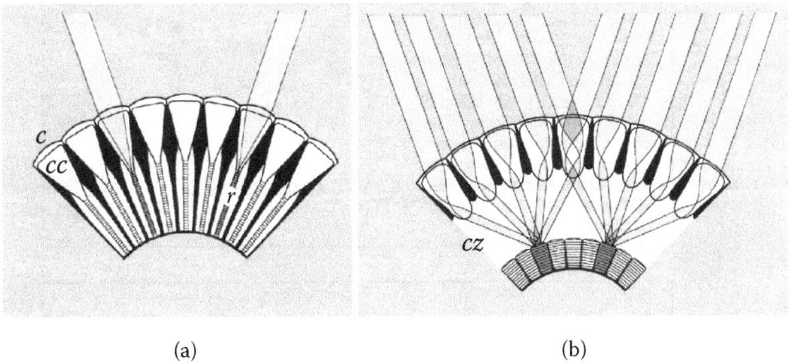

(a) (b)

FIGURE 6.2
Compound eyes are composed of identical units called ommatidia, each consisting of a lens element—the corneal lens (*c*) and crystalline cone (*cc*)—that focuses light incident from a narrow region of space onto the rhabdom (*r*), a photoreceptive structure composed of membranous microvilli. Each ommatidium is responsible for reading the average intensity, color, and (in some cases) plane of polarization within the small region of space that it views. Two neighboring ommatidia view two neighboring regions of space. Thus, each ommatidium supplies a pixel of information to a larger image of pixels that the entire compound eye constructs. Larger compound eyes with more ommatidia thus have the potential for greater spatial resolution. (a) A focal apposition compound eye (showing nine ommatidia in cross section). Light reaches the photoreceptors exclusively from the small corneal lens located directly above. This eye design is typical of day-active insects. (b) A refracting superposition compound eye (showing nine ommatidia in cross section). A large number of corneal facets and bullet-shaped crystalline cones collect and focus light—across the clear zone of the eye (*cz*)—toward single photoreceptors in the retina. Several hundred, or even thousands of, facets service a single photoreceptor. Not surprisingly, many nocturnal and deep-sea animals have refracting superposition eyes, and benefit from the significant improvement in sensitivity. (Diagrams courtesy of Dan-Eric Nilsson. Adapted from Warrant, E.J., *Current Biology*, 14:1309–1318, 2004.)

Apis (Greiner et al. 2004a, Warrant et al. 2004). These differences in rhabdom and facet size endow *Megalopta* with a roughly 27 times greater visual sensitivity than in *Apis*, albeit only at the cost of spatial resolution.

6.2 Neural Strategies for Reliable Nocturnal Vision in *Megalopta*

6.2.1 Photoreceptor Strategies

Recent experiments have revealed that nocturnal insects have photoreceptors that are uniquely suited to a life in dim light (Laughlin and Weckström 1993, Frederiksen et al. 2008). These adaptations were discovered by intracellularly recording the responses of photoreceptors to Gaussian-distributed white-noise light stimuli in closely related nocturnal and diurnal halictid

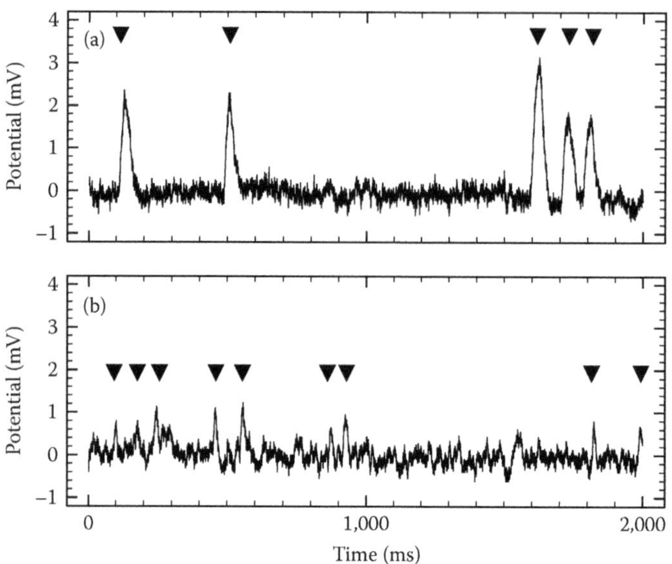

FIGURE 6.3
Responses to single photons (or photon bumps: arrowheads) recorded from photoreceptors in the nocturnal halictid bee *Megalopta genalis* (a) and the diurnal halictid bee *Lasioglossum leucozonium* (b). Note that the bump amplitude is larger, and the bump time course much slower, in *Megalopta* than in *Lasioglossum*. (Adapted from Frederiksen, R., *Current Biology*, 18:349–353, 2008.)

species: the nocturnal *Megalopta genalis* and the diurnal *Lasioglossum leucozonium* (Frederiksen et al. 2008). Two important differences in photoreceptor performance were found between these two species, each of which highlights an adaptation for vision in dim light, and both of which have importance for the new night vision algorithm.

First, the photoreceptor's dark-adapted voltage responses to single photons—so-called quantum bumps—are much larger in nocturnal *Megalopta* than in diurnal *Lasioglossum* (Figure 6.3). These larger bumps, which have also been reported from other nocturnal arthropods (e.g., crane flies, cockroaches, and spiders: Laughlin et al. 1980, Laughlin and Weckström 1993, Heimonen et al. 2006, Pirhofer-Walzl et al. 2007), indicate that the photoreceptor's gain of transduction is greater in *Megalopta* than in *Lasioglossum*. This higher transduction gain manifests itself as a higher "contrast gain," which means that the photoreceptor has a greater voltage response per unit change in light intensity (or contrast). At lower light levels the contrast gain is up to five times higher in *Megalopta* than in *Lasioglossum*, which results in greater signal amplification and improved visual reliability in dim light.

Second, like in many other nocturnal arthropods, the time response of dark-adapted photoreceptors of *Megalopta* is slow. Despite compromising temporal resolution, slower vision in dim light (analogous to having a longer exposure time on a camera) is beneficial because it increases the visual signal-to-noise ratio and improves contrast discrimination at lower temporal frequencies by suppressing photon noise at frequencies that are too high to be reliably resolved (van Hateren 1992, 1993). Visual speed can be measured in several ways, for instance, by measuring the ability of a photoreceptor to follow a light source whose intensity modulates sinusoidally over time: a photoreceptor that can follow a light source modulating at a high frequency is considered to be fast. Thus, in the frequency domain, slower vision is equivalent to saying that the temporal corner frequency is low—this is the frequency where the response has fallen to 50% of its maximum value, and lower values indicate slower vision. In *Megalopta* it is around 7 Hz in dark-adapted conditions. In diurnal *Lasioglossum* the dark-adapted corner frequency is around 20 Hz, a value that is nonetheless considerably less than that typical of the diurnal, highly maneuverable, and rapidly flying higher flies (50–107 Hz) (Laughlin and Weckström 1993). The difference in temporal properties between the two bee species is most likely due to different photoreceptor sizes, and different numbers and types of ion channels in the photoreceptor membrane (Laughlin 1996).

6.2.2 Higher-Order Strategies

Apart from having more sensitive eyes with photoreceptors that are slower and possess greater contrast gain, nocturnal animals have one further visual strategy at their disposal: the neural summation of light in space and time (Snyder et al. 1977a, 1977b, Snyder 1977c, Laughlin 1981, 1990, Warrant 1999). This summation strategy, thought to reside in the neural circuits processing the incoming visual signal, can potentially improve visual reliability significantly (Warrant 1999). We have already discussed the temporal summation of photons above: when light gets dim, the visual systems of nocturnal animals can improve visual reliability by integrating signals over longer periods of time (Laughlin 1990, van Hateren 1992, 1993). In the eye, this can be achieved by having slower photoreceptors. Even slower vision could be obtained by integrating signals at a higher level in the neural visual system. This temporal summation only comes at a price: it can drastically degrade the perception of fast-moving objects, potentially disastrous for a fast-flying nocturnal animal (like a nocturnal bee or moth) that needs to negotiate obstacles. Not surprisingly, temporal summation is more likely to be employed by slowly moving animals.

Summation of photons in space can also improve image quality. Instead of each visual channel collecting photons in isolation (a strategy used in bright light), the transition to dim light could activate specialized laterally spreading neurons that couple the channels together into groups. Each summed group—themselves now defining the channels—could collect considerably more photons over a much wider visual angle, albeit with a simultaneous and unavoidable loss of spatial resolution. Despite being much brighter, the image would become necessarily coarser.

Evidence for laterally spreading neurons has been found in the first optic ganglion (*lamina ganglionaris*), the first processing station in the brain for the analysis of signals arriving from the retina of the eye. These laterally spreading neurons, which have been found in nocturnal cockroaches (Ribi 1977), fireflies (Ohly 1975), and hawkmoths (Strausfeld and Blest 1970), have been interpreted as an adaptation for spatial summation (Laughlin 1981). The nocturnal bee *Megalopta genalis* also appears to have such laterally spreading neurons (Greiner et al. 2004b, 2005). In contrast, they are absent in diurnal bees such as *Apis* and *Lasioglossum* (Ribi 1981, Greiner et al. 2004b, 2005).

Spatial and temporal summation strategies have the potential to greatly improve the visual signal-to-noise ratio in dim light, and thereby the reliability of vision, for a narrower range of spatial and temporal frequencies (Warrant 1999, Theobald et al. 2006, Klaus and Warrant 2009). Thus, summation would ensure that nocturnal visual reliability is maximized for the slower and coarser features of the world. Those features that are faster and finer—and inherently noisy—would be filtered out. However, for a nocturnal bee struggling to find its way home in the dark, the ability to see a slow and coarse world, rather than nothing at all, would probably mean the difference between a successful return to the nest and becoming hopelessly lost.

Despite their tiny and relatively insensitive apposition eyes, nocturnal bees have successfully conquered the nocturnal niche, and taken advantage of the benefits that this niche provides for foraging and the avoidance of enemies, feats that require reliable vision in dim light. This reliability is the result of the combined action of a variety of adaptations within the eyes and visual system. Greatly enlarged corneal facet lenses and rhabdoms, and slow photoreceptors with high contrast gain, ensure that visual signal strength is maximal as it leaves the eye and travels to the brain. Visual signals are then spatially and temporally summed, resulting in an enhanced signal and reduced noise. The greatly improved signal-to-noise ratio that this strategy affords, while confined to a narrower range of spatial and temporal frequencies, would ensure that nocturnal visual reliability is maximized for the slower and coarser features of the world.

6.3 A Biologically Inspired Night Vision Algorithm

Like nocturnal eyes, artificial imaging systems operating in dim light also suffer from limitations in detection reliability. Hence, these systems should benefit from the same strategies for improving reliability. Thus, our idea was to take visual solutions present in nature and to incorporate them in a computer algorithm that radically improves the quality of video sequences collected in dim light. The remainder of this section summarizes the results.

6.3.1 Overview of Algorithm

We assume that the input to our algorithm is either color or grey-scale video data, collected under very poor lighting conditions. We assume nothing further about the data, and it could be recorded by any type of camera, be it a consumer grade or machine vision camera.

The first step in our image processing approach is to increase the contrast in order to make the structures in the darker areas of the images more visible. This can be done in several different ways. We have chosen to apply a modified version of the well-known procedure of histogram equalization (Gonzalez and Woods 1992). In its basic implementation this is a parameter-free approach that tries to flatten out the intensity histogram of the image as much as possible, and in this way widen the dynamic range and increase the contrast. A variant of this, where there is a limit on the slope of the function that transforms the intensities, is referred to as contrast-limited histogram equalization in the literature (Pizer et al. 1987). More information on the contrast enhancement step will be given in the next section.

Following contrast enhancement, amplification of the signal in the dark areas also results in an amplification of the noise in the image sequence. The next step therefore aims to reduce this noise significantly, while preserving as much real detail and structure in the sequence as possible. Inspired by the previously described higher-order summation principles discovered in the eyes of nocturnal animals, we propose to apply an adaptive spatiotemporal intensity smoothing. For every pixel in the image sequence a specific summation kernel (a small matrix of pixels) is constructed that adapts to the local spatiotemporal intensity structure. This kernel is then used to calculate a weighted mean in the neighborhood of the current pixel. Instead of applying a combination of pure spatial and pure temporal summation, as in our biological model (Warrant 1999), we view the three-dimensional spatiotemporal space (two spatial dimensions and one temporal dimension) symmetrically, and the summation kernels are directed and scaled according to the local three-dimensional intensity structure. This leads to an optimal way of smoothing the intensities, while preserving important object contours and preventing motion blur. The algorithm works very well on color image data and maximally restores the original chromaticity values that are seen in brighter lighting, along with the luminance. When dealing with raw color image data, obtained in a so-called Bayer pattern (Bayer 1976), the interpolation to a three-channel RGB image is done concurrently with the spatiotemporal smoothing.

The implementation of the algorithm is actually quite simple but requires a very large number of multiplications and summations at every pixel position. However, by performing spatiotemporal summation calculations on a graphics processing unit (GPU), found on a standard graphics card, the processing can be made parallel. In this way, the complete processing can reach real-time performance.

We have tested our image enhancement approach on a large amount of very challenging input data, with excellent results. The method outperforms other low-light-level methods that we have tested. We will now describe the different steps of the algorithm in greater depth. More details can be found in Malm and colleagues (Malm and Warrant 2006, Malm et al. 2007).

6.3.2 Contrast Enhancement

The very first step in our image enhancement approach is to apply an amplifying intensity transformation such that the dynamic range of the dark areas is increased while the structures in brighter areas are preserved, if such areas exist. We refer to this part of the algorithm as the contrast enhancement step. The term *tone mapping* is sometimes also used for this procedure. This is akin to the increased amplification typically found in nocturnal insect photoreceptors.

In the virtual exposures method of Bennett and McMillan (2005), a contrast enhancement procedure is applied where the intensity transformation is a logarithmic function, similar to the one proposed by Drago et al. (2003). The method also contains additional smoothing using spatial and temporal bilateral filters and an attenuation of details, found by the subtraction of a filtered image from a nonfiltered image. According to the processing pipeline suggested by the results from animal eyes, we instead choose to do all smoothing in a separate noise reduction stage and here concentrate on the contrast enhancement, i.e., the amplification step.

The method of Bennett involves several parameters, both for the bilateral smoothing filters and for changing the acuteness of two different mapping functions. These parameters have to be set manually and will not adapt if the lighting conditions change in the image sequence. Since we aim at an automatic procedure, we instead opt for a modified version of the well-known procedure of histogram equalization (Gonzalez and Woods 1992). Histogram equalization is parameter-free and increases the contrast in an image by finding an intensity transformation that is determined so that it flattens out the intensity histogram of the input image. However, for many images histogram equalization gives a too extreme transformation, which, for example, saturates the brightest intensities so that structure information there is lost. We therefore apply contrast-limited histogram equalization as presented by Pizer et al. (1987), but without the tiling that applies different transformations to different areas (tiles) in the image. In contrast-limited histogram equalization a parameter, known as the clip limit β, sets a limit on the derivative of the slope of the transformation function. If the function, found by histogram equalization, exceeds this limit, the increase in the critical areas is spread equally over the dynamic range.

A comparison of the different amplification methods is shown in Figure 6.4.

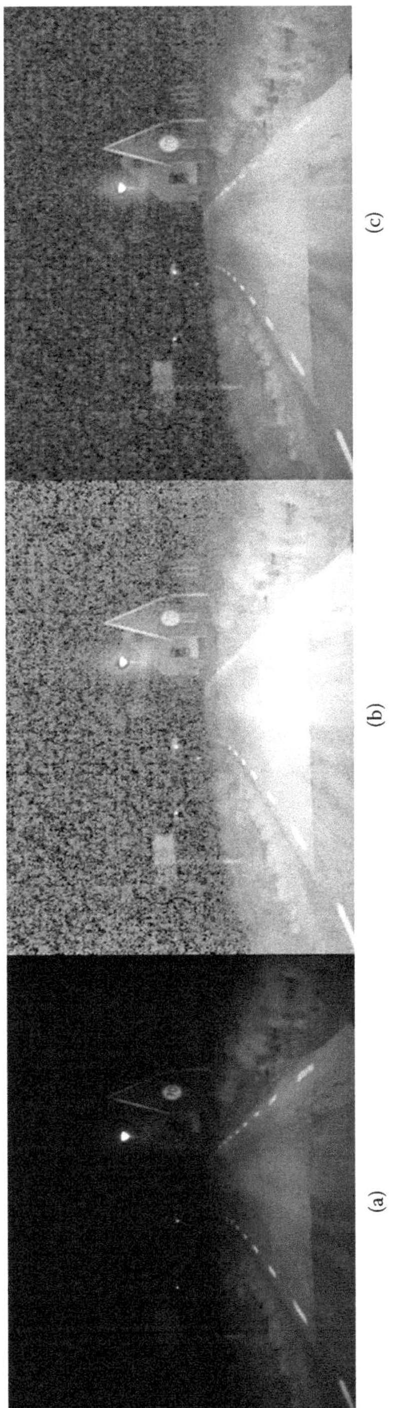

FIGURE 6.4 (See color figure at http://www.crcpress.com/product/isbn/9781439877463)
Contrast enhancement of a dark image. (a) The original low-light image. (b) The image after applying the transformation resulting from histogram equalization. (c) The image after contrast-limited histogram equalization. The clip limit was in this case set to 0.05.

6.3.3 Noise Reduction

After the contrast enhancement, the noise in the image sequence will be significant. This is especially apparent in the former dark areas, where the signal-to-noise ratio is low and the intensities have been heavily amplified. Spatial and temporal summation is a major strategy for improving vision in dim light for many nocturnal animals, and is also the central component of our algorithm. As mentioned above, we apply adaptive spatiotemporal averaging to reduce the noise while keeping the important edges and structures intact.

Surprisingly few published studies exist that especially target image enhancement of low-light-level video. Lee et al. (2005) combine very simple operations in a system presumably developed for easy hardware implementation, for example, in mobile phone cameras and other compact digital video cameras. In our tests of this method, it is evident that this method cannot handle high levels of noise.

The approach taken by Bennett and McMillan (2005) for low dynamic range image sequences is more closely connected to our technique. Their *virtual exposures* framework includes the bilateral ASTA-filter (adaptive spatiotemporal accumulation) and a contrast enhancement technique (also called tone mapping). The ASTA-filter, which changes from temporal filtering to relatively more spatial filtering when motion is detected, is in this way also related to our biological model (Warrant 1999). Since bilateral filters are applied, the filtering is edge sensitive and the temporal bilateral filter is additionally used for the local motion detection.

There exist a multitude of noise reduction techniques that apply spatiotemporal weighted averaging for noise reduction purposes. Many authors have additionally realized the benefit of reducing the noise by filtering the sequences along motion trajectories in spatiotemporal space. In this way they ideally avoid motion blur and unnecessary amounts of spatial averaging. These noise reduction techniques are usually collectively referred to as motion-compensated (spatio-)temporal filtering. Kalivas and Sawchuk (1990) calculated means along the motion trajectories, while Özkan et al. (1993) and Miyata and Taguchi (2002) applied weighted averages, dependent on the intensity structure in a small neighborhood. Sezan et al. (1991) use so-called linear minimum mean-square-error filtering along the trajectories. A drawback of motion-compensating methods is that the filtering relies on a good motion estimation to give good output without excessive blurring.

Another class of noise-reducing video processing methods uses a cascade of directional filters that analyze the spatiotemporal intensity structure in the neighborhood of each point. The filtering and smoothing is done primarily in the direction that corresponds to the filter that gives the highest output (Martinez and Lim 1985, Arce 1991, Ko and Forest 1993). These methods work well for directions that coincide with the fixed filter directions, but produce a pronounced degradation in the output in directions that fall in between the filter directions. For a review of spatiotemporal noise reduction methods see Brailean et al. (1995).

An interesting family of smoothing techniques for noise reduction is the ones that solve an edge-preserving anisotropic diffusion equation on the images. This approach was pioneered by Perona and Malik (1990) and has had many successors, including the work by Weickert (1998). These techniques have also been extended to three dimensions and spatiotemporal noise reduction in video by Uttenweiler et al. (2003) and Lee and Kang (1998). Uttenweiler et al., and also Weickert (1999), apply the so-called structure tensor or second moment matrix in a similar manner to our approach in order to analyze the local spatiotemporal intensity structure and steer the smoothing accordingly. The drawbacks of techniques based on diffusion equations include the fact that the solution has to be found using an often time-consuming iterative procedure. Moreover, it is very difficult to find a suitable stopping time for the diffusion, at least in a general and automatic manner. These drawbacks make these approaches in many cases unsuitable for video processing.

A better approach is to apply single-step structure-sensitive adaptive smoothing kernels. The bilateral filters introduced by Tomasi and Manduchi (1998) for two-dimensional images fall within this class. Here, edges are maintained by calculating a weighted average at every point using a Gaussian kernel, where the coefficients in the kernel are attenuated based on how different the intensities are in the corresponding pixels, compared to the center pixel. This makes the local smoothing very dependent on the correctness of the intensity in the center pixel, which cannot easily be estimated in images heavily disturbed by noise. This drawback can be somewhat reduced by a presmoothing before pixel comparisons, but this inevitably also leads to a higher degree of overall smoothing.

An approach that is closely connected to bilateral filtering, and also to anisotropic diffusion techniques based on the structure tensor, is the structure-adaptive anisotropic filtering method (Yang et al. 1996). In our search for an optimal strategy to apply spatiotemporal summation for noise reduction in low-light-level video, we naturally settled for this approach. Since our current methodology is an extension of this technique, it will be presented in detail. For a study of the connection between anisotropic diffusion, adaptive smoothing, and bilateral filtering, see Barash and Comaniciu (2004).

The method of Yang et al. (1996) computes a new image $f_{out}(x)$, by applying at each spatiotemporal point $x_0 = (x'_0, y'_0, t'_0)$, a kernel $k(x_0, x)$ to the original image $f_{in}(x)$ such that

$$f_{\text{out}}(x_0) = \frac{1}{\mu(x_0)} \iiint_{\Omega} k(x_0, x) f_{\text{in}}(x) dx \tag{6.1}$$

where

$$\mu(x_0) = \iiint_{\Omega} k(x_0, x)dx \tag{6.2}$$

is a normalizing factor. The normalization makes the sum of the kernel elements equal to 1 in all cases, so that the mean image intensity does not change. The area Ω over which the integration (or in the discrete case, summation) is taken, is chosen as a finite neighborhood centered on x_0.

Since we want to adapt the filtering to the spatiotemporal intensity structure at each point, in order to reduce blurring over spatial and temporal edges, we calculate a kernel $k(x_0, x)$ individually for each point x_0. The kernels should be wide in directions of homogeneous intensity and narrow in directions with important structural edges. To find these directions, the intensity structure is analyzed by the so-called structure tensor or second moment matrix. This object has been developed and applied in image analysis in numerous papers (for example, Jähne 1993). The tensor $J_\rho(x_0)$ is defined in the following way:

$$J_\rho(\nabla f(x_0)) = G_\rho * (\nabla f(x_0))(\nabla f(x_0))^T, \tag{6.3}$$

where

$$\nabla f(x_0) = \begin{bmatrix} \dfrac{\partial f}{\partial x_0'} & \dfrac{\partial f}{\partial y_0'} & \dfrac{\partial f}{\partial t_0'} \end{bmatrix}^T \tag{6.4}$$

is the spatiotemporal intensity gradient of f at the point x_0. G_ρ is the Gaussian kernel function

$$G_\rho = \frac{1}{\mu} e^{\frac{-(x'^2+y'^2+t'^2)}{2\rho^2}}, \tag{6.5}$$

where μ is the normalizing factor. The notation * means element-wise convolution with the matrix $(\nabla f(x_0))(\nabla f(x_0))^T$ in a neighborhood centered at x_0. It is this convolution that gives us the smoothing in the direction of gradients, and it is the key to the noise insensitivity of this method.

Eigenvalue analysis of J_ρ will now give us the structural information that we seek. The eigenvector v_1 corresponding to the smallest eigenvalue λ_1 will be approximately parallel to the direction of minimum intensity variation, while the other two eigenvectors are orthogonal to this direction. The magnitude of each eigenvalue will be a measure of the amount of intensity variation in the direction of the corresponding eigenvector. For a deeper discussion on eigenvalue analysis of the structure tensor see Haussecker and Spies (1999).

The basic form of the kernels $k(x_0, x)$ constructed at each point x_0 is a Gaussian function,

$$k(x_0, x) = e^{-\frac{1}{2}(x-x_0)^T R\Sigma^2 R^T (x-x_0)}, \tag{6.6}$$

including a rotation matrix R and a scaling matrix Σ. The rotation matrix is constructed from the eigenvectors v_i of J_ρ,

$$R = \begin{bmatrix} v_1 & v_2 & v_3 \end{bmatrix}, \tag{6.7}$$

while the scaling matrix has the following form:

$$\Sigma = \begin{bmatrix} \frac{1}{\sigma(\lambda_1)} & 0 & 0 \\ 0 & \frac{1}{\sigma(\lambda_2)} & 0 \\ 0 & 0 & \frac{1}{\sigma(\lambda_3)} \end{bmatrix}. \tag{6.8}$$

$\sigma(\lambda_i)$ is a decreasing function that sets the width of the kernel along each eigenvalue direction. The method of Yang et al. (1996) is mainly developed for two-dimensional images and measures of corner strength and of anisotropism, both involving ratios of the maximum and minimum eigenvalues, and calculated at every point x_0. An extension of this to the three-dimensional case is then discussed. However, we have not found these two measures to be adequate for the three-dimensional case, as they focus too much on singular corner points in the video input and largely disregard the linear and planar structures that need to be preserved in the spatiotemporal space. For example, a dependence of the kernel width in the temporal direction on the eigenvalues corresponding to the spatial direction does not seem appropriate in a static background area. Instead, an exponential function is constructed that depends directly on the eigenvalue λ_i in the current eigenvector direction v_i:

$$\sigma(\lambda_i, x_0) = \Delta\sigma e^{-\frac{\lambda_i}{d}} + \sigma_{min}, \tag{6.9}$$

where $\Delta\sigma = (\sigma_{max} - \sigma_{min})$, so that σ_i attains its maximum σ_{max} at $\lambda = 0$ and asymptotically approaches its minimum σ_{min} when $\lambda \to \infty$. The parameter d scales the width function along the λ-axis and has to be set in relation to the current noise level. Since the part of the noise that stems from the quantum nature of light (i.e., the photon shot noise) depends on the brightness level, it is signal dependent and the parameter d should ideally be set locally using a local noise measurement, but we will settle for a global setting in this presentation.

When the widths $\sigma(\lambda_i, x_0)$ have been calculated and the kernel subsequently constructed according to Equation (6.6), Equation (6.1) is used to calculate the output intensity f_{out} of the smoothing stage in the current pixel x_0.

6.4 Consideration for Color

The discussion so far has dealt with monochrome intensity images. Now, we will discuss some special aspects of the algorithm related to processing color images.

In applying the algorithm to RGB color image data one could envision a procedure where the color data in the images are first transformed to another color space including an intensity channel, for instance, the HSV (hue, saturation, value) color space (Gonzalez and Woods 1992). The algorithm could then be applied unaltered to the intensity channel, while smoothing of the other two channels could be performed either with the same kernel as in the intensity channel or by isotropic smoothing. The HSV image would then be transformed back to the RGB color space.

However, in very dark color video sequences there is often a significant difference in the noise levels in the different input channels: for example, the blue channel often has a relatively higher noise level. It is therefore essential that there is a possibility to adapt the algorithm to this difference. To this end, we chose to calculate the structure tensor J_ρ, and its eigenvectors and eigenvalues, in the intensity channel, which we simply define as the mean of the three color channels. The widths of the kernels are then adjusted separately for each color channel by using a different value of the scaling parameter d for each channel. This gives a clear improvement of the output with colors that are closer to the true chromaticity values and with less false color fluctuations than in the HSV approach mentioned above.

When acquiring raw image data from a charge coupled device (CCD) or complimentary metal oxide semiconductor (CMOS) sensor, the pixels are usually arranged according to the so-called Bayer pattern. It was shown by Hirakawa and Parks (2006) that it is efficient and suitable to perform the interpolation from the Bayer pattern to three separate channels, so-called demosaicing, at the same time as the denoising of the image data. We apply this approach here for each channel by setting to zero the coefficients in the summation kernel $k(x_0, x)$ corresponding to pixels where the intensity data are not available, and then normalizing the kernel. A smoothed output is then calculated for both the noisy input pixels and the pixels where data are missing.

6.5 Implementation on a Graphics Processing Unit

We have implemented the whole adaptive enhancement methodology as a combined CPU/GPU algorithm. The filtering using the summation kernels $k(x_0, x)$ is an inherently parallelizable task for which the graphics processing units (GPUs) of modern graphics cards are very well suited.

All image pre- and postprocessing is performed on the central processing unit (CPU). This includes image input/output and the contrast enhancement step. The histogram equalization, which implements the contrast enhancement, requires summation over all pixels. This computation is not easily adapted to a GPU, as the summation would have to be done in multiple passes. However, as these steps constitute a small amount of the execution time, a simpler CPU implementation is adequate here.

The most time-expensive parts of the complete algorithm are the calculation of the structure tensor, including the gradient calculation, the element-wise smoothing, and the actual filtering, or summation, all of which we perform entirely on the GPU. To calculate the gradients we upload n frames of the input sequence to the graphics card as floating-point two-dimensional textures, and perform spatial gradient computations on each of these. Next we take temporal differences of each successive frame and use the resulting gradient to compute the structure tensor for each pixel in each frame. Using the separability of the Gaussian kernel, we then compute the isotropically smoothed tensor for each frame. Typically these filtering steps are performed using filter sizes of up to $7 \times 7 \times 7$ pixels. For the normalization of the filter, the alpha channel is used to store the sum of the filter weights.

The smoothing kernels are then computed in a fragment program on the GPU. This involves finding the eigenvalues and eigenvectors of the structure tensor, which can be done efficiently in a single pass on modern GPUs. The kernel coefficients are temporarily stored as textures and the final spatiotemporal summation is performed, similarly to the isotropic prepass, in multiple two-dimensional passes. As the filter kernels are unique for each pixel, the filter weights are recomputed on the fly during the filtering process.

In summary, by exploiting the massively parallel architecture of modern GPUs, we obtain interactive frame rates on a single nVidia GeForce 8800-series graphics card. For some sample RGB sequences, we obtain the following computational times per image (in ms):

Resolution	CPU Computation	Tensor Computation	Spatiotemporal Smoothing
360 × 288	14	37	95
1,024 × 768	150	265	702

These timings assume an isotropic presmoothing with kernels of size 7^3 pixels, and adaptive spatiotemporal smoothing with filter kernels as large as 13^3 pixels.

6.6 Results and Comparison to Other Methods

We have tested our complete enhancement method on a variety of grey-scale and color image sequences, differing in resolution and light level, obtained by both consumer grade and machine vision cameras. The tested input sequences include movies obtained by stationary cameras as well as by moving cameras. Overall, the method gives strikingly clean and bright output, considering the bad quality of the input data, as shown in Figures 6.5 and 6.6.

We have compared our method mainly to three other related methods. First, we implemented the low-light-level enhancement method presented by Lee et al. (2005) In this method, Poisson noise is removed in stationary areas by calculation of the median in a small spatial neighborhood. So-called false color noise is removed by choosing either the intensity value of the current pixel value or the value of the preceding corresponding pixel, depending on a simple measure of the variance in the neighborhood of the pixel. We have implemented this method and tested it on the same sequences as presented in the last section. However, for the very low light levels that we are dealing with here, and the high noise levels that this implies, the output of the method of Lee et al. (2005) was visually very disappointing. For instance, the median calculations in a small neighborhood leave heavy flickering in the output.

For a related noise reduction approach, we have implemented the adaptive weighted averaging (AWA) filter of Özkan et al. (1993), where a weighted mean is calculated within a spatiotemporal neighborhood along the motion trajectories in the sequence. The motion trajectories are calculated using optical flow estimation. For this task, we used the optical flow method suggested by Bruhn et al. (2005). However, the heavy noise in our test sequences results in poor estimations of the optical flow, and this leads to excessive smoothing in the output of the AWA algorithm. Moreover, much of the noise is left unaffected at the contours of moving objects in the scene.

The competing method that gave the best output with our dark noisy test sequences was the virtual exposures method proposed by Bennett and McMillan (2005), described briefly previously. When the sequence is obtained with a stationary camera, the virtual exposure method results mostly in temporal filtering, which removes a large part of the heavy noise in the input sequence. However, the output is not as spatially smooth as the output of the method presented in this paper. When the camera is moving, an external motion estimation is needed. Moreover, the contrast enhancement achieved by the tone mapping is very disappointing for the cases we tested. We suspect that this is due to the fact that Bennett's tone mapping approach uses separate processing of the low- and high-frequency parts of the input data, which is not optimal for the extreme levels of noise that we are dealing with here.

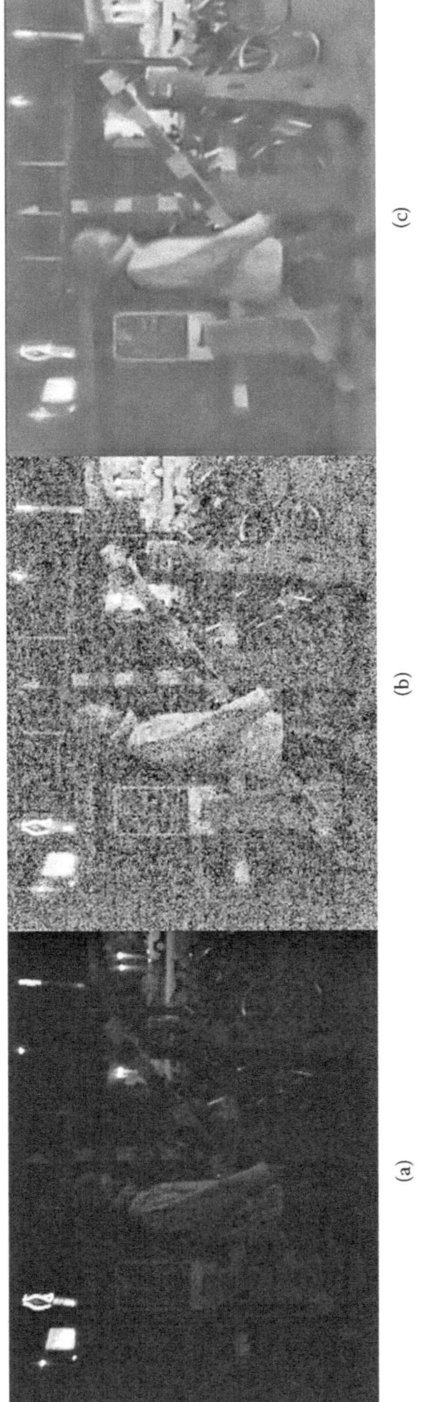

FIGURE 6.5 (See color insert.)
One frame of a color sequence obtained by a Sony DCR-HC23 handheld consumer camera. The resolution of this clip is 360 × 288. (a) An image from the original dark sequence. (b) The contrast-enhanced image, utilizing contrast-limited histogram equalization. (c) An image taken from the sequence after the adaptive smoothing approach has been applied.

FIGURE 6.6 (See color figure at http://www.crcpress.com/product/isbn/9781439877463)
The algorithm at work on a 640 × 480 color sequence taken by a moving camera. The camera used was an AVT Marlin machine vision camera. The input to the algorithm is the raw sensor data, and therefore the demosaicing of the Bayer pattern is done simultaneously to the denoising. As the camera moves, a spider moves from left to right in the scene. The algorithm has no problem in smoothing the intensity in the best possible way for these different motions, and the spider is well preserved.

6.7 Summary

We have presented here a methodology for adaptive enhancement and noise reduction for very dark image sequences with very low dynamic range. The method has been inspired by the spatiotemporal summation principles used by nocturnal animals in dim light and generally follows the same main processing steps employed by their visual systems. The approach is very general and adapts to the spatiotemporal intensity structure in order to prevent motion blur and smoothing across important structural edges.

Most parameters can be generally set for a very large group of input sequences. These parameters include the clip limit in the contrast-limited histogram equalization, the maximum and minimum widths of the filtering kernels, and the width of the isotropic smoothing used in the calculations of the gradients and the structure tensor. However, the scaling parameter for the width function has to be adjusted to the noise level in each sequence. We have also discussed the best approach for applying the method to color images, and this includes demosaicing from the Bayer pattern in raw input color data simultaneously to the noise reduction. We have implemented the method using a GPU and achieved interactive performance. For very noisy video input data, which are the result of filming in very low light levels, the method presented here outperforms (in terms of output quality) all competing methods that we have come across at the time of writing.

References

Aho, A.-C., Donner, K., Helenius, S., Larsen, L.O., and Reuter, T. (1993). Visual performance of the toad (*Bufo bufo*) at low light levels, retinal ganglion cell responses and prey-catching accuracy. *Journal of Comparative Physiology A*, 172:671–682.

Aho, A.-C., Donner, K., Hydén, C., Larsen, L.O., and Reuter, T. (1988). Low retinal noise in animals with low body temperature allows high visual sensitivity. *Nature*, 334:348–350.

Arce, G.R. (1991). Multistage order statistic filters for image sequence processing. *IEEE Transactions on Signal Processing*, 39:1146–1163.

Baird, M., Kreiss, E., Wcislo, W.T., Warrant, E.J., and Dacke, M. (2011). Nocturnal insects use optic flow for flight control. *Biology Letters*, 7:499–501.

Barash, D., and Comaniciu, D. (2004). A common framework for nonlinear diffusion, adaptive smoothing, bilateral filtering and mean shift. *Image and Vision Computing*, 22:73–81.

Barlow, H.B. (1956). Retinal noise and absolute threshold. *Journal of the Optical Society of America*, 46:634–639.

Bayer, B. (1976). Color imaging array. U.S. patent 397165.

Baylor, D.A., Matthews, G., and Yau, K.-W. (1980). Two components of electrical dark noise in toad retinal rod outer segments. *Journal of Physiology*, 309:591–621.

Bennett, E., and McMillan, L. (2005). Video enhancement using per-pixel virtual exposures. In Proceedings of ACM SIGGRAPH, Los Angeles, CA, pp. 845–852.

Brailean, J., Kleihorst, R., Efstratiadis, S., Katsaggelos, A., and Lagendijk, R. (1995). Noise reduction filers for dynamic image sequences: a review. *Proceedings of the IEEE*, 9:1272–1292.

Bruhn, A., Wieckert, J., and Schnörr, C. (2005). Lucas Kanade meets Horn/Schunk: combining local and global optic flow methods. *International Journal of Computer Vision*, 61:211–231.

De Vries, H. (1943). The quantum character of light and its bearing upon threshold of vision, the differential sensitivity and visual acuity of the eye. *Physica*, 10:553–564.

Drago, F., Myszkowski, K., Annen, T., and Chiba, N. (2003). Adaptive logarithmic mapping for displaying high contrast scenes. In *Proceedings of Eurographics*, 22:419–426.

Doujak, F.E. (1985). Can a shore crab see a star? *Journal of Experimental Biology*, 166:385–393.

Dubs, A., Laughlin, S.B., and Srinivasan, M.V. (1981). Single photon signals in fly photoreceptors and first order interneurons at behavioural threshold. *Journal of Physiology*, 317:317–334.

Frederiksen, R., Wcislo, W.T., and Warrant, E.J. (2008). Visual reliability and information rate in the retina of a nocturnal bee. *Current Biology*, 18:349–353.

Gonzalez, R., and Woods, R. (1992). *Digital image processing*. Addison-Wesley, Boston, MA.

Greiner, B., Ribi, W.A., and Warrant, E.J. (2004a). Retinal and optical adaptations for nocturnal vision in the halictid bee *Megalopta genalis*. *Cell and Tissue Research*, 316:377–390.

Greiner, B., Ribi, W.A., and Warrant, E.J. (2005). A neural network to improve dim-light vision? Dendritic fields of first-order interneurons in the nocturnal bee *Megalopta genalis*. *Cell and Tissue Research*, 323:313–320.

Greiner, B., Ribi, W.A., Wcislo, W.T., and Warrant, E.J. (2004b). Neuronal organisation in the first optic ganglion of the nocturnal bee *Megalopta genalis*. *Cell and Tissue Research*, 318:429–437.

Haussecker, H., and Spies, H. (1999). *Handbook of computer vision and applications*, 125–51. Academic Press.

Heimonen, K., Salmela, I., Kontiokari, P., and Weckström, M. (2006). Large functional variability in cockroach photoreceptors: optimization to low light levels. *Journal of Neuroscience*, 26:13454–13462.

Hirakawa, K., and Parks, T. (2006). Joint demosaicing and denoising. *IEEE Transactions of Image Processing*, 15:2146–2157.

Jähne, B. (1993). *Spatio-temporal image processing*. Springer, Berlin.

Kalivas, D., and Sawchuk, A. (1990). Motion compensated enhancement of noisy image sequences. In *Proceedings of the IEEE International Conference on Acoustic Speech Signal Processing*, Albuquerque, NM, pp. 2121–2124.

Kelber, A., Warrant, E.J., Pfaff, M., Wallén, R., Theobald, J.C., Wcislo, W., and Raguso, R. (2006). Light intensity limits the foraging activity in nocturnal and crepuscular bees. *Behavioural Ecology*, 17:63–72.

Klaus, A., and Warrant, E.J. (2009). Optimum spatiotemporal receptive fields for vision in dim light. *Journal of Vision*, 9:1–16.

Ko, S.-J., and Forest, T. (1993). Image sequence enhancement based on adaptive symmetric order statistics. *IEEE Tran. Circ. Sys. II: Analog and Digital Signal Processing*, 40:504–509.

Land, M.F. (1981). Optics and vision in invertebrates. In *Handbook of sensory physiology*, ed. H. Autrum, 471–592. Vol. VII/6B. Berlin: Springer.

Laughlin, S.B. (1981). Neural principles in the peripheral visual systems of invertebrates. In *Handbook of sensory physiology*, ed. H. Autrum, 133–280. Vol. VII/6B. Berlin: Springer.

Laughlin, S.B. (1990). Invertebrate vision at low luminances. In *Night vision*, ed. R.F. Hess, L.T. Sharpe, and K. Nordby, 223–50. Cambridge: Cambridge University Press.

Laughlin, S.B. (1996). Matched filtering by a photoreceptor membrane. *Vision Research*, 36:1529–1541.

Laughlin, S.B., Blest, A.D., and Stowe, S. (1980). The sensitivity of receptors in the posterior median eye of the nocturnal spider *Dinopis*. *Journal of Comparative Physiology*, 141:53–65.

Laughlin, S.B., and Lillywhite, P.G. (1982). Intrinsic noise in locust photoreceptors. *Journal of Physiology*, 332:25–45.

Laughlin, S.B., and Weckström, M. (1993). Fast and slow photoreceptors: a comparative study of the functional diversity of coding and conductances in the Diptera. *Journal of Comparative Physiology A*, 172:593–609.

Lee, S., and Kang, M. (1998). Spatio-temporal video filtering algorithm based on 3D anisotropic diffusion equation. *IEEE Proceedings of the International Conference on Image Processing*, 2:447–450.

Lee, S.W., Maik, V., Jang, J., Shin, J., and Paik, J. (2005). Noise-adaptive spatio-temporal filter for real-time noise removal in low light level images. *IEEE Transactions of Consumer Electronics*, 51:648–653.

Lillywhite, P.G. (1977). Single photon signals and transduction in an insect eye. *Journal of Comparative Physiology*, 122:189–200.

Lillywhite, P.G., and Laughlin, S.B. (1979). Transducer noise in a photoreceptor. *Nature*, 277:569–572.

Malm, H., Oskarsson, M., Warrant, E., Clarberg, P., Hasselgren, J., and Lejdfors, C. (2007). Adaptive enhancement and noise reduction in very low light-level video. In *Proceedings of the 11th International Conference on Computer Vision*, Rio de Janiero, pp. 1395–1402.

Malm, H., and Warrant, E. (2006). Motion dependent spatiotemporal smoothing for noise reduction in very dim light image sequences. In *Proceedings of the 18th International Conference on Pattern Recognition*, Hong Kong, pp. 954–959.

Martinez, D., and Lim, J. (1985). Implicit motion compensated noise reduction of motion video scenes. In *Proceedings of the IEEE International Conference on Acoustic Speech Signal Processing*, pp. 375–378.

Miyata, K., and Taguchi, A. (2002). Spatio-temporal separable data-dependent weighted average filtering for restoration of the image sequences. In *Proceedings of the IEEE International Conference on Acoustic Speech Signal Processing*, Albuquerque, NM, pp. 3696–3699.

Ohly, K.P. (1975). The neurons of the first synaptic regions of the optic neuropil of the firefly, *Phausius splendidula* L. (Coleoptera). *Cell and Tissue Research*, 158:89–109.

Özkan, M., Sezan, M., and Tekalp, A. (1993). Adaptive motion-compensated filtering of noisy image sequences. *IEEE Transactions on Circuits, Systems and Video Techniques*, 3:277–290.

Perona, P., and Malik, J. (1990). Scale-space and edge detection using anisotropic diffusion. *IEEE Transactions of Pattern Analysis and Machine Intelligence*, 12:629–639.

Pirhofer-Walzl, K., Warrant, E.J., and Barth, F.G. (2007). Adaptations for vision in dim light: impulse responses and bumps in nocturnal spider photoreceptor cells (*Cupiennius salei* Keys). *Journal of Comparative Physiology A*, 193:1081–1087.

Pizer, S., Amburn, E., Austin, J., Cromartie, R., Geselowitz, A., Geer, T., teer Haar Romeny, B., Zimmerman, J., and Zuiderveld, K. (1987). Adaptive histogram equalization and its variations. *Computer Vision, Graphics and Image Processing*, 39:355–368.

Ribi, W.A. (1977). Fine structure of the first optic ganglion (lamina) of the cockroach *Periplaneta americana*. *Tissue Cell*, 9:57–72.

Ribi, W.A. (1981). The first optic ganglion of the bee. IV. Synaptic fine structures and connectivity patterns of receptor cell axons and first order interneurones. *Cell and Tissue Research*, 215:443–464.

Rose, A. (1942). The relative sensitivities of television pickup tubes, photographic film and the human eye. *Proceedings of the Institute of Radio Engineers New York*, 30:293–300.

Sezan, M., Özkan, M., and Fogel, S. (1991). Temporally adaptive filtering of noisy image sequences. In *Proceedings of the IEEE International Conference on Acoustic Speech Signal Processing*, Toronto, Canada, pp. 2429–2432.

Snyder, A.W. (1977c). Acuity of compound eyes, physical limitations and design. *Journal of Comparative Physiology*, 116:161–182.

Snyder, A.W., Laughlin, S.B., and Stavenga, D.G. (1977b). Information capacity of eyes. *Vision Research*, 17:1163–1175.

Snyder, A.W., Stavenga, D.G., and Laughlin, S.B. (1977a). Spatial information capacity of compound eyes. *Journal of Comparative Physiology*, 116:183–207.

Somanathan, H., Borges, R.M., Warrant, E.J., and Kelber, A. (2008). Nocturnal bees learn landmark colours in starlight. *Current Biology*, 18:R996–97.

Strausfeld, N.J., and Blest, A.D. (1970). Golgi studies on insects. I. The optic lobes of Lepidoptera. *Philosophical Transactions of the Royal Society of London B*, 258:81–134.

Theobald, J.C., Greiner, B., Wcislo, W.T., and Warrant, E.J. (2006). Visual summation in night-flying sweat bees: a theoretical study. *Vision Research*, 46:2298–2309.

Tierney, S.M., Sanjur, O., Grajales, G.G., Santos, L.M., Bermingham, E., and Wcislo, W.T. (2012). Photic niche invasions: phylogenetic history of the dim-light foraging augochlorine bees (Halictidae). *Proceedings of the Royal Society of London B* 279:794–803.

Tomasi, C., and Manduchi, R. (1998). Bilateral filtering for gray and color images. In *Proceedings of the 6th International Conference on Computer Vision*, pp. 839–846.

Uttenweiler, D., Weber, C., Jähne, B., Fink, R., and Scharr, H. (2003). Spatiotemporal anisotropic diffusion filtering to improve signal-to-noise ratios and object restoration in fluorescence microscopic image sequences. *Journal of Biomedical Optics*, 8:40–47.

van Hateren, J.H. (1992). Real and optimal neural images in early vision. *Nature*, 360:68–70.

van Hateren, J.H. (1993). Spatiotemporal contrast sensitivity of early vision. *Vision Research*, 33:257–267.
Warrant, E.J. (1999). Seeing better at night: life style, eye design and the optimum strategy of spatial and temporal summation. *Vision Research*, 39:1611–1130.
Warrant, E.J. (2008a). Nocturnal vision. In *The senses: A comprehensive reference: Vision II*, ed. T. Albright and R.H. Masland, 53–86. Vol. 2. Oxford: Academic Press.
Warrant, E.J. (2008b). Seeing in the dark: vision and visual behaviour in nocturnal bees and wasps. *Journal of Experimental Biology*, 211:1737–1746.
Warrant, E.J., Kelber, A., Gislén, A., Greiner, B., Ribi, W., and Wcislo, W.T. (2004). Nocturnal vision and landmark orientation in a tropical halictid bee. *Current Biology*, 14:1309–1318.
Wcislo, W.T., Arneson, L., Roesch, K., Gonzalez, V., Smith, A., and Fernandez, H. (2004). The evolution of nocturnal behaviour in sweat bees, *Megalopta genalis* and *M. ecuadoria* (Hymenoptera, Halictidae): an escape from competitors and enemies? *Biological Journal of the Linnean Society*, 83:377–387.
Wcislo, W.T., and Tierney, S.M. (2009). Behavioural environments and niche construction: the evolution of dim-light foraging in bees. *Biological Reviews*, 84:19–37.
Weickert, J. (1998). *Anisotropic diffusion in image processing*. Stuttgart: Teubner-Verlag.
Weickert, J. (1999). Coherence-enhancing diffusion filtering. *International Journal of Computer Vision*, 31:111–127.
Yang, G., Burger, P., Firmin, D., and Underwood, S. (1996). Structure adaptive anisotropic image filtering. *Image and Vision Computing*, 14:135–145.

7

Modeling and Simulation of Structural Colors

Shuichi Kinoshita, Dong Zhu, and Akira Saito

CONTENTS

7.1 Modeling and FDTD Calculations ... 192
7.1.1 Introduction ... 192
7.1.2 FDTD Method ... 193
7.1.2.1 General Descriptions ... 193
7.1.2.2 Formulations of FDTD Method ... 193
7.1.2.3 Nonstandard FDTD Method ... 200
7.1.2.4 Demonstration of NS-FDTD Method ... 205
7.1.3 FDTD Method Applied to the *Morpho* Butterfly Scale ... 207
7.1.3.1 Introduction ... 207
7.1.3.2 Microscopic Structure and Optical Measurements ... 207
7.1.3.3 FDTD Analyses ... 211
7.1.4 Summary ... 222
References ... 223
7.2 Simulation Analysis on the Optical Role of the Random Arrangement in *Morpho* Butterfly Scales ... 226
7.2.1 Optical Properties ... 227
7.2.1.1 Aspect of the Randomness in Structures ... 227
7.2.1.2 Measurement ... 228
7.2.2 Software and Computer Platform ... 228
7.2.3 Origin of the Smooth Angular Dependence without Fringes ... 229
7.2.3.1 Effect of Randomness in the Structure ... 229
7.2.3.2 Prevention of Interference Fringes ... 230
7.2.3.3 Consideration on the Optical Coherence ... 231
7.2.4 Role of Randomness in the Structure under Incoherent Light ... 234
7.2.4.1 Role of Lateral Randomness ... 234
7.2.4.2 Role of Vertical Randomness ... 234
7.2.4.3 Role of the Number of Components ... 236
7.2.5 Further Future Applications ... 239
7.2.6 Summary ... 239
Acknowledgments ... 241
References ... 241

7.1 Modeling and FDTD Calculations

Shuichi Kinoshita and Dong Zhu

7.1.1 Introduction

Colors produced by light phenomena, such as interference, diffraction, and scattering of light, are generally called structural colors [1–8]. In recent years, structural colors have attracted considerable attention, particularly in the field of photonics, where sophisticated next-generation photonic materials are of interest, and in industries where novel, visually appealing materials are in demand, for example, for use in fibers, paints, cosmetics, and jewels. In nature, structural colors are widely distributed in living things, accompanying surprisingly elaborate nanostructures, and have been a subject of extensive studies for more than a century [9–15].

However, most of the structural color materials, especially those found in living beings, have so complicated nanostructures that it is difficult to find a physical approach to determine the mechanism responsible for the structural colors. We can take the *Morpho* butterfly as such an example. Its wings show an extremely glossy and strongly blue reflection and have been known as the most challenging example of a structural color material [16–23]. Owing to the multilayered shelf structures on the scales on the wings, a multilayer interference model has often been employed to estimate the spectral properties of the reflected light [24–27]. However, recent minute angle-resolved reflection measurements have shown that the simple multilayer interference model cannot explain the spectral shape of the reflection and its angular dependence [24–38]. It has further been clarified that there is a considerable amount of irregularities in the shelf structure and also in the array of shelves [29,30].

Thus, it is quite difficult to elucidate the physical mechanism responsible for structural colors in such complicated structures. Although various theoretical models and a variety of calculation methods have been applied so far, a method to directly solve Maxwell's equations for nanostructures seems to be the most fundamental and to have wide applicability. A computational approach including the finite-difference time-domain (FDTD) method [39–42] has become popular. Although the FDTD calculation cannot directly provide us with the physical mechanism, we can compare the computed result with experimental results, analyze the difference between the results using a simple model, and finally determine the true mechanism. Thus, FDTD calculations have become an indispensable tool from observation for determining the mechanism. In this chapter, we describe the fundamentals of the FDTD method and show how the reflection properties of the *Morpho* butterfly can be analyzed using this method.

7.1.2 FDTD Method

7.1.2.1 General Descriptions

FDTD [39–42] is a universal method to simulate an electromagnetic field in the presence of a complex dielectric material. The FDTD algorithm can be directly derived from the basic equations of electromagnetism, Maxwell's equations, by discretizing both the space of interest and time using a finite-difference approximation. Thus, it can, in principle, offer accurate solutions to general light scattering problems.

The FDTD method usually involves algorithms to calculate the total field (TF), scattered field (SF), and total field/scattered field (TF/SF). The method involves the use of additional algorithms: one is for determining the boundary condition. Since the FDTD calculation is performed for a limited space, the boundary condition should always be considered to remove the influence of the boundary. Usually, the absorbing boundary condition (ABC) is considered. Several ABC methods have been proposed so far, for example, the Mur [43], Higdon [44], and perfectly matched layer (PML) [45]. Another is to consider the incident source, which can be chosen arbitrarily, for example, a point source, line source, plane wave, and current source. Yet another algorithm is concerned with the conjecture of electromagnetic fields at a distant place. Since the FDTD simulation is performed for a limited space, the electromagnetic fields obtained are for a small space involving scatterers, and they are usually called near fields. To evaluate the reflectance and transmittance at a distance sufficiently far from the scatterer (far field), near-to-far-field (NTFF) transformation [46,47] is necessary after completing the FDTD simulation.

Further, an improved version of the FDTD method, called nonstandard FDTD (NS-FDTD) method [48–50], has been proposed recently. It has a much higher accuracy than the standard FDTD method for a monochromatic incident source, and has been shown to be effective even for a coarse lattice. Thus, there are many methods to perform the FDTD calculation. However, in the present section, we describe only the essential part of the FDTD method, from which readers can understand the fundamental concept of the FDTD method. We touch on the highly accurate NS-FDTD method, which is expected to be useful for analyzing scattering problems associated with complex structures in nature. Owing to space limitations, we omit the FDTD algorithms for pure SF and TF/SF, the treatment of incident sources, and the methods of ABCs and NTFF transformation. Readers interested in these algorithms can refer to papers and books listed in the references [41,42,51,52].

7.1.2.2 Formulations of FDTD Method

The general FDTD algorithm starts by considering the following Maxwell's equations:

$$\frac{\partial \mathbf{H}}{\partial t} = -\frac{1}{\mu}\nabla \times \mathbf{E}, \tag{7.1}$$

$$\frac{\partial \mathbf{E}}{\partial t} = \frac{1}{\varepsilon}\nabla \times \mathbf{H} - \frac{\sigma}{\varepsilon}\mathbf{E}, \tag{7.2}$$

where **E** and **H** are the electric and magnetic field vectors and ϵ, μ, and σ denote the electric permittivity, magnetic permeability, and electric conductivity, respectively. The FDTD method is based on the fact that the spatiotemporal partial derivatives of arbitrary functions can be approximated by the following finite differences:

$$\frac{\partial f(x_j)}{\partial x_j} \approx \frac{f(x_j + \Delta x_j / 2) - f(x_j - \Delta x_j / 2)}{\Delta x_j}, \tag{7.3}$$

$$\frac{\partial g(t)}{\partial t} \approx \frac{g(t + \Delta t / 2) - g(t - \Delta t / 2)}{\Delta t}, \tag{7.4}$$

where $x_j = x, y,$ and z, and we have employed the central difference with Δx_j and Δt being the discretized steps with respect to space and time. Thus, a spatiotemporal lattice point in three dimensions (3D) is generally expressed by $(x, y, z, t) = (i\Delta x, j\Delta y, k\Delta z, n\Delta t)$, where i, j, k, and n are all integer-valued indices. Here, let us denote a function f of position and time evaluated at a discrete lattice point by

$$f(i\Delta x, j\Delta y, k\Delta z, n\Delta t) \equiv f^n(i, j, k). \tag{7.5}$$

To solve Equations (7.1) and (7.2) by substituting the finite differences, we should provide the fields at half-integer lattice points such as $(i + 1/2)\Delta x$ or $(n + 1/2)\Delta t$. The problem of finding the fields at half-integer lattice points was first solved by Yee in 1966, and it is now known as the Yee algorithm [39]. As shown in Figure 7.1, the Yee lattice in a 3D spatial domain consists of lattice points such that every **E** vector is surrounded by four circulating **H** vectors and every **H** vector is surrounded by four circulating **E** vectors. That is, neither **E** nor **H** has all-integer indices. For instance, E_x has the set of indices $(i + 1/2, j, k)$, while H_x has the set $(i, j + 1/2, k + 1/2)$. In the temporal domain, the Yee algorithm solves the problem by defining leapfrog time steps: **E** is given at steps of $n\Delta t$ and **H** at $(n + 1/2)\Delta t$. Thus, **E** always has an integer time index n, while **H** has a half-integer time index $n + 1/2$.

Now, we can apply the above calculation procedures to numerically evaluate Maxwell's equations. First, Equation (1.1) is approximated at a temporal lattice point $t = n\Delta t$ as

$$\frac{\mathbf{H}^{n+\frac{1}{2}} - \mathbf{H}^{n-\frac{1}{2}}}{\Delta t} = -\frac{1}{\mu}\mathbf{D}_1 \times \mathbf{E}^n, \tag{7.6}$$

where $\mathbf{D}_1$ denotes the finite difference corresponding to ∇. We will not write down the explicit form of $\mathbf{D}_1$ right now because each component of **H** will

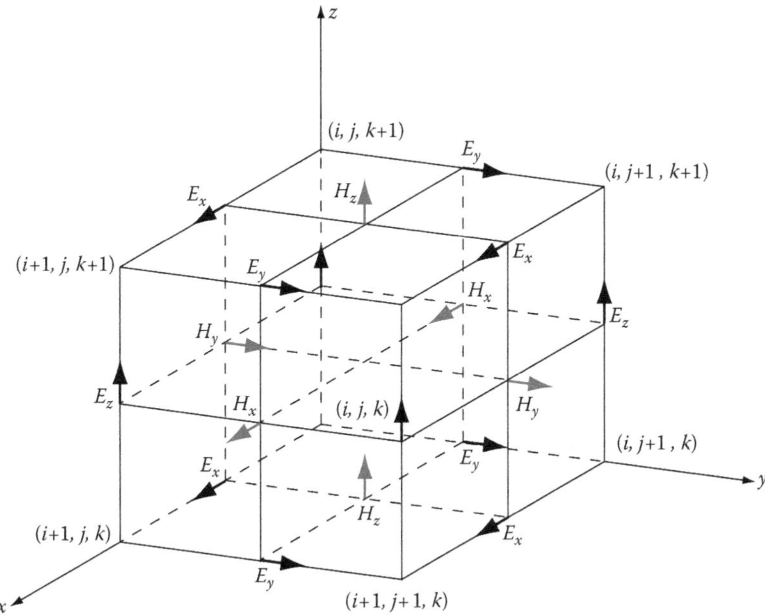

FIGURE 7.1
Yee lattice in a 3D spatial domain.

have a different expression for the position index, so that $\mathbf{D}_1$ will have three complicated expressions. In a manner similar to the case of Equation (1.1), Equation (1.2) is approximated at $t = (n - 1/2)\Delta t$ as

$$\frac{\mathbf{E}^n - \mathbf{E}^{n-1}}{\Delta t} = -\frac{\sigma}{\varepsilon}\mathbf{E}^{n-\frac{1}{2}} + \frac{1}{\varepsilon}\mathbf{D}_1 \times \mathbf{H}^{n-\frac{1}{2}}.$$

Since the electric field is defined only for an integer-valued temporal lattice, it is necessary to estimate $\mathbf{E}^{n-\frac{1}{2}}$ in the first term on the right-hand side. Usually, this is obtained by the semi-implicit approximation

$$\mathbf{E}^{n-\frac{1}{2}} \approx \frac{\mathbf{E}^n + \mathbf{E}^{n-1}}{2}.$$

Thus, we obtain

$$\frac{\mathbf{E}^n - \mathbf{E}^{n-1}}{\Delta t} = -\frac{\sigma}{\varepsilon}\frac{\mathbf{E}^n + \mathbf{E}^{n-1}}{2} + \frac{1}{\varepsilon}\mathbf{D}_1 \times \mathbf{H}^{n-\frac{1}{2}}. \tag{7.7}$$

Rearranging Equations (7.6) and (7.7), we obtain the difference equations corresponding to Maxwell's equations as

$$\mathbf{H}^{n+\frac{1}{2}} = \mathbf{H}^{n-\frac{1}{2}} - \frac{\Delta t}{\mu}\mathbf{D}_1 \times \mathbf{E}^n, \tag{7.8}$$

$$\mathbf{E}^n = \frac{1-\sigma\Delta t/(2\varepsilon)}{1+\sigma\Delta t/(2\varepsilon)}\mathbf{E}^{n-1} + \frac{\Delta t/\varepsilon}{1+\sigma\Delta t/(2\varepsilon)}\mathbf{D}_1 \times \mathbf{H}^{n-\frac{1}{2}}. \tag{7.9}$$

Hereafter, we will confine ourselves to a two-dimensional (2D) case, where the electromagnetic waves are assumed to propagate within a 2D plane spanned by the *xy* coordinate system, because most of the natural structure-based colors reported so far have been analyzed within a 2D framework. The 2D electromagnetic field can be classified into two modes according to the polarization employed: (1) transverse magnetic (TM) mode, where the electric field is perpendicular to the 2D plane, and (2) transverse electric (TE) mode, where the magnetic field is perpendicular to the 2D plane. Thus, the TM mode has only E_z, H_x, and H_y components, while the TE mode has only H_z, E_x, and E_y. An arbitrary electromagnetic field can be expressed by a linear combination of these two modes.

The Yee lattice of the TM mode can be easily obtained by extracting a plane given by $z = (k + \frac{1}{2})\Delta z$ from the Yee lattice shown in Figure 7.1, while that of the TE mode can be obtained by extracting a plane given by $z = k\Delta z$. Actually, both TM and TE modes have an identical lattice [53], as shown in Figure 7.2, as long as E_z is surrounded by four circulating **H** vectors and H_z is

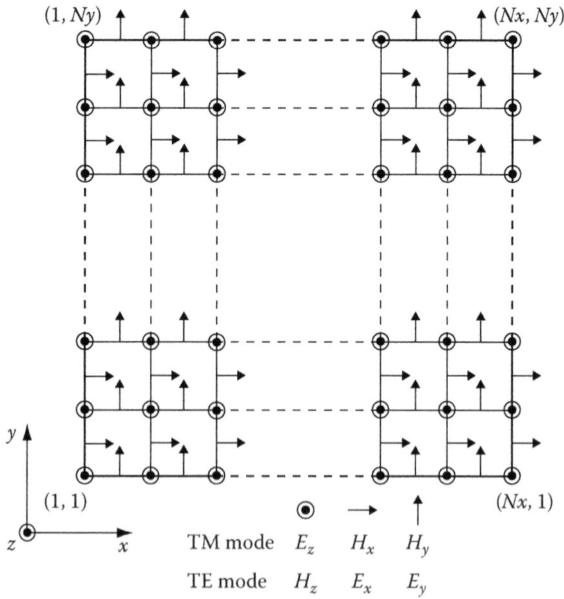

FIGURE 7.2
Yee lattice in a 2D spatial domain.

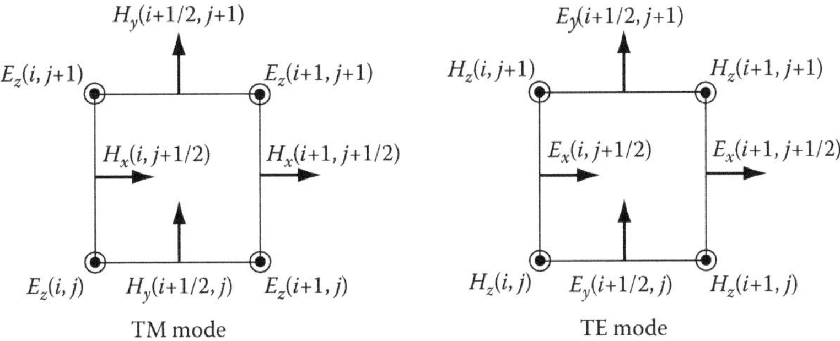

FIGURE 7.3
Yee cell with position indices in the 2D spatial domain.

surrounded by four circulating **E** vectors. We highly recommend this identical lattice because it makes it easier to program the Yee algorithm, especially in the case of an absorbing boundary condition. Position indices in one Yee cell for the TM and TE modes are shown in Figure 7.3.

7.1.2.2.1 Yee Algorithm for TM Mode

Now, let us transform Equations (7.8) and (7.9) into the final expressions containing spatial differences. First, we explicitly rewrite each component of Equation (7.8) in terms of the following two formulas to conform to the TM mode:

$$H_x^{n+\frac{1}{2}} = H_x^{n-\frac{1}{2}} - \frac{\Delta t}{\mu}\frac{\partial E_z^n}{\partial y}, \tag{7.10}$$

$$H_y^{n+\frac{1}{2}} = H_y^{n-\frac{1}{2}} + \frac{\Delta t}{\mu}\frac{\partial E_z^n}{\partial x}. \tag{7.11}$$

Replacing $\partial E_z^n / \partial y$ and $\partial E_z^n / \partial x$ with the finite differences at the spatial lattice points of $(i, j + 1/2)$ and $(i + 1/2, j)$, respectively, we obtain

$$\frac{\partial E_z^n}{\partial y} \approx \frac{E_z^n(i, j+1) - E_z^n(i, j)}{\Delta y},$$

$$\frac{\partial E_z^n}{\partial x} \approx \frac{E_z^n(i+1, j) - E_z^n(i, j)}{\Delta x}.$$

We then obtain the final expressions for H_x and H_y as

$$\begin{aligned} H_x^{n+\frac{1}{2}}\left(i, j+\frac{1}{2}\right) &= H_x^{n-\frac{1}{2}}\left(i, j+\frac{1}{2}\right) \\ &\quad - C_{hx}\left(i, j+\frac{1}{2}\right)\left\{E_z^n(i, j+1) - E_z^n(i, j)\right\}, \end{aligned} \tag{7.12}$$

$$H_y^{n+\frac{1}{2}}\left(i+\frac{1}{2},j\right)=H_y^{n-\frac{1}{2}}\left(i+\frac{1}{2},j\right)+C_{hy}\left(i+\frac{1}{2},j\right)\left\{E_z^n(i+1,j)-E_z^n(i,j)\right\},\tag{7.13}$$

where we have defined the following two new variables:

$$C_{hx}\left(i,j+\frac{1}{2}\right)\equiv\frac{\Delta t}{\mu\left(i,j+\frac{1}{2}\right)\Delta y},\quad C_{hy}\left(i+\frac{1}{2},j\right)\equiv\frac{\Delta t}{\mu\left(i+\frac{1}{2},j\right)\Delta x}.$$

In the same way, Equation (7.9) is transformed into the final expression

$$\begin{aligned}E_z^n(i,j)&=C_{ez}(i,j)E_z^{n-1}(i,j)\\&+C_{ezdx}(i,j)\left\{H_y^{n-\frac{1}{2}}\left(i+\frac{1}{2},j\right)-H_y^{n-\frac{1}{2}}\left(i-\frac{1}{2},j\right)\right\}\\&-C_{ezdy}(i,j)\left\{H_x^{n-\frac{1}{2}}\left(i,j+\frac{1}{2}\right)-H_x^{n-\frac{1}{2}}\left(i,j-\frac{1}{2}\right)\right\},\end{aligned}\tag{7.14}$$

with

$$C_{ez}(i,j)\equiv\left\{1-\frac{\sigma(i,j)\Delta t}{2\varepsilon(i,j)}\right\}\Big/\left\{1+\frac{\sigma(i,j)\Delta t}{2\varepsilon(i,j)}\right\},$$

$$C_{ezdx}(i,j)\equiv\frac{\Delta t}{\varepsilon(i,j)\Delta x}\Big/\left\{1+\frac{\sigma(i,j)\Delta t}{2\varepsilon(i,j)}\right\},$$

$$C_{ezdy}(i,j)\equiv\frac{\Delta t}{\varepsilon(i,j)\Delta y}\Big/\left\{1+\frac{\sigma(i,j)\Delta t}{2\varepsilon(i,j)}\right\}.$$

7.1.2.2.2 Yee Algorithm for TE Mode

Since the derivation of the expressions for the TE mode is similar to that for the TM mode, we skip the derivation and show only the final results:

$$E_x^n\left(i,j+\frac{1}{2}\right)=C_{ex}\left(i,j+\frac{1}{2}\right)E_x^{n-1}\left(i,j+\frac{1}{2}\right)+C_{exdy}\left(i,j+\frac{1}{2}\right)\left\{H_z^{n-\frac{1}{2}}(i,j+1)-H_z^{n-\frac{1}{2}}(i,j)\right\},\tag{7.15}$$

$$E_y^n\left(i+\frac{1}{2},j\right)=C_{ey}\left(i+\frac{1}{2},j\right)E_y^{n-1}\left(i+\frac{1}{2},j\right)$$
$$-C_{eydx}\left(i+\frac{1}{2},j\right)\left\{H_z^{n-\frac{1}{2}}(i+1,j)-H_z^{n-\frac{1}{2}}(i,j)\right\}, \tag{7.16}$$

$$H_z^{n+\frac{1}{2}}(i,j)=H_z^{n-\frac{1}{2}}(i,j)$$
$$-C_{hzdx}(i,j)\left\{E_y^n\left(i+\frac{1}{2},j\right)-E_y^n\left(i-\frac{1}{2},j\right)\right\} \tag{7.17}$$
$$+C_{hzdy}(i,j)\left\{E_x^n\left(i,j+\frac{1}{2}\right)-E_x^n\left(i,j-\frac{1}{2}\right)\right\}.$$

Here,

$$C_{ex}\left(i,j+\frac{1}{2}\right)\equiv\left\{1-\frac{\sigma\left(i,j+\frac{1}{2}\right)\Delta t}{2\varepsilon\left(i,j+\frac{1}{2}\right)}\right\}\Bigg/\left\{1+\frac{\sigma\left(i,j+\frac{1}{2}\right)\Delta t}{2\varepsilon\left(i,j+\frac{1}{2}\right)}\right\},$$

$$C_{exdy}\left(i,j+\frac{1}{2}\right)\equiv\frac{\Delta t}{\varepsilon\left(i,j+\frac{1}{2}\right)\Delta x}\Bigg/\left\{1+\frac{\sigma\left(i,j+\frac{1}{2}\right)\Delta t}{2\varepsilon\left(i,j+\frac{1}{2}\right)}\right\},$$

$$C_{ey}\left(i+\frac{1}{2},j\right)\equiv\left\{1-\frac{\sigma\left(i+\frac{1}{2},j\right)\Delta t}{2\varepsilon\left(i+\frac{1}{2},j\right)}\right\}\Bigg/\left\{1+\frac{\sigma\left(i+\frac{1}{2},j\right)\Delta t}{2\varepsilon\left(i+\frac{1}{2},j\right)}\right\},$$

$$C_{eydx}\left(i+\frac{1}{2},j\right)\equiv\frac{\Delta t}{\varepsilon\left(i+\frac{1}{2},j\right)\Delta y}\Bigg/\left\{1+\frac{\sigma\left(i+\frac{1}{2},j\right)\Delta t}{2\varepsilon\left(i+\frac{1}{2},j\right)}\right\},$$

with

$$C_{hzdx}(i,j)\equiv\frac{\Delta t}{\mu(i,j)\Delta y},\quad C_{hzdy}(i,j)\equiv\frac{\Delta t}{\mu(i,j)\Delta} \tag{7.18}$$

Usually, the final expressions for both TM and TE modes can be further simplified by replacing $\mu(i, j)$, $\mu(i + 1/2, j)$, and $\mu(i, j + 1/2)$ with μ_0, where μ_0 is the magnetic permeability of vacuum. Thus, it is unnecessary to allocate arrays to $C_{hx}(i, j + 1/2)$, $C_{hy}(i + 1/2, j)$, $C_{hzdx}(i, j)$, and $C_{hzdx}(i, j)$. Furthermore, we normally construct a uniform Yee lattice, where we put $\Delta x = \Delta y$, so that we can reduce C_{hx}, C_{hy}, C_{hzdx}, and C_{hzdx} to only one constant.

7.1.2.3 Nonstandard FDTD Method

The NS-FDTD method, which was first proposed by Mickens [54] in 1989, is an improved version of the standard FDTD (S-FDTD) method. The S-FDTD method is generally easy to program and has wide applicability, but its accuracy is sometimes not high when the cell size is not sufficiently small. Actually, the error in the S-FDTD method is generally given by $\varepsilon \sim h^2$, where h is the cell size. In contrast, the NS-FDTD method, which we will describe here, has a much smaller error of $\varepsilon \sim h^6$. This method promises a result with much higher accuracy for nearly the same computational cost as that in the S-FDTD method [50]. On the other hand, the NS-FDTD method is only applicable when the functional form of the fields is expressed by a special form such as an exponential function. The method is, however, quite useful for the present purpose—to simulate an electromagnetic wave interacting with complicated structures.

Since the NS-FDTD method was originally derived from a wave equation [55], let us first consider a 1D wave equation:

$$(\partial_t^2 - v^2\partial_x^2)\psi(x,t) = 0, \tag{7.19}$$

where v is the velocity of the wave; we have abbreviated the differential operators such that $\partial_t \equiv \partial/\partial t$ and $\partial_x \equiv \partial/\partial x$. The central finite-difference approximation for the first derivative of a function $\psi(x)$ with respect to x is generally given by

$$\psi'(x) \approx \frac{d_x\psi(x)}{h}, \tag{7.20}$$

where the difference operator d_x is defined by

$$d_x\psi(x) = \psi(x + h/2) - \psi(x - h/2). \tag{7.21}$$

Replacing ψ with ψ' in Equation (7.20), we obtain the second derivative:

$$\psi''(x) \approx \frac{d_x^2\psi(x)}{h^2}, \tag{7.22}$$

where $d_x^2 = d_x d_x$ and

$$d_x^2\psi(x) = \psi(x+h) + \psi(x-h) - 2\psi(x). \tag{7.23}$$

The finite difference for the second derivative with respect to t is expressed in a similar form by considering $\psi = \psi(t)$:

$$d_t^2\psi(t) = \psi(t+\Delta t) + \psi(t-\Delta t) - 2\psi(t). \tag{7.24}$$

Thus, replacing the derivatives in the wave equation of Equation (7.19) with the above finite differences, we obtain

$$\left(d_t^2 - \bar{v}^2 d_x^2\right)\psi(x,t) = 0, \tag{7.25}$$

where $\bar{v}$ is defined as $v\Delta t/h$. Rearranging Equation (7.25) and expanding d_t^2, we obtain

$$\psi(x,t+\Delta t) = -\psi(x,t-\Delta t) + 2\psi(x,t) + \bar{v}^2 d_x^2\psi(x,t), \tag{7.26}$$

which corresponds to the expression used in the S-FDTD algorithm for the 1D wave equation.

The NS-FDTD method is based on the following concept: if we already know a functional form of ψ beforehand, we can obtain the following identity, in a strict sense, only by choosing a proper function of $s(h)$:

$$\psi'(x) = \frac{d_x\psi(x)}{s(h)}. \tag{7.27}$$

Backward- and forward-propagating monochromatic solutions of the wave equation with angular frequency ω and wavenumber k are generally given by the functional form $\psi_{\pm}(x, t) = \exp[i(kx \mp \omega t)]$. A general monochromatic solution is a linear combination of ψ_+ and ψ_-, and it is given as $\psi = \alpha_+\psi_+ + \alpha_-\psi_-$, where $\alpha_{\pm}$ are constants. Using this monochromatic solution and computing $d_x\psi$ using Equation (7.21), we can obtain

$$d_x\psi = 2i\sin(kh/2)\psi. \tag{7.28}$$

On the other hand, the mathematical expression for $\psi'(x)$ can be given as $\psi'(x) = ik\psi(x)$. Therefore, substituting these expressions into Equation (7.27), we obtain the simple result $s_k(h) = (2/k)\sin(kh/2)$, where $s_k(h)$ is related to the derivative with respect to x. Consequently, we can obtain the following finite difference for the second derivative:

$$\psi''(x) = \frac{d_x^2\psi(x)}{s_k^2(h)}, \tag{7.29}$$

with $s_k(h)$ as given above.

In the same way, the second derivative with respect to t is given by $\psi''(t) = d_t^2\psi(t)/s_\omega^2(\Delta t)$, where $s_\omega(\Delta t) = (2/\omega)\sin(\omega\Delta t/2)$. Substituting the two new finite differences into the 1D wave equation, we obtain

$$\left(\frac{d_t^2}{4\sin^2(\omega\Delta t/2)/\omega^2} - v^2\frac{d_x^2}{4\sin^2(kh/2)/k^2}\right)\psi(x,t) = 0. \tag{7.30}$$

Using the relation $k = \omega/v$ and defining a new parameter u_1, we obtain

$$\left(d_t^2 - u_1^2 d_x^2\right)\psi(x,t) = 0, \tag{7.31}$$

where

$$u_1 = \sin(\omega\Delta t/2)/\sin(kh/2). \tag{7.32}$$

Expanding d_t^2 using Equation (7.24), we obtain the expression of the NS-FDTD algorithm for the 1D wave equation:

$$\psi(x,t+\Delta t) = -\psi(x,t-\Delta t) + 2\psi(x,t) + u_1^2 d_x^2\psi(x,t). \tag{7.33}$$

Thus, the 1D NS-FDTD algorithm has an expression analogous to the S-FDTD one, and hence S-FDTD users can easily switch to the NS-FDTD calculation by the simple replacement $\bar{v} \rightarrow u_1$. Further, the 1D NS-FDTD algorithm has exactly zero numerical dispersion, which is essentially unattainable in the S-FDTD method, although it is only applicable to 1D dielectric structures such as multilayered structures.

The wave equation in 2D is generally given as

$$\left(\partial_t^2 - v^2\nabla^2\right)\psi(\mathbf{r},t) = 0, \tag{7.34}$$

where $\nabla^2 = \partial_x^2 + \partial_y^2$ and $\mathbf{r} = (x, y)$. The S-FDTD expression becomes

$$\left(d_t^2 - \bar{v}^2\mathbf{D}_1^2\right)\psi(\mathbf{r},t) = 0, \tag{7.35}$$

where $\mathbf{D}_1^2 = d_x^2 + d_y^2$ and $\bar{v} = v\Delta t/h$. Following the method used for deriving the 1D NS-FDTD algorithm, we can get

$$\nabla^2\psi(\mathbf{r},t) = -k^2\psi(\mathbf{r},t),$$

and

$$\mathbf{D}_1^2\psi(\mathbf{r},t) = -4\left\{\sin^2(k_x h/2) + \sin^2(k_y h/2)\right\}\psi(\mathbf{r},t),$$

where it is assumed that $\psi(\mathbf{r}, t)$ can be expressed as the monochromatic plane wave $\psi_p(\mathbf{r}, t) = exp[\mathrm{i}\,(\mathbf{k} \cdot \mathbf{r} - \omega t)]$. Thus, ∇^2 is given as

$$\nabla^2\psi(\mathbf{r},t) = \frac{k^2}{4\sin^2(k_x h/2) + \sin^2(k_y h/2)}\mathbf{D}_1^2\psi(\mathbf{r},t). \tag{7.36}$$

Substituting this expression into the 2D wave equation leads to

$$\left(\frac{d_t^2}{4\sin^2(\omega\Delta t/2)/\omega^2} - v^2\frac{\mathbf{D}_1^2}{4\sin^2(k_x h/2)/k^2 + 4\sin^2(k_y h/2)/k^2}\right)\psi(\mathbf{r},t) = 0. \tag{7.37}$$

By defining $u_2^2 = \sin^2(\omega\Delta t/2)/\{\sin^2(k_x h/2) + \sin^2(k_y h/2)\}$, the NS-FDTD algorithm for the 2D wave equation seems to be easily obtained:

$$\left(d_t^2 - u_2^2\mathbf{D}_1^2\right)\psi(\mathbf{r},t) = 0, \tag{7.38}$$

which is analogous to the 1D case.

Unfortunately, Equation (7.38) is satisfied only for one particular direction of $\mathbf{k} = (k_x, k_y)$ because the parameter u_2 involves k_x and k_y. In a 2D computational space, electromagnetic waves usually change their propagation directions owing to reflection, diffraction, or scattering after interaction with dielectric materials. Therefore, in addition to the incident wave vector $\mathbf{k}_i$, wave vectors with arbitrary directions should exist in the computational space. What we really need is a parameter that is independent of direction and that minimizes the error caused by the finite-difference approximation.

To solve this problem, Cole [55] proposed a new method in which another finite difference was phenomenologically introduced in the expression for ∇^2 such that

$$\begin{aligned} 2\mathbf{D}_2^2\psi(x,y) = {} & \psi(x+h, y+h) + \psi(x+h, y-h) \\ & + \psi(x-h, y+h) + \psi(x-h, y-h) - 4\psi(x,y). \end{aligned} \tag{7.39}$$

A new difference operator $\mathbf{D}_0^2$ was then defined as a superposition of $\mathbf{D}_1^2$ and $\mathbf{D}_2^2$:

$$\mathbf{D}_0^2 = \gamma\mathbf{D}_1^2 + (1-\gamma)\mathbf{D}_2^2, \tag{7.40}$$

where γ is a free parameter that should be determined to minimize the error. This parameter was obtained as

$$\gamma = 1 - \frac{\sin^2(k_{x'}h/2) + \sin^2(k_{y'}h/2) - \sin^2(kh/2)}{2\sin^2(k_{x'}h/2)\sin^2(k_{y'}h/2)}, \tag{7.41}$$

with $(k_{x'}, k_{y'}) = k\left(2^{-1/4}, \sqrt{1-2^{-1/2}}\right)$.

By expanding $\mathbf{D}_0^2\psi_\mathrm{p} / \psi_\mathrm{p}$ in a Taylor series around $\sin\theta = 0$, Cole showed

$$\begin{aligned}\frac{\mathbf{D}_0^2\psi_\mathrm{p}}{\psi_\mathrm{p}} &= -4\sin^2(kh/2) \\ &\quad - \frac{1}{24192}(kh)^8\left[\left(\sqrt{2}-1\right)\sin^2 2\theta - \frac{1}{2}\sin^4 2\theta\right] + \cdots,\end{aligned} \tag{7.42}$$

which indicated that $\mathbf{D}_0^2$ had very small angular dependence. Thus, a highly accurate finite-difference expression for $\nabla^2\psi_\mathrm{p}$ was obtained as

$$\nabla^2\psi_\mathrm{p} \cong \frac{\mathbf{D}_0^2\psi_\mathrm{p}}{s_k^2(h)}, \tag{7.43}$$

by inserting the above expression into $s_k^2(h)$.

By making the replacements $\mathbf{D}_1^2 \to \mathbf{D}_0^2$ and $u_2 \to u_0$ in Equation (7.38), we obtain

$$\left(d_t^2 - u_0^2\mathbf{D}_0^2\right)\psi(\mathbf{r},t) = 0, \tag{7.44}$$

where

$$u_0 = \frac{\sin(\omega\Delta t/2)}{\sin(kh/2)}. \tag{7.45}$$

In Equation (7.44), neither k_x nor k_y is explicitly involved. The final expression for the NS-FDTD method with respect to the 2D wave equation thus becomes

$$\psi(\mathbf{r},t+\Delta t) = -\psi(\mathbf{r},t-\Delta t) + 2\psi(\mathbf{r},t) + u_0^2\mathbf{D}_0^2\psi(\mathbf{r},t) \tag{7.46}$$

which is again analogous to the S-FDTD expression.

Before closing this section, we shall briefly comment on the NS-FDTD method. For computational purposes, the algorithm based on a wave equation generally has clear advantages compared to that based on Maxwell's equations: the former algorithm is easier to program and has a lower computational cost and lower memory consumption. However, the above derivation from a wave function is actually applicable only to the TM mode because in the TE mode, the E_x and E_y components are combined in a complicated manner to form coupled wave equations. Recently, Okada and Cole [56] have proposed a method to solve the coupled wave equations and have presented an NS-FDTD algorithm applicable even to the TE mode, which shortens the computational time considerably. On the other hand, since ω is involved in u_0, the NS-FDTD algorithm has high accuracy only for one particular frequency. Hence, to calculate the reflection spectrum in the frequency domain, the NS-FDTD simulation has to be repeated for different frequencies. Recently, a wideband version of the NS-FDTD method [57] was proposed to overcome this drawback.

7.1.2.4 Demonstration of NS-FDTD Method

To demonstrate the superiority of the NS-FDTD method, we had previously calculated the Mie scattering [58,59] by an infinitely long circular cylinder illuminated by a plane light wave propagating along the positive x-axis, as shown in Figure 7.4. In Figure 7.5, we show the near-field result calculated by the NS-FDTD method and compare it with those obtained by the S-FDTD method and also with an analytical solution. In these calculations, both the incident wavelength and the radius of the cylinder are set at 400 nm, while the refractive index of the cylinder is set at 7.6. The cylinder is embedded in a medium with a refractive index of 1.0. In both FDTD calculations, we dare to set the cell size h at a comparably large value of 50 nm to clearly bring out the advantage of the NS-FDTD method, and the grid size is taken to be 60×60. It is clear that the NS-FDTD method gives a much better result than the S-FDTD method.

Theoretically, if the near-field solution is correct, the far-field pattern calculated by performing the near-to-far-field (NTFF) transformation [46,47] should also be correct. In order to confirm this quantitatively, we have

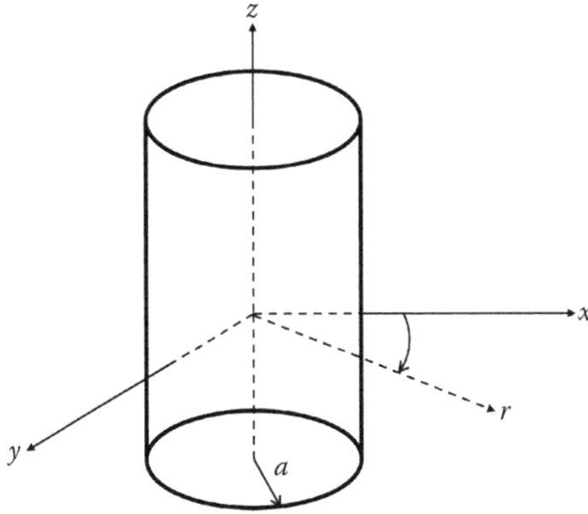

FIGURE 7.4
Infinitely long circular cylinder with radius a illuminated by a plane light wave propagating along the x-axis.

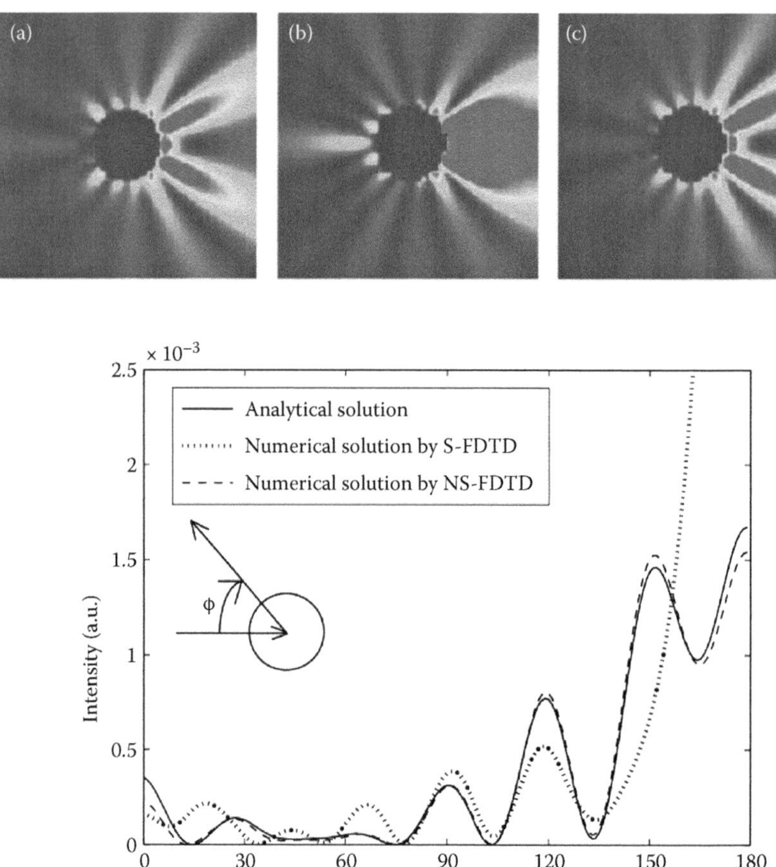

FIGURE 7.5
Upper: E_z of the scattered near fields by an infinitely long cylinder with a radius of 400 nm and a refractive index of 1.6. The cylinder is illuminated, from the left side of this figure, by a plane light wave with a wavelength of 400 nm. The cylinder is embedded in a medium with a refractive index of 1.0. (a) Analytical solution, and numerical solutions obtained by (b) S-FDTD and (c) NS-FDTD method. Lower: Angle-dependent scattering intensities by the same cylinder. The horizontal axis represents the scattering angle φ, while the vertical axis indicates the scattering intensity. The solid line denotes the analytical solution, while the dotted and dashed lines represent the numerical solutions for the S-FDTD and NS-FDTD methods. The numerical solutions are obtained by performing the NTFF transformations. (From Zhu, D., Modeling and Simulation of *Morpho* Butterfly Scale Using NS-FDTD Method, PhD thesis, University of Tsukuba, 2009. With permission.)

calculated the angular dependence of the scattered field in the far-field region and have compared it with the analytical solution. The comparison is shown in the lower part of Figure 7.5; the angle-dependent scattering intensities, which are calculated from the scattering amplitude in the near field by performing the NTFF transformation, are shown for both the S-FDTD and NS-FDTD methods, together with the analytical solution. As expected, the scattering intensity calculated by the NS-FDTD method gives a much better result, which is actually very close to the analytical solution.

7.1.3 FDTD Method Applied to the *Morpho* Butterfly Scale

7.1.3.1 Introduction

The glossy and strongly blue-colored wings of *Morpho* butterflies are one of the most exciting examples of structural color material and have been attracting scientists' attention for a long time [9–15]. It was just after the invention of the electron microscope that their nanostructures were discovered [16,17]. Since then, microscope observations of their structures have been repeatedly reported by many researchers; the structures are among the most challenging that nature has ever produced [18–23].

In contrast to the progress made on the basis of the microscopic observations, theoretical analyses concerning its reflection properties relied mostly on multilayer interference theory [24–27]. However, recent detailed spectral and angular reflection measurements of a wing and even individual scales [24–38] have clarified that the *Morpho* wing shows a strong reflection band in the blue region with extremely broad and anisotropic angular dependence; this band differs drastically from that for multilayer reflection.

The FDTD method is found to be effective when the nanostructure is so complicated that no analytical solution can be used. The nanostructure found on a scale of the *Morpho* butterfly wing is just in accordance with this criterion. In the present section, we will elucidate the mechanism responsible for the peculiar optical properties of the *Morpho* butterfly, which might not have been understood without the FDTD method. Since the nanostructure in the *Morpho* butterfly is known to vary from species to species, we will focus on *Morpho rhetenor*, which is known as a species that displays the most glossy blue wing.

7.1.3.2 Microscopic Structure and Optical Measurements

First, we show the nanostructure and its optical properties of *M. rhetenor* obtained through electron microscope observations and angle-dependent reflection measurements. In Figure 7.6, we show the nanostructure in a scale of *M. rhetenor* [38]. The nanostructure is characterized by well-developed ridges arranged with a spacing of 600–670 nm, and both sides of each ridge show multilayered shelf structures with an interval of 150–240 nm and

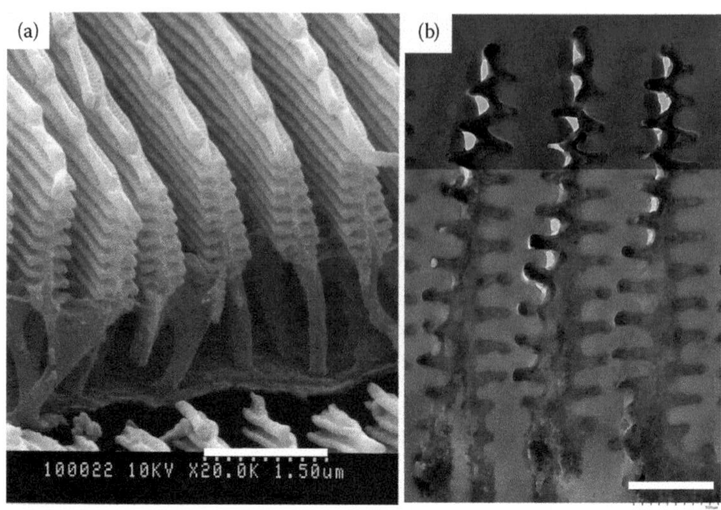

FIGURE 7.6
(a) Scanning and (b) transmission electron microscope images of the cross section of a ground scale of the *M. rhetenor*. Scale bars are (a) 1.5 μm and (b) 0.5 μm. (From Kambe, M., et al., *J. Phy. Soc. Jpn.* 80:054801, 2011. With permission.)

a thickness of 50–90 nm. The shelves on one side alternate with those on the other side of a ridge and extend over an identical vertical distance. It is known that the shelves are not parallel to the base plane of the scale, but slightly inclined, which causes the reflection direction to be toward the inner side of the wing. Further, as reported previously [29,30], the ends of the shelves observed at the top of the ridges seem to be randomly distributed, which disturbs the interference of light reflected at adjacent ridges. Thus, the blue reflection from this species of the *Morpho* butterfly originates from interference of light resulting from regularly arranged shelves and not from interference from regularly arranged ridges. In addition, light-absorbing pigments are known to exist in the lower part of the scale.

More recently, our group has reported detailed angle-resolved reflection spectra of the *M. rhetenor* wing and have clarified that the *Morpho* wing possesses strong retroreflection (or backscattering) properties [38]. In this work, we have compared three typical angle-resolved reflection spectra, which have been frequently employed so far: (1) detector and (2) sample rotations and (3) $\theta - 2\theta$ scan. Here, we will define the angular parameters for the angle-resolved measurements, which are summarized in Figure 7.7. We define the angles of incidence and reflection as θ and φ, respectively. Further, since we adjust the direction of the scale to be along the axis of rotation, the incident light beams with electric fields vertical and horizontal to the laboratory table automatically correspond to those parallel and perpendicular to the average directions of the ridges on the scale, which are called TM and TE modes, respectively.

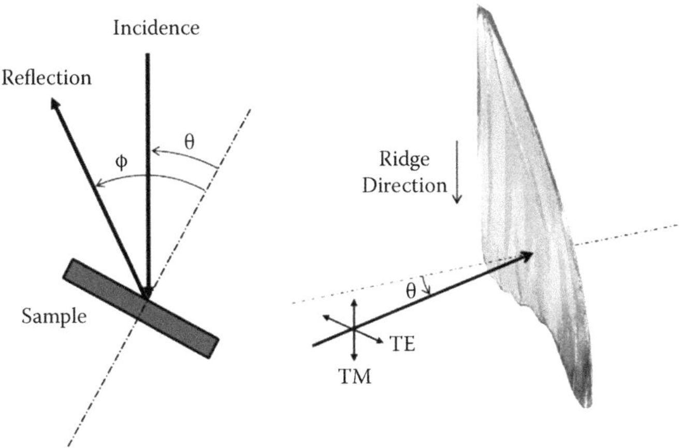

FIGURE 7.7
Definitions of angular parameters and polarization directions. (From Kambe, M., et al., *J. Phy. Soc. Jpn.* 80:054801, 2011. With permission.)

Many past studies have used the detector rotation method to obtain the angular dependence of the reflection; in this method, θ is fixed while φ is varied (see Figure 7.8(a)). On the other hand, in the sample rotation method, only the sample is rotated about a rotation axis, while the angle between the incident beam and the detector is fixed so that θ and φ will vary simultaneously (see Figure 7.8(b)). If we define $\eta \equiv \theta - \varphi$, $\eta = 0°$ corresponds to the so-called retroreflection (or backscattering) geometry. When the sample is rotated by a step of $\Delta\theta$ and the detector by a step of $2\Delta\theta$, the method is called $\theta - 2\theta$ scan and is the most commonly employed method in x-ray diffractometry (see Figure 7.8(c)).

The results of these three kinds of measurements are summarized in Figure 7.8. In the detector rotation method, under normal incidence, the reflection spectrum (Figure 7.8(a)) is characterized by a broad hump around 480 nm, which extends over the spectral range 450–530 nm and an angular range from −45° to +45°. It is found that the reflection in the normal direction is weakly suppressed, and accordingly faint twin peaks appear at $\varphi \approx \pm 20°$. On the other hand, the sample rotation method under an almost retroreflection geometry of $\eta \sim 0°$ displays a considerably different pattern (Figure 7.8(b)). The twin peaks are clearly separated and the suppression in the normal direction is more conspicuous. It is also found that the absolute reflectivity is much higher than that in the detector rotation method. The result of a $\theta - 2\theta$ scan is shown in Figure 7.8(c). In this method, a peak is observed only around $\varphi \sim 0°$ in the wavelength range 430–520 nm, which broadens only within ± 20°. It is found that the absolute reflectivity is much smaller than those obtained in the other two.

The completely different appearances of the spectra in the three methods can be well understood, if we employ an inclusive method for the $\theta - \varphi$ scan, in which both the sample and detector are rotated exhaustively. Then, an

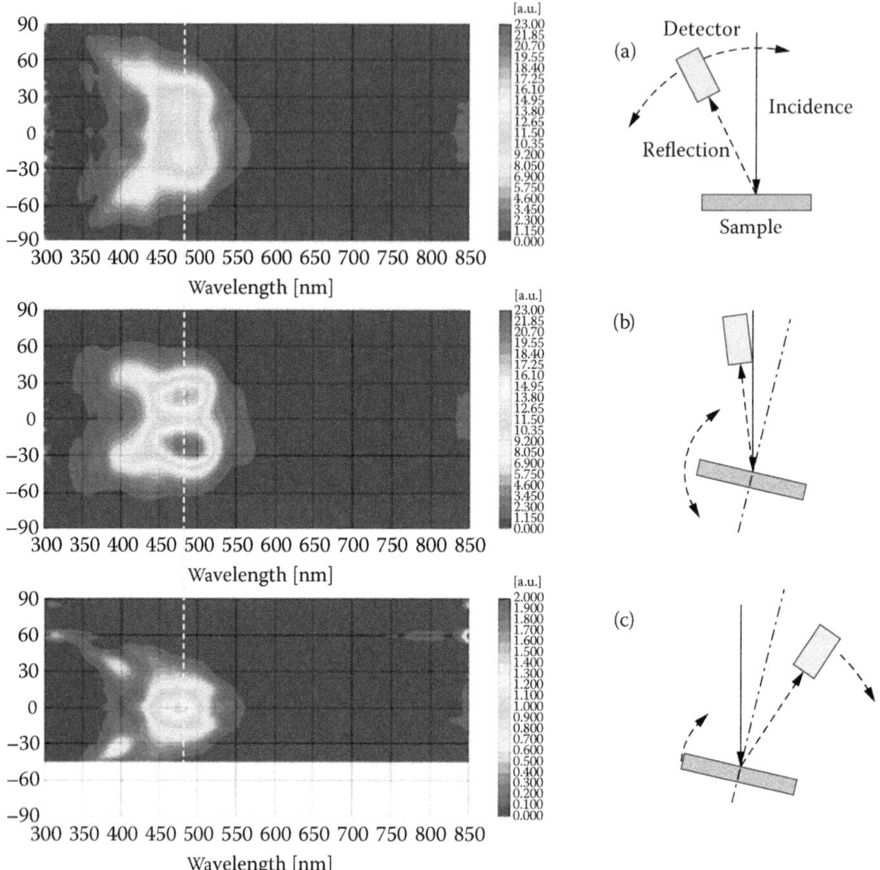

FIGURE 7.8 (See color figure at http://www.crcpress.com/product/isbn/9781439877463) Angle-resolved reflection spectra obtained by the (a) detector rotation method, (b) sample rotation method, and (c) $\theta - 2\theta$ scan for the *M. rhetenor* wing for TE polarization. The experimental geometry employed in each measurement is shown on the right side. (From Kambe, M., et al., *J. Phy. Soc. Jpn.* 80:054801, 2011. With permission.)

intensity map for all θ and φ can be obtained. We call this method the $\theta - \varphi$ scan, and it is essentially similar to the well-known bidirectional reflectance distribution function (BRDF) measurement; further, we call the resulting map the $\theta - \varphi$ map. In Figure 7.9, we show a typical result for the *M. rhetenor* wing, measured at 480 nm. It is surprising that clear twin spots appear along the 45° inclined line corresponding to $\theta = \varphi$. The locations of the twin spots are somewhat asymmetric with respect to the origin and located at the angular coordinates (17°, 17°) and (−27°, −27°).

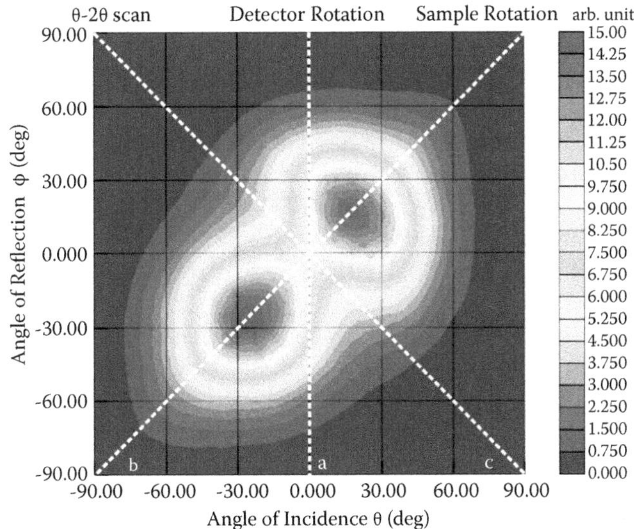

FIGURE 7.9 (See color figure at http://www.crcpress.com/product/isbn/9781439877463) $\theta - \varphi$ map of the *M. rhetenor* wing at 480 nm for TE polarization. The dashed lines (a), (b), and (c) indicate scanning lines corresponding to the detector rotation method, sample rotation method, and $\theta - 2\theta$ scan method, respectively. (From Kambe, M., et al., *J. Phy. Soc. Jpn.* 80:054801, 2011. With permission.)

We can analyze the results of the above three experiments by using the $\theta - \varphi$ map. Since the detector rotation method involves scanning φ, while θ is fixed, it is generally expressed by a vertical line in this map. When $\theta = 0°$, it is coincident with the φ-axis, which is shown as a dashed line a in Figure 7.9. On the other hand, the sample rotation method is characterized by a 45° inclined line, and for the retroreflection geometry of $\eta \sim 0°$, it is expressed by a straight line corresponding to $\theta = \varphi$, shown as a dashed line (b). Further, the $\theta - 2\theta$ scan is represented by the −45° inclined line, which is shown as a dashed line (c).

Comparing these with the $\theta - \varphi$ map measured at 480 nm, it is quite easy to understand their apparent patterns. Since the scattering properties of the *Morpho* butterfly wing are characterized by the twin spots lying on a 45° inclined line, the detector rotation method only probes the foot parts of the twin spots, while the $\theta - 2\theta$ scan probes only a valley between the spots. In contrast, the sample rotation method probes the most important parts of the reflection characteristics by examining the centers of each of the twin spots.

7.1.3.3 FDTD Analyses

The above experiments have given an interesting result: the wing of *M. rhetenor* displays strong retroreflection capability in the blue region. To find the reason for this, we reinvestigate the electron microscope image of the *M. rhetenor* scale in detail. It appears that the retroreflection shown in Figure 7.9 is somehow related to the alternating shelf structure. The presence of this structure was noticed in the 1980s [20], and was often taken into account in various computational models [33,61–63]. Upon inspecting the shelf structure more carefully, we have noticed that the shelves on the right and left sides have a common base, which is flattened and slightly inclined to form the alternating shelves (see Figure 7.6(a)). Thus, the flattened shelf leans to the left so that the right shelf is always higher than the left shelf when one faces the base of the scale. Thus, the shelf structure is essentially asymmetric in itself and at the scale level.

In order to investigate the relation between these structural features and the peculiar retroreflection properties, an analytical approach is unsuitable because the nanostructure is quite complicated and involves a considerable amount of irregularities. It is of great interest to determine which part of the above complicated structure actually contributes to the production of *Morpho* blue and to the retroreflection capability. The FDTD method is quite effective for analyzing this type of complicated problem.

First, we have attempted to investigate the angle-dependent reflection from an intact structure obtained directly from a transmission electron micrograph, as shown in Figure 7.10(a) [60]. The result is shown in Figure 7.10(b), where we show the angular dependence obtained for normal incidence at several wavelengths between 400 and 550 nm. Although the overall intensity is actually increased around the blue region of 480 nm, the large irregularity found in the angular dependence disturbs any quantitative interpretation. Since the intact structure shown in Figure 7.10(a) seems to assume considerable irregularities in both the ridge structure and its arrangement, we select four ridges labeled 1–4 and calculate the angular dependence for each ridge. The angular dependence is shown in Figure 7.10(c). It is clear that the angular dependence is largely dependent on the ridge selected so that a large amount of irregularities is clearly involved in each ridge. Thus, even if the FDTD method is believed to fairly effective, the sampling number is quite important to investigate the nanostructure involving irregularities. In the present case, it is clear that the sampling number of ridges is so small that we cannot obtain the statistical average of optical properties of the ridges.

However, it is natural that as the sampling number is increased, the computational load increases rapidly. Thus, it is necessary to construct a simple model that does not have any irregularity and yet contains the main characteristics of the *Morpho* blue. After inspecting the nanostructure in detail, we hypothesize that the retroreflection originates from an alternating stacked shelf structure. To confirm this hypothesis, we consider the following two simple models: a flat multilayer structure and an alternating structure. The former consists of nine flat shelves with a width of 300 nm and a thickness of 55 nm, and the shelves are equidistantly arrayed with an interval of 205 nm. The refractive index of the shelf is assumed to be 1.55, while that of the surrounding medium is set at 1.00. In the latter model, each shelf

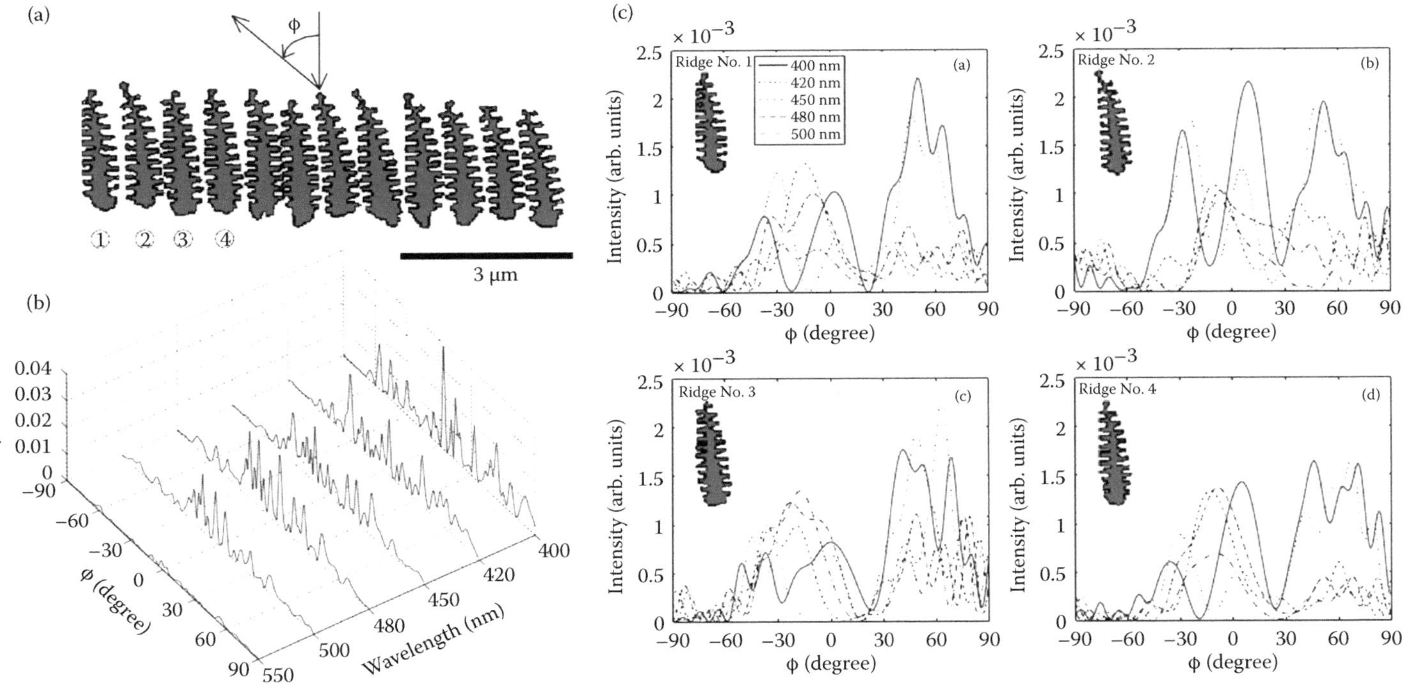

FIGURE 7.10
(a) Actual nanostructure [27] obtained by binarizing the TEM data for the transverse cross section of a *M. rhetenor* ground scale. The scale bar is 3 μm. (b) Angular dependence of the reflection intensities at wavelengths of 400, 420, 450, 480, 500, and 550 nm calculated for the structure shown in (a) for normal incidence. (c) Angular dependence of the reflection for various wavelengths from 400 to 500 nm in the TM mode. The ridge number shown in each figure corresponds to the ridge number in (a). The calculations were performed with a mesh size of 10 × 10 nm^2. (From Zhu, D., et al., *Phys. Rev. E* 80:051924, 2009. With permission.)

in the former model is exactly cut into two halves, and one half is shifted down by 118 nm to mimic the alternating shelf structure; hereafter, we call this model alternating multilayer model. The amount of the shift is chosen to agree with a half of the optical path length of one shelf and the adjacent spacing.

The $\theta - \varphi$ maps at the peak positions of the reflection bands calculated by the NS-FDTD method are shown in Figure 7.11. It is clear that the flat model shows only a single spot in the $\theta - \varphi$ map, which is located at the origin, while in the alternating multilayer model, twin spots are clearly seen. The twin spots are not exactly identical, and the spot on the left side is somewhat elongated. This is natural since the alternating structure considered here is

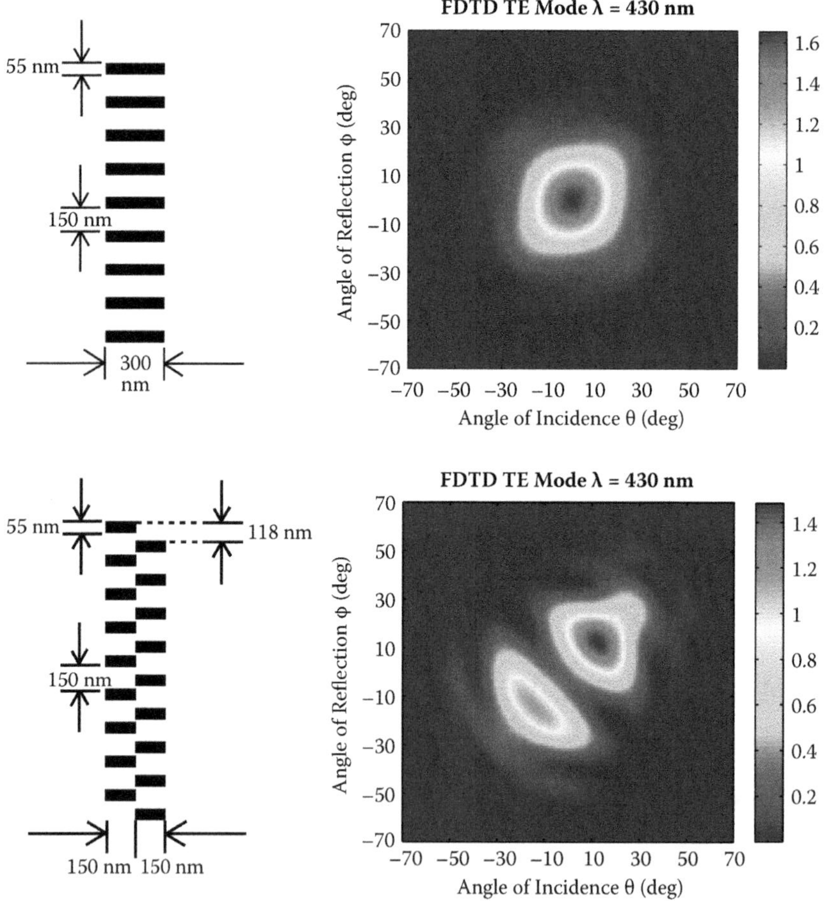

FIGURE 7.11 (See color figure at http://www.crcpress.com/product/isbn/9781439877463) Flat (upper) and alternating (lower) multilayer models and $\theta - \varphi$ maps at 430 nm for TE polarization, calculated by the NS-FDTD method. (From Kambe, M., et al., *J. Phy. Soc. Jpn.* 80:054801, 2011. Wih permission.)

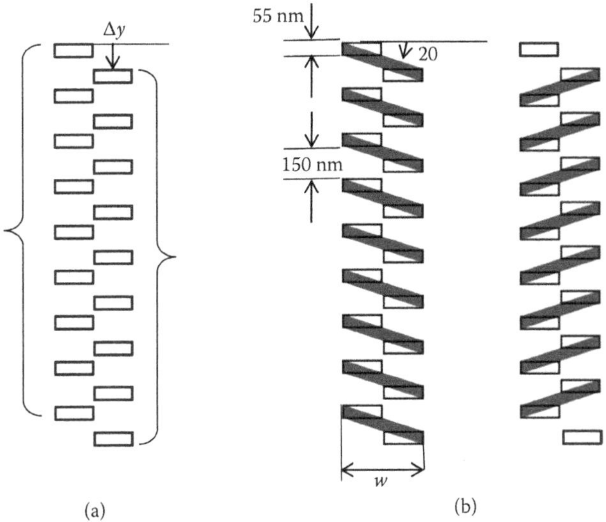

FIGURE 7.12
Comprehensive models for the alternating shelf structure. (a) The alternating multilayer model, which is regarded as two multilayer structures with a finite width, with one of the structures shifted vertically by an amount of Δy. (b) The model regarded as an ensemble of oblique shelves formed by connecting the shelves at both ends.

not symmetric with respect to the center line of the structure. The above calculation strongly suggests that the alternating structure is responsible for the twin spots and strong suppression of reflection in the normal direction.

We will further consider the origin of this retroreflection phenomenon by using two intuitive models (Figure 7.12). One is based on the fact that the alternating multilayer model can be regarded as two multilayer structures with a finite width, with one of the structures being shifted vertically by an amount Δy. If Δy is equal to $\lambda/4$, where λ is the wavelength of the incident light, this type of structure will strongly suppress the reflection in the normal direction for normal incidence because the light reflected from both multilayer structures will eventually disappear owing to destructive interference. In the other model, the alternating multilayer model is regarded as an ensemble of oblique shelves formed by connecting both ends of the right and left shelves.

In the case of the first model, according to the theory of light interference, the scattered/reflected light intensity from two exactly identical structures separated by Δx and Δy in Cartesian coordinates is generally expressed by

$$\begin{aligned} I(\theta,\phi) &= I_s(\theta,\phi)\left|1+e^{ik(u\Delta x - v\Delta y)}\right|^2 \\ &= 4I_s(\theta,\phi)\cdot\cos^2\left[\frac{k(u\Delta x - v\Delta y)}{2}\right], \end{aligned} \tag{7.47}$$

where $u = \sin\theta + \sin\varphi$ and $v = \cos\theta + \cos\varphi$, $I_s(\theta, \varphi)$ denotes the scattering intensity for scattering by a single structure, and k represents the wave vector of the incident light. From this relation, it is expected that when $k(u\Delta x - v\Delta y)/2 = (2m + 1)\pi/2$, that is, $u\Delta x - v\Delta y = (2m + 1)\lambda/2$, the scattering intensity vanishes owing to destructive interference, disregarding $I_s(\theta, \varphi)$. For normal incidence and reflection, $u = 0$ and $v = 2$ hold, so that the above condition is reduced to $\Delta y = -(2m + 1)\lambda/4$. Thus, for $\Delta y = \pm\lambda/4, \pm 3\lambda/4, \cdots$, strong suppression in the normal direction will occur.

From the above consideration, the optical property of the alternating multilayer model is well expressed by the product of a one-sided shelf structure and the interference term expressed by the $\cos^2$ function. Using the values of $\Delta x = 300/2$ (nm) and $\Delta y = -118$ (nm), we have calculated each of the terms at wavelengths of 370, 400, and 430 nm. The calculation results are summarized in Figure 7.13. The first and second columns, (a) and (b), are the results for the one-sided shelf structure and the $\cos^2$ term, respectively, while the last column (d) is the result for the alternating multilayer model. It appears that the pattern in column (a) is sampled by pattern (b), which results in pattern (d). In fact, as shown in column (c), the product of columns (a) and (b) is somewhat similar to column (d).

Thus, the $\theta - \varphi$ map for the alternating multilayer model seems to be expressed at least partly by the product of the optical properties of a one-sided multilayer and the $\cos^2$ term. Since this treatment is justified only when multiple scattering between the structural units is neglected, a clear discrepancy is also observed at 400 nm. The retroreflection in this case results from an appropriate angular broadening associated with a single structure and the condition $\cos^2[k(u\Delta x - v\Delta y)/2] \sim 1$. The latter condition further leads to $k(u\Delta x - v\Delta y)/2 \sim m\pi$. The conditions $m = 0$ and $m = 1$ contribute to the pattern in the present case for $\Delta x = 150$ (nm) and $\Delta y = \lambda/4$. Hence, the retroreflection is interpreted in terms of the zeroth- and first-order interference of reflected light from two multilayer structures at different positions.

In the second model, if both ends of the alternating layers are connected to each other, an obliquely inclined multilayer is formed. Further, if both ends with different combinations are connected, two inversely inclined multilayer models will be virtually created (see Figure 7.12(b)), and these models will contribute to retroreflection in two directions normal to the inclined layers. However, we should be careful while making a decision about the above intuitive speculation because if the width of the inclined layers w is too small, such a multilayer model will become a simple diffraction grating aligned vertically. Thus, the direction of reflection is no longer along the normal to the inclined multilayer. It is considered that an obliquely inclined multilayer model marks the limit for an infinitely large width, while a diffraction grating corresponds to limit for an infinitely small width. Here, we investigate their difference by calculating the reflection properties for various widths by means of the NS-FDTD method.

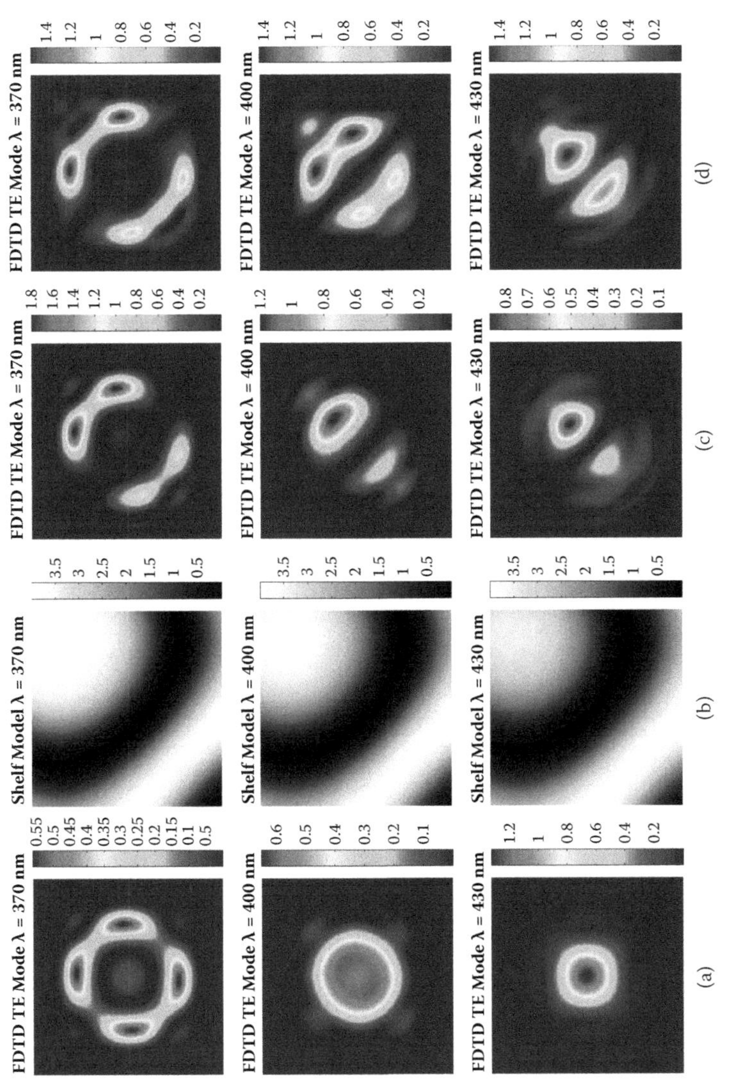

FIGURE 7.13 (See color figure at http://www.crcpress.com/product/isbn/9781439877463)
$\theta - \varphi$ maps calculated for (a) a one-sided shelf structure, (b) $\cos^2[k(u\Delta x - v\Delta y)/2]$, and (d) the alternating multilayer model. The vertical and horizontal axes in each map correspond to φ and θ, respectively, with the angular range being from −70 to 70° for p (TE) polarization. The third column (c) is obtained by multiplying (a) and (b). The wavelengths employed are 370, 400, and 430 nm from the top to the bottom. (From Kambe, M., et al., *J. Phy. Soc. Jpn.* 80:054801, 2011. With permission.)

The results for the right-inclined multilayer model (the left type in Figure 7.12(b)) are shown in Figure 7.14. In this figure, the top three diagrams are $\theta - \varphi$ maps for $w = 50$ nm, while the middle and lower ones are those for $w = 300$ and 2,000 nm, respectively. The top row is representative of a diffraction grating, while the lowest one is representative of the inclined multilayer; the middle row pertains to our model corresponding to actual alternating layers. It is natural that for a large width of w = 2,000 nm, the dominant reflection is specular, which satisfies the interference condition

$$2\{nd\cos(\theta_1+\alpha)+(a-d)\cos(\theta+\alpha)\}\cos\alpha = m\lambda. \quad (7.48)$$

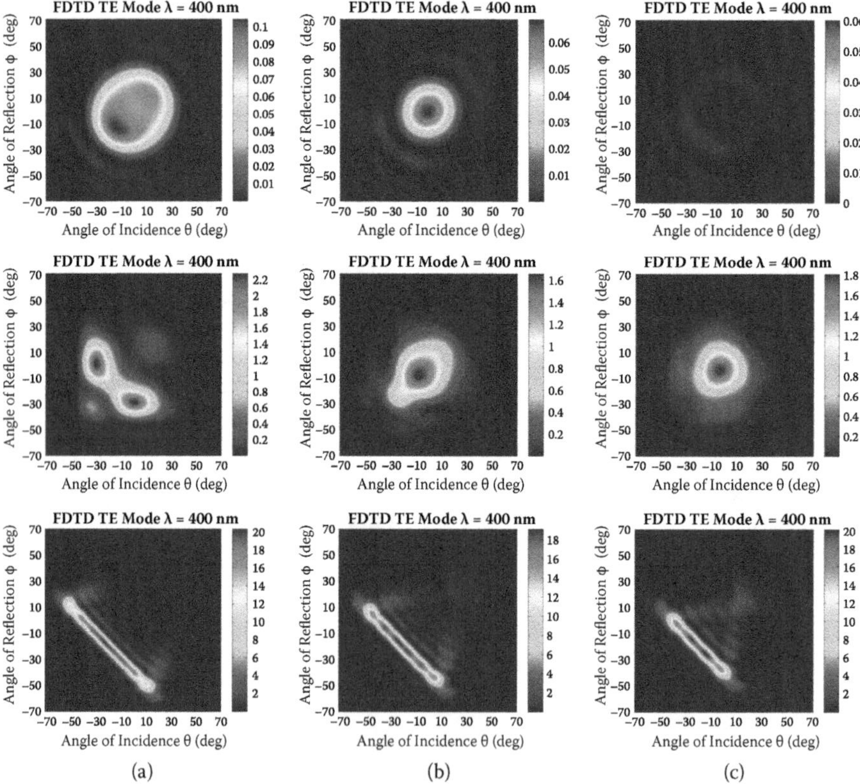

FIGURE 7.14 (See color figure at http://www.crcpress.com/product/isbn/9781439877463) $\theta - \varphi$ maps calculated for obliquely inclined shelves (the left type in Figure 7.12(b)) with widths of 50, 300, and 2,000 nm for p (TE) polarization. The inclined shelves have a thickness of 55 nm and an interval of 150 nm, both of which are measured vertically. The inclination angle is basically determined by connecting both ends of the alternating shelves, and it is approximated to an angle of 20°. The wavelengths are (a) 400, (b) 430, and (c) 460 nm. The vertical and horizontal axes in each map correspond to φ and θ, respectively, with the angular range being from –70 to 70°. (From Kambe, M., et al., *J. Phy. Soc. Jpn.* 80:054801, 2011. With permission.)

It is also natural that φ should satisfy the relation $\varphi = -\theta - 2\alpha$, where n, d, a, θ_1, and α are the refractive index, the thickness of an inclined layer measured vertically, the interval of layers measured vertically, the angle of refraction, and the inclination angle of the layers, respectively, with m being an integer. Further, under the conditions of $n = 1.55$, $d = 55$ nm, $a = 205$ nm, and $\alpha = 20°$, the first-order multilayer reflection is expected to occur at 442 nm for $\theta = \varphi = -\alpha$, and below this wavelength, twin spots indicating the occurrence of specular reflection will appear. However, owing to the large number of layers and the large difference between the refractive indices of the shelves, instead of twin spots, a slender segment lying on the straight line $\varphi = -\theta - 2\alpha$ is observed. Its maximum angular range is further restricted to $\theta = -77 \sim 37°$ owing to Brewster's law.

On the other hand, if the width becomes too small, the inclined multilayer model approaches to a vertically aligned diffraction grating, whose reflection properties are characterized by a factor of $\{\sin(Mkav/2)/\sin(kav/2)\}^2$, with M being the number of layers. Hence, spots are observed at angles satisfying $kav/2 = m\pi$, that is, $\cos\theta + \cos\varphi = m\lambda/a$, and they form a quasi-circle in the $\theta - \varphi$ map, centered at the origin. In the case of a diffraction grating of an infinite size with a period of 205 nm, the longest wavelength to generate the first-order diffraction spot occurs at 410 nm under the retroreflection configuration of $\theta = \varphi = 0°$. However, when the size of the grating is not infinite, a weak diffraction spot appears around the origin in the $\theta - \varphi$ map even at wavelengths longer than 410 nm. The calculated result justifies this speculation.

In view of the behavior at each limit, the reflection properties for $w = 300$ nm are considered to be intermediate between those at the two extremes because clear twin spots, which are characteristic of a multilayer, that appear at 400 nm, are not on the line $\varphi = -\theta - 2\alpha$, but on a line slightly shifted toward the origin. Further, the retroreflection takes place around 460 nm and tends to approach the origin as the wavelength increases. Thus, the obliquely inclined multilayer model partly explains the occurrence of retroreflection but is not complete when the layer width is not very large. From a biological viewpoint, however, this half-finished work will be beneficial to the considerable extension of the angular range of reflection, which may be indispensable in the struggle for existence of this butterfly.

So far, the analysis has been restricted to a single ridge having an alternating shelf structure and also to the case of p (TE) polarization of incident light. We will proceed to a more general case in which many ridges are arrayed almost equidistantly and consider the polarization dependence of this case. However, we will soon encounter difficulties: If the ridges are arrayed regularly, this structure will act as a simple diffraction grating, and diffraction spots corresponding to the spacing of ridges will be clearly seen in the angular dependence of the reflection. On the other hand, if the ridges are regularly arrayed but their heights are irregular, as in an actual scale, an irregular reflection pattern will be obtained.

These speculations are easily confirmed by considering a simple model in which 10 sets of alternating multilayer models are arrayed equidistantly

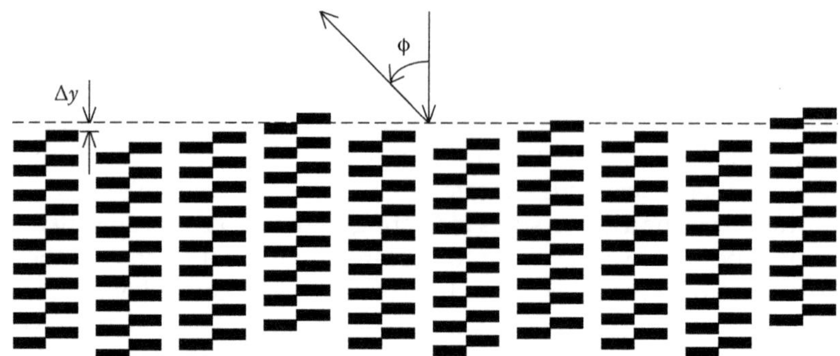

FIGURE 7.15
Model for an array of alternating multilayer models with random heights. The deviation from the mean value of the ridge height is denoted as Δy. (From Zhu, D., et al., *Phys. Rev. E* 80:051924, 2009. With permission.)

with an interval of 700 nm, with their heights being distributed randomly within a range of $\pm\Delta y_{max}$, as shown in Figure 7.15. The result is shown in Figure 7.16 for *s* (TM) polarization, and it confirms that the angular dependence of reflection can be expressed as an ensemble of sharp peaks when no height distribution is present, while with increasing irregularity, noise is introduced in the form of spikes. Thus, extracting information from the reflection pattern in either case appears to be difficult.

To overcome this difficulty, we calculate the intensity of light reflected from an array of ridges, and then integrate the intensity over all reflection angles. This approach considerably reduces the noise in the spectrum. The results are shown in the right column of Figure 7.16. These figures are obtained by generating random numbers for the ridge heights five times in each case and by plotting the corresponding results as well as their average. In spite of adding the irregularity to the ridge height, the noise appearing in the reflection spectrum is considerably suppressed and is confined to the range of 300–400 nm. It is also noticeable that a peak appearing in a regular array at 340 nm gradually decreases with increasing irregularity.

We compare the reflection spectrum obtained for $\Delta y_{max} = 205$ (nm) with that for a single ridge for *s* (TM) polarization. The result is shown in Figure 7.17. It is quite interesting that the ultraviolet region of 220–380 nm in the reflection spectrum for a single ridge is largely enhanced compared to the result for the array of ridges. We perform calculations for the case of *p* (TE) polarization for making a similar comparison. It is surprising that the spectra for a single ridge and the array of ridges almost completely agree with each other. Further, the spectra calculated for the array of ridges for the *s* and *p* polarizations are quite similar to each other with respect to the peak position and the spectral line shape. Thus, the array formation only affects the optical properties for *s*-polarized incident light, which manifests as the strong suppression of the reflection intensity in the ultraviolet region.

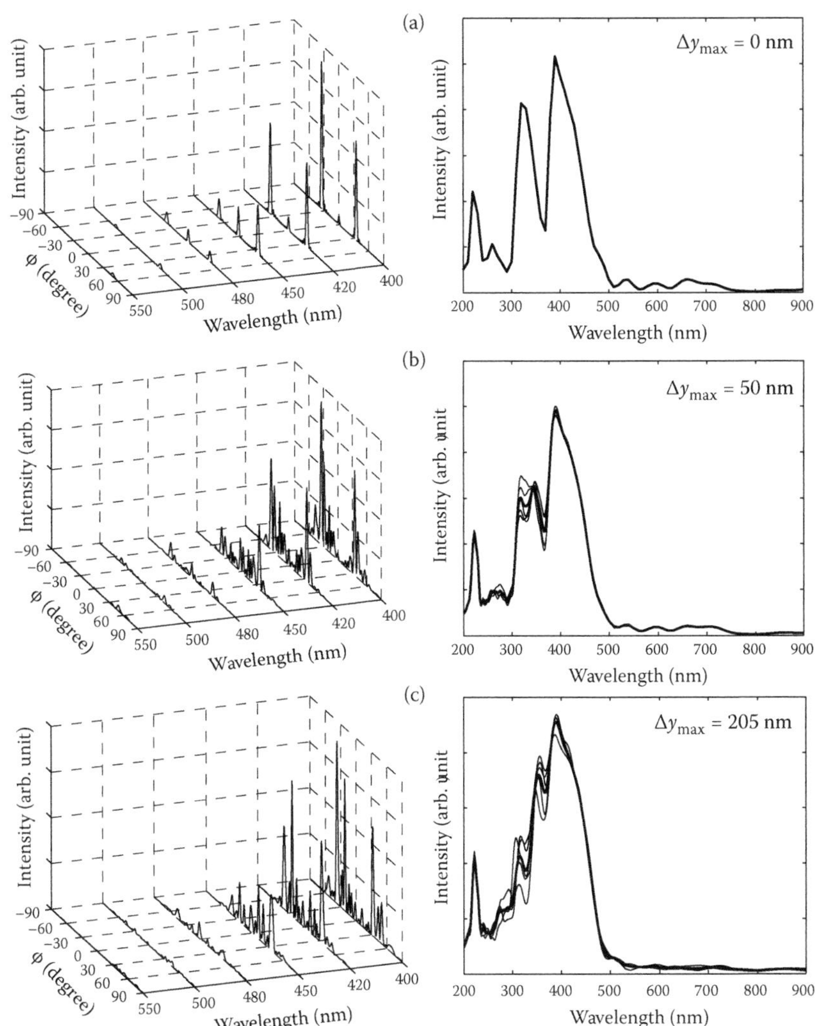

FIGURE 7.16
Angular dependence of the reflection intensities (left) and angle-integrated reflection spectra (right) for an array of alternating multilayer models with irregularity in the ridge height. The irregularity is introduced by giving a uniform random number within a maximum range of $\pm \Delta y_{max}$, where Δy_{max} = (a) 0, (b) 50, and (c) 205 (nm). In (b) and (c), the angle-integrated spectra were calculated for five series. The five curves and their average are plotted as thin and thick curves, respectively. The calculations were performed with a mesh size of 10 × 10 nm^2. (From Zhu, D., et al., *Phys. Rev. E* 80:051924, 2009. With permission.)

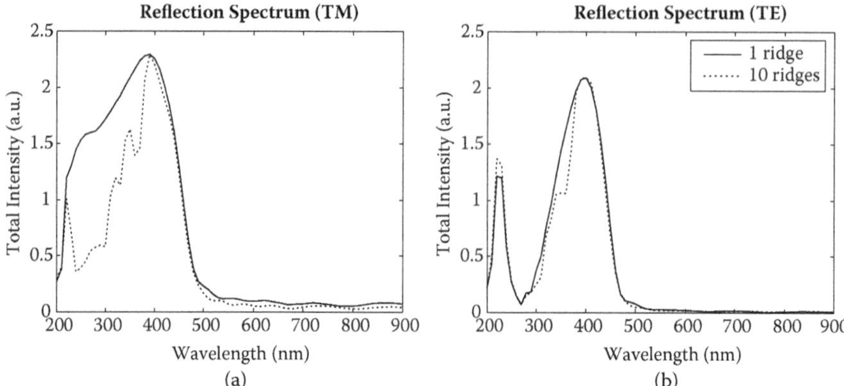

FIGURE 7.17
Angle-integrated reflection spectra for the (a) TM and (b) TE modes for a single alternating multilayer model (solid line) and an array of 10 alternating multilayer models with the irregularity in height for Δy_{max} = 205 nm (dotted line). The reflection spectrum for the array was obtained by averaging five series of calculations. (From Zhu, D., et al., *Phys. Rev. E* 80:051924, 2009. With permission.)

This suppression comes from the scattering of the reflected light by adjacent ridges, and it is particularly remarkable for *s* polarization since the direction of emission in *s* polarization is uniformly distributed over the entire angular range, while that in *p* polarization is restricted to specific angles about the normal direction. The suppression is greater at shorter wavelengths because the angle of reflection tends to be large when shorter wavelengths are employed for normal incidence. As a consequence, the reflection spectrum for *s* polarization is considerably affected by scattering from adjacent ridges, which prevents reflection at shorter wavelengths. On the other hand, the reflection spectrum for *p* polarization is rather unaffected by adjacent ridges, and it seems to be solely determined by a single alternating multilayer. Thus, a clear difference in the mechanisms between the two polarizations is found, even though the obtained spectra are similar to each other.

7.1.4 Summary

We have described the fundamentals of the standard and nonstandard FDTD methods and have shown the effectiveness of the latter by applying it to the most complex nanostructure found in the *Morpho* butterfly scale. The full ability of the FDTD method can be seen when the electromagnetic field interacts with extremely complicated nanostructures. Analysis of such interaction is far beyond the analytical approach. However, a major problem with these types of sophisticated calculation methods is that while they immediately provide the electromagnetic fields and their scattering properties, they do not provide us with any physical meaning. Thus, it is absolutely necessary to construct a proper model to elucidate the mechanism without losing the essence of the structure.

References

1. M. Srinivasarao. (1999). Nano-optics in the biological world: beetles, butterflies, birds, and moths. *Chem. Rev.* 99:1935–1961.
2. P. Vukusic and J.R. Sambles. (2003). Photonic structures in biology. *Nature* 424:852–855.
3. S. Kinoshita and S. Yoshioka. (2005). *Structural colors in biological systems—principles and applications*. Osaka University Press, Osaka.
4. S. Kinoshita and S. Yoshioka. (2005). Structural colors in nature: the role of regularity and irregularity in the structure. *ChemPhysChem* 6:1442–1459.
5. A.R. Parker and N. Martini. (2006). Structural colour in animals—simple to complex optics. *Opt. Laser Technol.* 38:315–322.
6. S. Berthier. (2006). *Iridescences—the physical colors of insects*. Springer Science + Business, New York.
7. S. Kinoshita, S. Yoshioka, and J. Miyazaki. (2008). Physics of structural colors. *Rep. Prog. Phys.* 71:076401.
8. S. Kinoshita. (2008). *Structural colors in the realm of nature*. World Scientific, Singapore.
9. A.A. Michelson. (1911). On metallic colouring in birds and insects. *Phil. Mag.* 21:554–566.
10. O.M. Rayleigh, F.R.S. (1917). On the reflection of light from a regularly stratified medium. *Proc. R. Soc. Lond. A* 93:565–577.
11. O.M. Rayleigh. (1919). On the optical character of some brilliant animal colours. *Phil. Mag.* 37:98–111.
12. H. Onslow. (1923). On a periodic structure in many insect scales, and the cause of their iridescence colours. *Phil. Trans.* 211:1–74.
13. F. Süffert. (1924). Morphologie und Optik der Schmetterlingsschuppen, Insbesondere die Schillerfarben der Schmetterlinge. *Z. Morphol. Ökol. Tiere* 1:171–308.
14. E. Merritt. (1925). A spectrophotometric study of certain cases of structural color. *J. Opt. Soc. Am. Rev. Sci. Instrum.* 11:93–98.
15. C.W. Mason. (1927). Structural colors in insects. II. *J. Phys. Chem.* 31:321–354.
16. T.F. Anderson and A.G. Richards. (1942). An electron microscope study of some structural colors of insects. *J. Appl. Phys.* 13:748–758.
17. K. Gentil. (1942). Elektronenmikrosckopische Untersuchung des Feinbaues schillernder Leisten von Morpho-schuppen. *Z. Morphol. Ökol. Tiere* 38:344–355.
18. W. Lippert and K. Gentil. (1959). Uber lamellare Feinstruckturen bei den Schillerschuppen der Schmetterlinge vom Urania- und Morpho-Typ. *Z. Morphol. Ökol. Tiere* 48:115–122.
19. H. Ghiradella and W. Radigan. (1976). Development of butterfly scales. II. Struts, lattices and surface tension. *J. Morphol.* 150:279–298.

20. H. Ghiradella. (1984). Structure of iridescent Lepidopteran scales: variations on several themes. *Ann. Entomol. Soc. Am.* 77:637–645.
21. H. Ghiradella. (1994). Structure of butterfly scales: patterning in an insect cuticle. *Microsc. Res. Tech.* 27:429–438.
22. L. Bingham, I. Bingham, S. Geary, J. Tanner, C. Driscoli, B. Cluff, and J.S. Gardner. (1995). SEM comparison of Morpho butterfly dorsal and ventral scales. *Microsc. Res. Tech.* 31:93–94.
23. H. Ghiradella. (1998). *Microscopic anatomy of invertebrates: Insecta*, 257. Vol. 11A. Wiley-Liss, New York.
24. P.K.C. Pillai. (1968). Spectral reflection characteristics of Morpho butterfly. *J. Opt. Soc. Am.* 58:1019–1022.
25. A.M. Young. (1971). Wing coloration and reflectance in Morpho butterflies as related to reproductive behavior and escape from avian predators. *Oecologia* 7:209–222.
26. H. Tabata, K. Kumazawa, M. Funakawa, J. Takimoto, and M. Akimoto. (1996). Microstructures and optical properties of scales of butterfly wings. *Opt. Rev.* 3:139–145.
27. P. Vukusic, J.R. Sambles, C.R. Lawrence, and R.J. Wootton. (1999). Quantified interference and diffraction in single Morpho butterfly scale. *Proc. R. Soc. Lond. B* 266:1403–1411.
28. S.E. Mann, I.N. Miaoulis, and P.Y. Wong. (2001). Spectral imaging, reflectivity mesurements, and modeling of iridescent butterfly scale structures. *Opt. Eng.* 40:2061–2068.
29. S. Kinoshita, S. Yoshioka, and K. Kawagoe. (2002). Mechanisms of structural colour in the Morpho butterfly: cooperation of regularity and irregularity in an iridescent scale. *Proc. R. Soc. Lond. B* 269:1417–1421.
30. S. Kinoshita, S. Yoshioka, Y. Fujii, and N. Okamoto. (2002). Photophysics of structural color in the Morpho butterflies. *Forma* 17:103–121.
31. S. Berthier, E. Charron, and A. Da Silva. (2003). Determination of the cuticle index of the scales of the iridescent butterfly *Morpho menelaus*. *Opt. Commun.* 228:349–356.
32. S. Yoshioka and S. Kinoshita. (2004). Wavelength-selective and anisotropic light-diffusing scale on the wing of the Morpho butterfly. *Proc. R. Soc. Lond. B* 271:581–587.
33. L. Plattner. (2004). Optical properties of the scales of *Morpho rhetenor* butterflies: theoretical and experimental investigation of the back-scattering of light in the visible spectrum. *J. R. Soc. Interface* 1:49–59.
34. S. Wickham, M.C.J. Large, L. Poladian, and L.S. Jermiin. (2005). Exaggeration and suppression of iridescence: the evolution of two-dimensional butterfly structural colours. *J. R. Soc. Interface* 3:99–109.
35. S. Yoshioka and S. Kinoshita. (2006). Structural or pigmentary? Origin of the distinctive white stripe on the blue wing of a Morpho butterfly. *Proc. R. Soc. Lond. B* 273:129–134.
36. S. Berthier, E. Charron, and J. Boulenguez. (2006). Morphological structure and optical properties of the wings of Morphidae. *Insect Sci.* 13:145–157.
37. R.O. Prum, T. Quinn, and R.H. Torres. (2006). Anatomically diverse butterfly scales all produce structural colours by coherent scattering. *J. Exp. Biol.* 209:748–765.
38. M. Kambe, D. Zhu, and S. Kinoshita. (2011). Origin of retroreflection from a wing of the Morpho butterfly. *J. Phy. Soc. Jpn.* 80:054801.

39. K.S. Yee. (1966). Numerical solution of initial boundary value problems involving Maxwell's equations in isotropic media. *IEEE Trans. Anten. Prop.* 14:302–307.
40. K.S. Kunz and R.J. Luebbers. (1993). *The finite difference time domain method for electromagnetics*. CRC Press, Boca Raton, FL.
41. D.M. Sullivan. (2000). *Electromagnetic simulation using the FDTD method*. IEEE Press, Piscataway.
42. A. Taflove and S.C. Hagness. (2005). *Computational electrodynamics: the finite-difference time-domain method*. Artech House, Boston.
43. G. Mur. (1981). Absorbing boundary conditions for the finite-difference approximation of the time-domain electromagnetic-field equations. *IEEE Trans. Electromagn. Compat.* 23:377–382.
44. R.L. Higdon. (1986). Absorbing boundary conditions for difference approximations to the multi-dimensional wave equation. *Math. Comput.* 47:437–459.
45. J.P. Berenger. (1994). A perfectly matched layer for the absorption of electromagnetic waves. *J. Comput. Physics* 114:185–200.
46. R.J. Luebbers, K.S. Kunz, M. Schneider, and F. Hunsberger. (1991). A finite-difference time-domain near zone to far zone transformation. *IEEE Trans. Anten. Prop.* 39:429–433.
47. R.J. Luebbers, D. Ryan, and J. Beggs. (1992). A two-dimensional time-domain near-zone to far-zone transformation. *IEEE Trans. Anten. Prop.* 40:848–851.
48. J.B. Cole. (1997). A high-accuracy realization of the Yee algorithm using non-standard finite differences. *IEEE Trans. Micro. Theory Tech.* 45:991–996.
49. J.B. Cole. (2000). In *Applications of nonstandard finite difference schemes*, ed. R.E. Mickens, 109–153. World Scientific, Singapore.
50. J.B. Cole. (2002). High-accuracy Yee algorithm based on nonstandard finite differences: new developments and verifications. *IEEE Trans. Anten. Prop.* 50:1185–1191.
51. J.B. Schneider. (2004). Plane waves in FDTD simulations and a nearly perfect total-field/scattered-field boundary. *IEEE Trans. Anten. Prop.* 52:3280–3287.
52. R. Holland and J.W. Williams. (1983). Total-field versus scattered-field finite-difference codes: a comparative assessment. *IEEE Trans. Nucl. Sci.* 30:4583–4588.
53. D. Zhu. (2009). Modeling and simulation of Morpho butterfly scale using NS-FDTD method. PhD thesis, University of Tsukuba.
54. R.E. Mickens (ed.). (2005). *Advances in the applications of nonstandard finite difference schemes*. World Scientific, Singapore.
55. J.B. Cole. (1994). A nearly exact second-order finite-difference time-domain wave propagation algorithm on a coarse grid. *Computers in Physics* 8:730–734.
56. N. Okada and J.B. Cole. (2010). High-accuracy finite-difference time domain algorithm for the coupled wave equation. *J. Opt. Soc. Am. B* 27:1409–1413.
57. T. Ohtani, K. Taguchi, T. Kashiwa, Y. Kanai, and J.B. Cole. (2009). Nonstandard FDTD method for wideband analysis. *IEEE Trans. Anten. Prop.* 57:2386–2396.
58. G. Mie. (1908). Beiträge zur Optik trüber Medien, speziell kolloidaler Metallösungen. *Ann. Physik* 25:377–445.
59. M. Born and E. Wolf. (1999). *Principles of optics*. 7th ed. Cambridge University Press, Cambridge.
60. D. Zhu, S. Kinoshita, D. Cai, and J.B. Cole. (2009). Investigation of structural colors in Morpho butterflies using the nonstandard-finite-difference time-domain method: effects of alternately stacked shelves and ridge density. *Phys. Rev. E* 80:051924.

61. S. Banerjee, J.B. Cole, and T. Yatagai. (2007). Colour characterization of a Morpho butterfly wing-scale using a high accuracy nonstandard finite-difference time-domain method. *Micron* 38:97–103.
62. S. Banerjee and D. Zhu. (2007). Optical characterization of iridescent wings of Morpho butterflies using a high accuracy nonstandard finite-different time-domain algorithm. *Opt. Rev.* 14:359–361.
63. R.T. Lee and G.S. Smith. (2009). Detailed electromagnetic simulation for the structural color of butterfly wings. *Appl. Opt.* 48:4177–4190.

7.2 Simulation Analysis on the Optical Role of the Random Arrangement in *Morpho* Butterfly Scales

Akira Saito

The shining blue of some *Morpho* butterflies in South America is a typical example of a structural color (see Figure 3.32(a) in Chapter 3). The brilliant color has long attracted both scientific [1–5] and general interests, because metallic coloration in animals is a hot research topic. However, the blue coloration contains a physically mysterious feature. The butterfly wing scales are composed of an almost transparent cuticle protein and do not have any blue pigment. The principle of this coloration was first attributed to interference, which is also consistent with the high reflectivity of blue (>60%). However, the *Morpho* blue optical reflection properties cannot be explained solely by grating or multilayer interference: the color appears blue in too wide an angular range (more than ±40° from the normal) (Figure 3.31). The homogeneity of the color in such a wide angular range contradicts the concept of interference. Thus, *Morpho* blue, which is one of the most well-known structural colors, is a peculiar example.

The mystery of the high reflectivity of a single color has been explained by a model with a peculiar optical nanostructure [6]. This model is based on a skillful combination of two contrasting characters: ordered (regular) and random (irregular) structures. We have recently proven the theoretical model by an experimental and engineering approach by emulating and extracting the essence of the specific structures [7,8].

The experiments to reproduce *Morpho* color showed that the disorder in the structure was essential, especially to prevent multicolor interference. However, the randomness itself has not been defined precisely, nor estimated in view of the optical dependence on the spatial parameters. Thus, the role of randomness is still ambiguous. Approaches to fabricate nanostructures by varying the spatial parameters may provide direct evidence of their optical properties [9], but studies by empirical trials only are not efficient. Thus, simulation studies are necessary to obtain the guiding principles of coloration. Although the optical properties of random structures have been hard to simulate analytically, the recent development in numerical simulation methods to treat electromagnetic fields has enabled us to approach such nonanalytical objects, especially using the finite-difference time-domain (FDTD) analysis.

A few examples of the theoretical analyses on the optical effect of *Morpho* butterfly's wings have been reported by either conventional analytical calculation [6] or FDTD analysis [10–12]. Both were in principle based on the analysis of a unit or a limited number of shelf-like structures (Figure 3.32(b)), apart from the effect of random arrangement in the whole structure, except for one example of the FDTD simulation by Lee and Smith [13], who revealed the effect of taper and offset in each shelf-like unit, but did not focus on the randomness in the array of the shelves. They succeeded in realizing the wide angular distribution of the *Morpho* color, but the role of random parameters was not yet clarified.

Here, we report results of the quantitative analysis on the optical role of randomness to realize the specific *Morpho* color that prevents the sharp reflective angular dependence. The results were derived by considering different directions of randomness, i.e., lateral and vertical randomness, the number of random components at the nanometer scale, and randomness (incoherence) in the incident light.

7.2.1 Optical Properties

7.2.1.1 Aspect of the Randomness in Structures

A magnified scanning electron microscope (SEM) image of a typical *Morpho* butterfly's (*Morpho didius*) scale shows ridges in a distorted stripe structure in planar view (Figure 3.32(c)), but shows a shelf-like multilayer structure (Figure 3.32(b)). These features suggest an interference effect based on a multilayer or grating as an origin of the coloration, which, however, produces multicolor interference. The mystery of the high reflectivity that lacks the multicoloration can be explained with a model including randomness [6], which has been proven by artificially reproducing the optical properties by extracting the principles [7,8], which are briefly summarized in the following section, although the details and processes are presented in our past reports [7–9,14,15].

As shown in Figure 3.32(b) and (c), the surface of a scale has a fine and complex structure. The specific structure is characterized as one-dimensionally arrayed shelves, each of which is composed of a multilayer. These features allow us to depict some principles on the specific blue, as shown in Figure 3.33. In brief, the principles can be categorized in five fundamental qualities:

1. The wavelength of the blue color is determined by interference in a single shelf, which is composed of alternating layers of high refractive index protein (r.i. ~ 1.5) and air with an r.i. of 1.0.
2. The blue is diffracted into a wide angular range because the width of each shelf has a limited size (~300 nm), which is slightly smaller than the wavelength of the reflected light.

3. The arrangement of the discrete pieces (shelves) has a certain amount of randomness. This randomness prevents the multicolor interference because it prohibits interference of the light scattered from neighboring discrete shelves.
4. A high density of the discrete shelves leads to high reflectivity. Otherwise, a gap much wider than the wavelength will allow the incident light to penetrate the bottom of the multilayer, allowing it to be absorbed or transmitted.
5. Quasi 1D anisotropy, which can be seen in Figure 3.32(c): the shelves extend along the y-axis. Anisotropy plays an important role in generating high reflectivity in an angular range limited to the y-axis. Otherwise, if the patterns have an isotropic randomness such as pixels, the reflection will scatter everywhere isotropically, critically decreasing the reflectivity at a certain viewing angle.

7.2.1.2 Measurement

The measured optical property of the *Morpho* butterfly wing is shown in Figure 3.39, in which an angular dependence of the reflectivity was obtained along the lateral axis in Figure 3.33(b) in Chapter 3. This axis showing a particular wide angular dispersion that corresponds to the x-axis in Figure 3.32(c). Figure 3.39(a) shows a schematic image of the measurement setup, in which a white incident light enters a sample surface from a normal direction. By scanning an optical fiber along the lateral axis, the angular profile of the reflectivity was obtained. Also, dependence of the reflectivity on wavelengths was measured using a spectrometer. Figure 3.39(b)–(d) shows a comparison of the profiles from the real *Morpho* butterfly's wing, the reproduced *Morpho* film (described in detail in Chapter 3), and a usual (not discrete, but continuous) flat multilayer deposited on a flat substrate, respectively.

As foreseen, the usual multilayer shows a sharp peak because of the strong interference effect. On the other hand, the real *Morpho* butterfly's profile (Figure 3.38(b)) is broad due to the diffraction from the discrete shelves, and smooth (without fringes), which can be attributed to randomness [6]. The reproduced *Morpho* film shows basically common characteristics with the real *Morpho* butterfly's one, as mentioned in Section 3.3, far from that of the continuous flat multilayer (Figure 3.39(d)).

7.2.2 Software and Computer Platform

In our present research, the FDTD method [16], which is one of the most convincing numerical techniques to treat electromagnetic and optical problems, was used to study the reflection from the *Morpho* structure described above. For advantages of simplicity and wide generality in the analysis, we used the tools on the market both for the FDTD algorithm (FDTD Solutions, a software package provided by Lumerical Solutions, Inc.) and for computer resources (a standard personal computer equipped with a multicore CPU (Core 2 Quad, Intel, Inc.) with a memory size of 4 GB). Before starting to investigate the role of structural randomness, we calibrated the setting in the FDTD analysis prudently using a simple analytical object to check the accordance of the simulated results on the FDTD analysis with that in a conventional analytical calculation.

Finally, we computed the reflective angular dependence from the disordered array of the shelf structures for monochromatic illumination at normal incidence, corresponding to the far-field scattered intensities. The grid spacing to solve the model of Maxwell's equations was set to 1/30 wavelength for high accuracy. The light source of the total field/scattered field (TF/SF) was applied, which can separate the center area calculating the total field (scattered field + incident field) from the area treating solely the scattered field surrounding the center area. Perfectly matched layer (PML) absorbing boundary conditions [17] were applied to the frame of the simulated domain, which prevents the reflection of external fields generated at the surfaces. Details of the FDTD analysis have been presented widely in the literature [16].

Considering the cross-sectional feature of the nanostructure (Figure 3.32(b)) and the specific reflective property (Figure 3.39(b)), the cross-sectional 2D structure was treated as a base of the simulation. Our purpose was not to make a precise 3D mathematical model for the *Morpho* butterfly, but to clarify the role of randomness in the scale. To exclusively regard randomness in the array of the shelves, the structure of each shelf was supposed to be as simple as possible, neglecting the offset and taper [13] in each shelf. Thus, the parameter of a single shelf component was fixed at a uniform shape as shown in Figure 7.18(a), which has five layers with a numerical index of 1.5, width of 300 nm, thickness of 60 nm, and interval of 145 nm.

7.2.3 Origin of the Smooth Angular Dependence without Fringes

7.2.3.1 Effect of Randomness in the Structure

One of the characteristics in the *Morpho* color is the smooth reflective angular profile without sharp fringes, despite the interference effect producing high reflectivity. To realize the smoothness in the profile by simulation, a degree of randomness in the arrangement of the shelf arrays was estimated. Figure 7.18(b) shows a structural model to introduce randomness into the *Morpho* butterfly's scale structure, where randomness in the z-direction was given by a normal distribution with a standard deviation of σ_z = 50 nm, whereas the lateral distance d was set to 400 nm, referring to the real butterfly's scale (σ_z ~110 nm, d ~400 nm). The number (M) of shelves was given as a parameter to define the degree of randomness. The

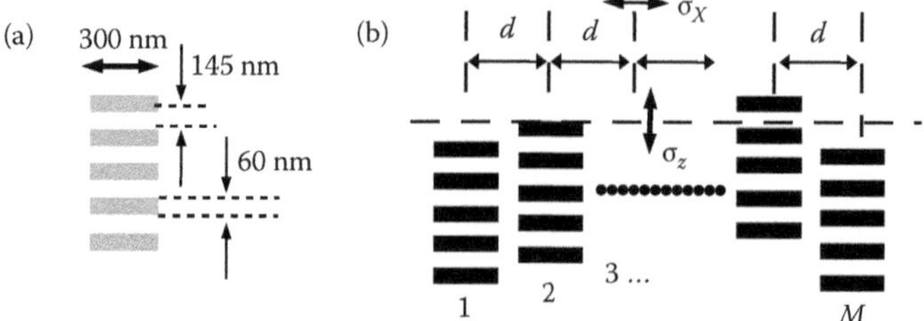

FIGURE 7.18
(a) Structural parameters of a single component in the shelf array on the *Morpho* scale. (b) Structural model to introduce randomness in the *Morpho* scale structure. The lateral distance d = 400 nm was set analogous to the butterfly's scale.

relatively smaller σ_z (50 nm) in the calculation than in reality (110 nm) was provided to consider the role of the parameter M as purely as possible without the influence of another parameter. On the other hand, a too small σ_z (<50 nm) gives an essentially wrong and different result from the reality with large sharp, peaks which loses the meaning to estimate the role of M. The simulated reflective angular profile for the case in which M = 50 (Figure 7.19) shows many sharp fringes differently from the reality (Figure 3.39(b)).

Since the result shown in Figure 7.19 seems to suggest a lack of randomness, much larger values of M than 50 were tested. However, surprisingly, the result did not show an improvement, even when M was 2,000 (Figure 7.19). Thus, we tried to introduce additional randomness in other directions.

7.2.3.2 Prevention of Interference Fringes

The incident light in our FDTD platform was defined as a perfectly coherent light, whereas the experiments were performed under an incoherent light irradiation. Thus, we mixed and averaged several (number = N) kinds of different results that were independently calculated for randomly arranged shelf structures, and compared the effect of the number N. Also, independent results for transverse electric (TE) and transverse magnetic (TM) fields were averaged [13]. Taking into consideration the power of our computer, M was fixed at 50 for all trials. Other parameters such as the lateral distance (d = 400 nm) and standard deviation (σ_z = 50 nm) were kept at the same value as in Figure 7.19 for comparison.

Figure 7.20 shows the dependence of a simulated reflective angular profile on the parameter N for (a) N = 20, (b) N = 50, and (c) N = 100. In comparison with the result shown in Figure 7.19 (N = 1), the mixture of N times the number of independent simulations shows a dramatic effect, even when the other parameters, d, M, and σ_z, have not been changed. As the value of N increased, a smoother profile was obtained.

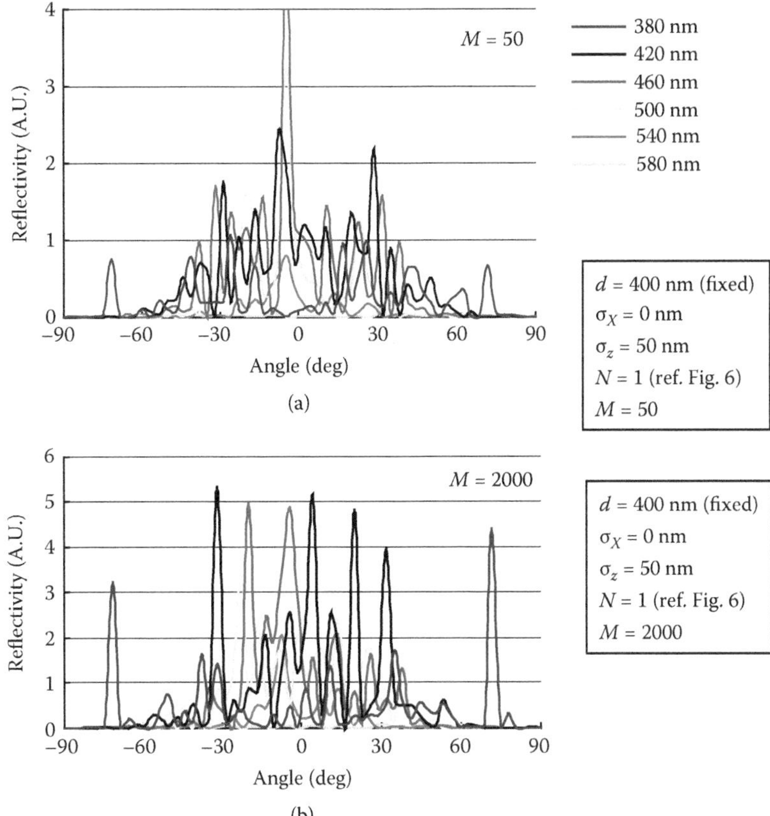

FIGURE 7.19 (See color insert.)
Comparison of the optical effects of the structural randomness. Simulated angular dependence of reflectivity for the model shown in Figure 7.18(b) for M = 50 and M = 2,000.

7.2.3.3 Consideration on the Optical Coherence

Figure 7.21 shows an interpretation of the meaning of the value N that reflects the concept of incoherence of light. By dividing a simulation space by N, an overabundant interference effect from the whole structure (Figure 7.21a) can be suppressed (Figure 7.21b), which leads to a smooth profile without sharp fringes. Thus, the division parameter N has a role providing incoherence to light in the simulation, which limits the size of the coherent block in the calculation. Hereafter, we refer to the value N as the incoherence parameter.

With reference to traditional understanding in optics, Figure 7.21a can be referred to as a situation where the summation of the electric field is taken in advance and then squared to obtain the intensity. In contrast, Figure 7.21b shows the condition where the squared intensity for small parts is obtained

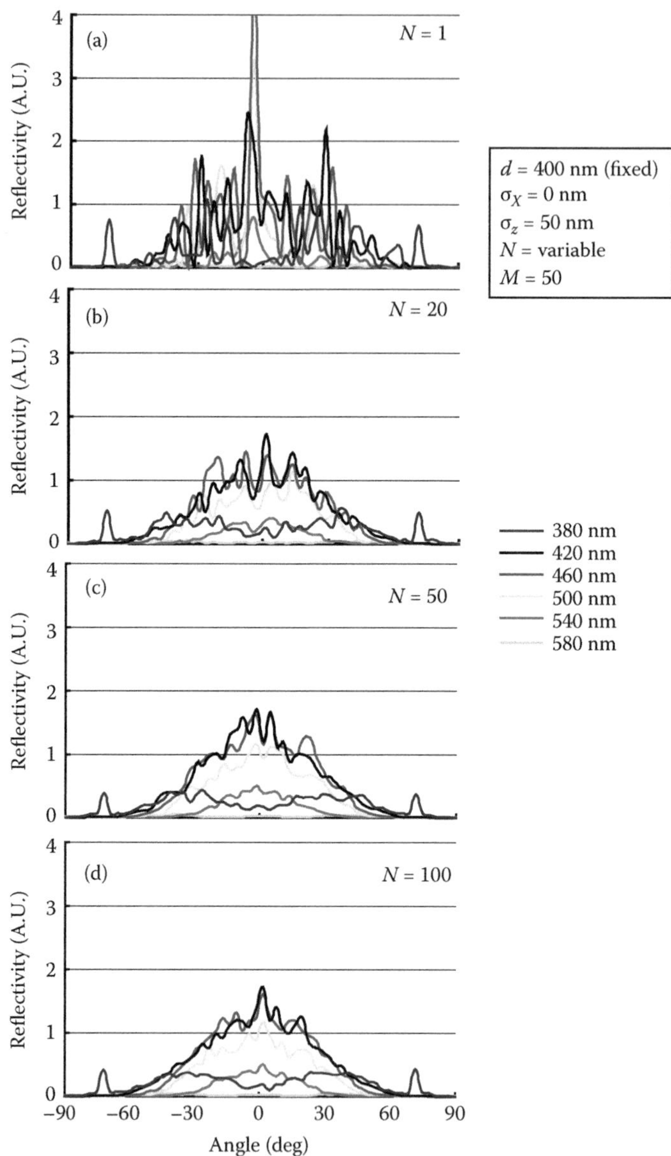

FIGURE 7.20 (See color insert.)
Optical effect of the incoherence (division) parameter N. Dependence of simulated angular profile of reflectivity on N. (a) N = 20, (b) N = 50, (c) N = 100. The results on the independent N blocks (one block is modeled in Figure 7.18(b), where M = 50, σ_z = 50 nm, d = 400 nm) were averaged.

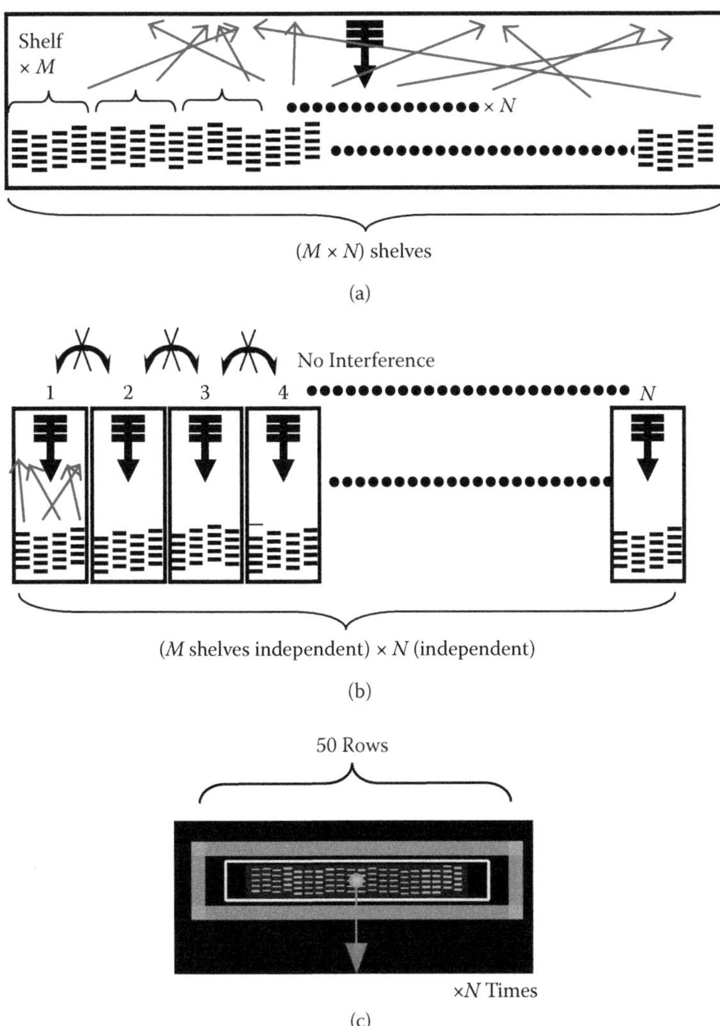

FIGURE 7.21
Concept of the incoherence (division) parameter N to give incoherence to the light in simulation. The arrows represent scattered light. (a) Whole number (M × N) shelves are simply taken in account. (b) For each of N times, the components of number M are randomly and independently arranged, and the calculated results are finally averaged. (c) Real image in the FDTD software used in our simulation.

in advance and then added. Finally, a criterion to treat the incoherence in incident light by changing the parameter N has been clarified. In other words, our trials derived the practical criterion of N to obtain sufficient incoherence.

According to the results shown in Figure 7.20, the incoherence value N of 50 gave a smoothing effect to the angular profile. Because we still needed to perform a variety of simulations by changing the other parameters, we thus fixed the incoherence value N at 50. Figure 7.21(c) shows the actual image in the FDTD software used in our simulation to help readers to image our real operations.

7.2.4 Role of Randomness in the Structure under Incoherent Light

7.2.4.1 Role of Lateral Randomness

Since the effect of the value N was clarified as an incoherence parameter that can solve the difficulty of "multifringes" shown in Figure 7.19, we start to estimate the role of nanorandomness in the structure. There is still a discord with the reality of the angular profile, because sharp peaks appeared at ±75° in Figure 7.20(d). This difficulty was solved by the addition of lateral randomness σ_x (Figure 7.22(a)), which was given with a mean value of the lateral distance d = 400 nm. Figure 7.22(b)–(d) show the simulated angular profile for the models in which lateral randomness σ_x was introduced and varied from 0 to 30 nm, where M = 50, N = 50, and σ_z = 50 nm were maintained constant for comparison with previous simulations. The sharp peaks at ±75° (Figure 7.20(d)) disappeared, and the smoothness in the profile increased. Thus, the role of lateral randomness can be summarized as being to prevent side peaks.

The origin of the side peaks can be attributed to a grating effect, an example of which is shown in Figure 7.23 (lateral distance d = 600 nm, σ_x = 0 nm) more clearly than in Figure 7.22. Figure 7.23 shows an effect of the missing lateral randomness σ_x, where the side peaks appear clearly, shifting by an angle corresponding to the wavelength. According to the principle of interference, the peak should appear at the angle corresponding to the condition $d \cdot \sin\theta = \lambda$, where θ is the angle measured from the normal (z) to the sample. Actually, the angle θ of the peak for each λ can be derived as (θ, λ) = (56°, 500 nm), (50°, 460 nm), (44°, 420 nm), which agrees with data shown in Figure 7.23.

Other parameters M = 50, N = 50, and σ_z = 50 nm were kept at the same values as in Figure 7.19 for comparison. Thus, the role of lateral randomness σ_x can be assigned to an antigrating effect.

7.2.4.2 Role of Vertical Randomness

Next, we independently estimated the role of vertical randomness σ_z. In comparison with the case shown in Figure 7.22(d), where both the specular peak and side peaks disappeared, all parameters other than σ_z (i.e., N = 50, M = 50, d = 400 nm, and σ_x = 30 nm) were kept the same as in Figure 7.22(d). Here, only the vertical randomness σ_z was removed, as depicted in Figure 7.24(a). Thus, Figure 7.24(b) shows the effect of vertical randomness in the reflective angular profile. The result shows a strong specular reflection as presented for a usual continuous multilayer (Figure 3.39(d)), and side peaks as shown

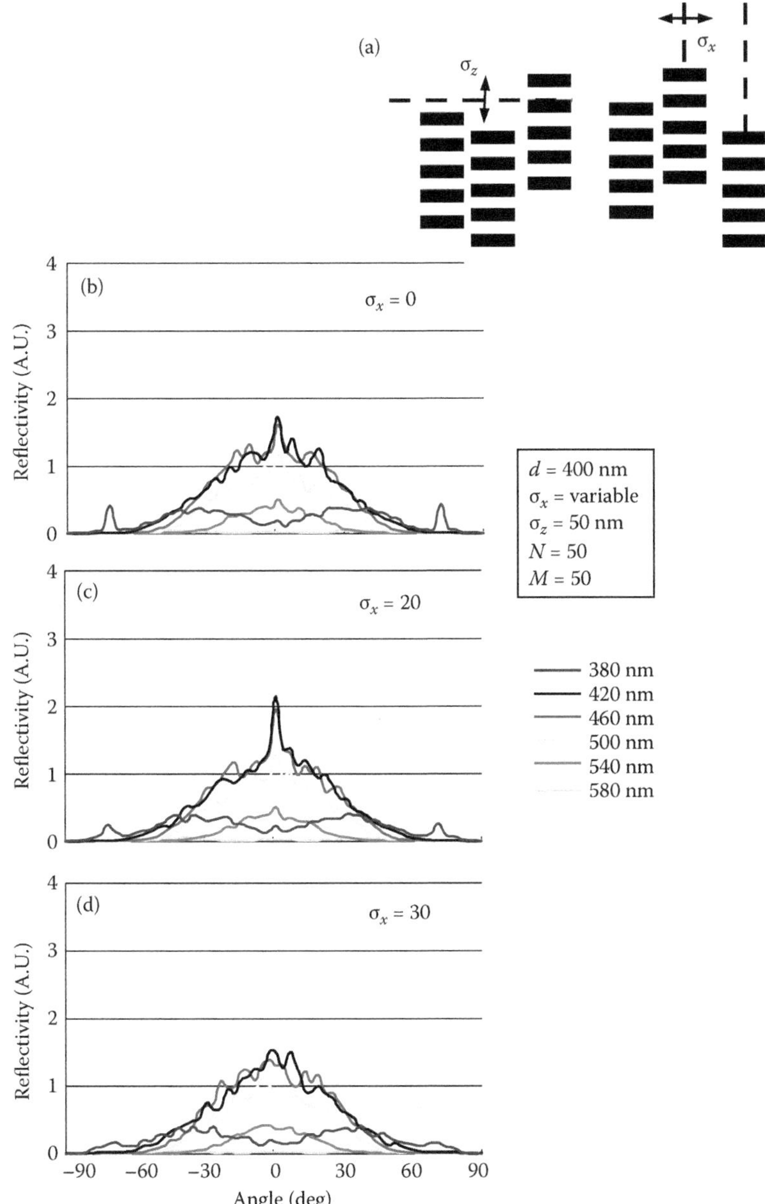

FIGURE 7.22 (See color insert.)
(a) Concept of the lateral randomness σ_x (b)–(d). Optical effect of the lateral randomness σ_x. Simulated angular profile of reflectivity for the models that have common averaged lateral distance d = 400 nm and the different lateral randomness σ_x: (b) σ_x = 0 nm, (c) σ_x = 20 nm, (d) σ_x = 30 nm. N = 50, M = 50, and σ_z = 50 nm were used.

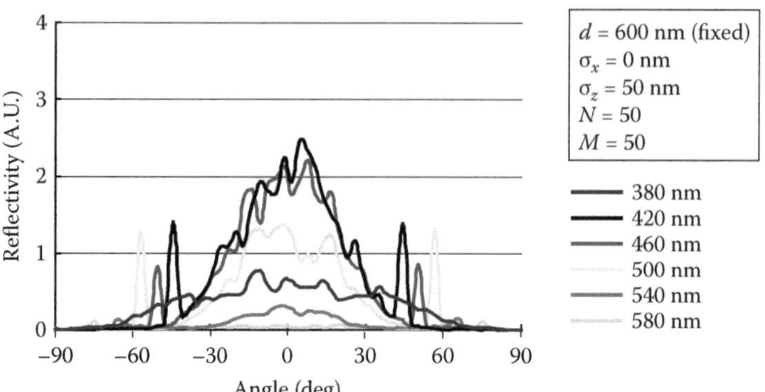

FIGURE 7.23 (See color insert.)
Example of the grating effect under the condition that $\sigma_x = 0$ nm, where side peaks appear dependent on wavelength. Simulated angular profile of reflectivity for the model in which the averaged lateral distance d = 600 nm, M = 50, N = 50, and $\sigma_z = 50$ nm.

in Figures 7.20(d) and 7.23. This result indicates that the lack of σ_z gives an effect identical to the usual multilayer and partially to the grating as well, even if the incoherent parameter N = 50, the number of random components M = 50, and lateral randomness $\sigma_x = 30$ nm are satisfied. To show the effect more clearly, the change of the lateral randomness $\sigma_x = 30$ nm into 10 nm (all other parameters are kept the same as in Figure 7.24(b)) does not change the specular peak, but strongly sharpens the grating peak at ±75°, as shown in Figure 7.24(c). This result shows that the role of the lateral randomness σ_x is focused on the antigrating.

Thus, the role of vertical randomness σ_z is to provide a wide angular range specific to the *Morpho* color, and the antigrating effect as well (because the grating peak does not disappear even if the lateral randomness $\sigma_x = 30$ nm is given as shown in Figure 7.24(b)).

7.2.4.3 Role of the Number of Components

After the role of the variables σ_x and σ_z was revealed, finally, the role of the number M of random components in the structure (originally shown in Figure 7.18(b)) was estimated. Other parameters such as the lateral distance (d = 400 nm) in the shelf array, the incoherent parameter N = 50 and the standard deviation $\sigma_z = 50$ nm were maintained at the same value as in Figure 7.20(c) for comparison.

Figure 7.25 shows the dependence of the simulated reflective angular profile on M ((a) M = 20, (b) M = 30, (c) M = 50). The results show that a small M provides a large interference effect, i.e., a large specular reflection, such as that given by the continuous multilayer (Figure 3.39(d)) due to the lack of randomness. On the other hand, as M increases, the wide angular

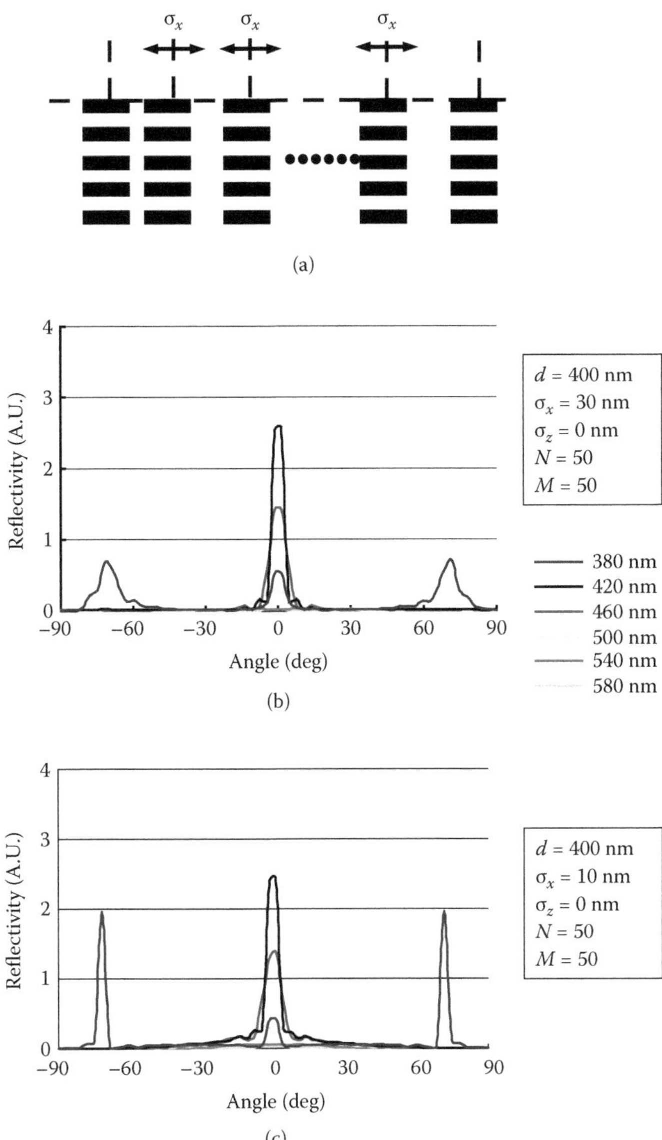

FIGURE 7.24 (See color insert.)
Optical effect of the vertical randomness σ_z. (a) Concept of the model that considers only the lateral randomness σ_x (average lateral distance d = 400 nm), whereas vertical randomness σ_z = 0 nm. (b, c) Simulated angular profiles of reflectivity for the model shown in (a), with the values of (b) σ_x = 30 nm and (c) σ_x = 10 nm. M = 50, N = 50 for both.

dispersion specific to the *Morpho* butterfly appears, because interference is suppressed by increased randomness in the structure. Thus, the number M works as a suppressor of interference to realize a wide reflective angular range. Although the fringes are not perfectly prevented, the number M was fixed at 50 to take into consideration the cost performance of the calculation.

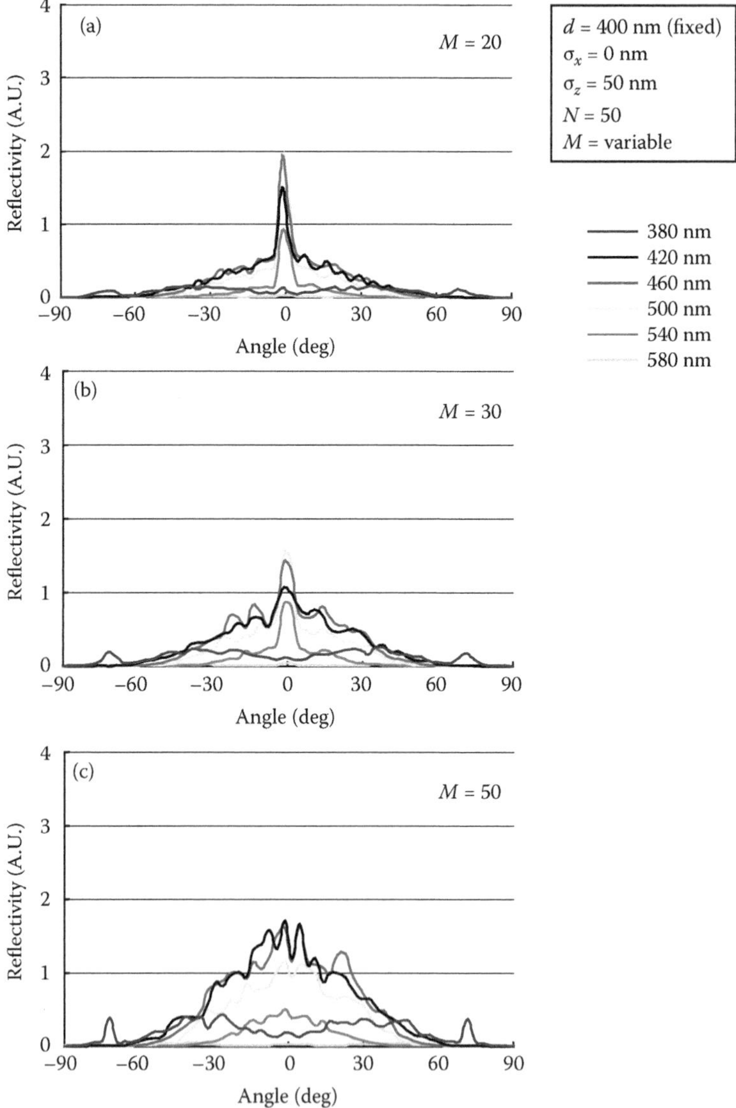

FIGURE 7.25 (See color insert.)
Optical effect of the number M of the random components (Figure 7.18(b)). Dependence of the simulated angular profile of reflectivity on M. (a) M = 20, (b) M = 30, (c) M = 50. N = 50, $\sigma_x = 0$ nm, $\sigma_z = 50$ nm, and d = 400 nm, for all.

7.2.5 Further Future Applications

Since the optical effect provided by randomness has recently been found to involve essentially new scientific occurrences [18,19], the direction of research shown here would contain a future potential importance. A criterion to treat the incoherence in incident light by changing the parameter N will serve to analyze various random structures, providing a guideline for simulation analysis.

In our past works to reproduce the specific angular-independent *Morpho* color, the minimum essence of the nanostructure in the *Morpho* butterfly's scale was extracted and emulated by a dielectric multilayer on stepped quartz (Section 3.3). The structures had been fabricated by conventional techniques in the semiconductor industry, namely, electron beam (EB) lithography and dry etching.

After successful reproduction of the fundamental optical properties of *Morpho* blue, the simple fabrication process opens the possibility to a variety of applications, beyond the experimental proof of purely scientific principles. In fact, in general argument, the structural color produces color without pigment, and makes a tone that is qualitatively impossible by pigments (brilliant luster with high reflectivity, speckle-like aspects, etc.). It has an ecological merit too, because it does not need chemical pigments, but the color can be controlled by changing the composition or thickness of only two different materials of the multilayer. Moreover, this color has lifetime advantages in comparison with conventional pigments, because it is resistant to discoloration due to chemical changes over time, as long as the structure is maintained [20]. Thus, a variety of applications are relevant to the structural color: cosmetics, decorations, textures, paint, etc. Especially in the case of the *Morpho* type color, it realizes both a wide reflective angular range and high reflectivity that are usually not compatible with each other. In addition, it provides a single color by an anti-interference effect that prevents multicoloration. These characters are fit for use in posters or displays.

Initially, to reproduce *Morpho* blue, randomness was simply and intuitively thought to be important to prevent multicolor interference, whereas here, the roles of randomness were estimated and quantitatively revealed. This analytical direction will serve to design precisely the optical properties, such as coloration (wavelength distribution) and spatial reflective distribution of the artificial materials [15], as depicted schematically in Figure 7.26.

7.2.6 Summary

Using FDTD analysis, the optical role of different forms of randomness in the structure of the *Morpho* butterfly's scale was studied, which has been previously impossible by conventional analytical methods because of complexity. It became possible to quantitatively estimate the optical properties that had previously been intuitively considered. The simulated results revealed two different parts: First, consideration of the incoherence of the light source is essential to obtain a specific smooth reflective angular profile

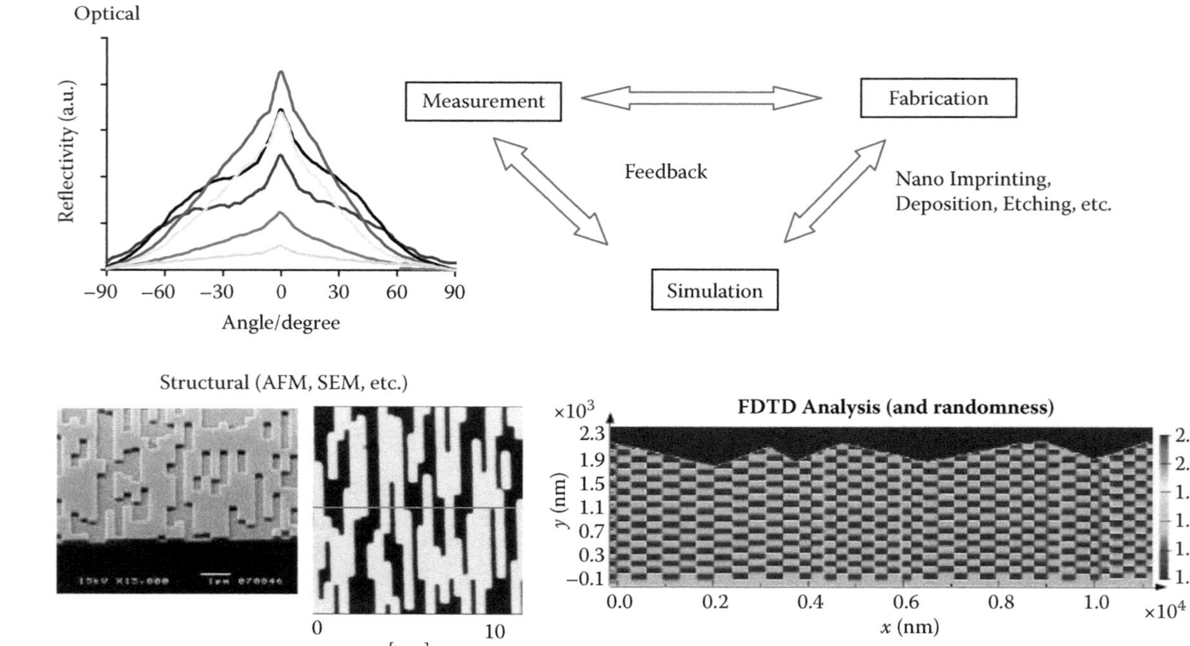

FIGURE 7.26 (See color figure at http://www.crcpress.com/product/isbn/9781439877463)
Schematic future direction of the application of FDTD analysis of random systems that will serve to design precisely the optical properties such as coloration (wavelength distribution) and spatial reflective distribution of the artificial materials.

without fringes. Smoothness in the angular profiles is not realized only by structural randomness (Figure 7.25). This effect is expressed by the number N that divides the entire simulated space into the individual small parts (Figures 7.20 and 7.21) to break the phase relationship. Second, different kinds of structural randomness (σ_x, σ_z, number M) were discussed quantitatively and their own roles were revealed. Under incoherent illumination, the wide angular reflective distribution was found to have originated from σ_z, which prevents sharp specular reflection of the multilayer interference (Figure 7.24). To activate the effect of σ_z, the number of random components M should be sufficiently large (>50); otherwise, the specular reflection becomes dominant (Figure 7.25). On the other hand, the grating-base side peaks (Figures 7.22(b) and 7.23) are prevented by σ_x (Figure 7.22(d)), whereas this antigrating effect is supported by σ_z (comparison between Figures 7.22(d) and 7.24(b)) too. The smoothness in the profile is enhanced by randomness in both σ_x and σ_z.

Acknowledgments

This study was achieved by collaborations with the members in Professor Y. Kuwahara's laboratory (Osaka University), especially M. Yonezawa, J. Murase, and Professor S. Juodkazis (Swinburne University of Technology) and Professor V. Mizeikis (Shizuoka University). This study was supported by Kakenhi (Grand-in-Aid from the Ministry of Education, Sports, Culture, Science and Technology, no. 18360039 and no. 21360033) and Japanese Science and Technology Agency (Seeds Innovation Project bridging academics and industry). We thank Drs. M. Kawano (Lumerical Solutions, Inc.) and O. Kudo (Cornes-Dodwell, Inc.) for their support for FDTD simulations.

References

1. Michelson, A.A. (1911). On metallic colouring in birds and insects. *Phil. Mag.* 21:554–566.
2. Rayleigh, J.W.S. (1918). On the optical character of some brilliant animal colours. *Phil. Mag.* 37:98–111.
3. Mason, C.W. (1923). Structural colors in feathers, part I. *J. Phys. Chem.* 27:201–251.
4. Anderson, T.F., and Richards Jr., A.G. (1942). An electron microscope study of some structural colors in insects. *J. Appl. Phys.* 13:48–58.
5. Ghiradella, H. (1998). Hairs, bristles, and scales. In Harrison, F.N., and Locke, M. (eds.), *Microscopic anatomy of invertebrates: Insecta*. Vol. 11A. Wiley-Liss, New York.

6. Kinoshita, S., Yoshioka, S., and Kawagoe, K. (2002). Mechanisms of structural colour in the *Morpho* butterfly: cooperation of regularity and irregularity in an iridescent scale. *Proc. R. Soc. Lond. B* 269:1417–21.
7. Saito, A., Yoshioka, S., and Kinoshita, S. (2004). Reproduction of the *Morpho* butterfly's blue: arbitration of contradicting factors. *Proc. SPIE* 5526:188–194.
8. Saito, A. (2005). Reproduction of *Morpho*-blue by artificial substrate. In Kinoshita, S., and Yoshioka, S. (ed.), *Structural colors in biological systems*. Osaka University Press, Osaka, pp. 287–295.
9. Saito, A., Ishikawa, Y., Miyamura, Y., Akai-Kasaya, M., and Kuwahara, Y. (2007). Optimization of reproduced *Morpho*-blue coloration. *Proc. SPIE* 6767:1–9.
10. Gralak, B., Tayeb, G., and Enoch, S. (2001). *Morpho* butterflies wings colour modeled with lamellar grating theory. *Optics Express* 9:567–578.
11. Plattner, L.J. (2004). Optical properties of the scales of *Morpho* rhetenor butterflies: theoretical and experimental investigation of the back-scattering of light in the visible spectrum. *J. R. Soc. Interface* 1:49–59.
12. Banerjee, S., Cole, J.B., and Yatagai, T. (2007). Colour characterization of a *Morpho* butterfly wing-scale using a high accuracy nonstandard finite-difference time-domain method. *Micron* 38:97–103.
13. Lee, R.T., and Smith, G.S. (2009). Detailed electromagnetic simulation for the structural color of butterfly wings. *Appl. Optics* 48:4177–4190.
14. Saito, A., Miyamura, Y., Nakajima, M., Ishikawa, Y., Sogo, K., Akai-Kasaya, M., Kuwahara, Y., and Hirai, Y. (2006). Reproduction of the *Morpho* blue by nano casting lithography. *J. Vac. Sci. Technol. B* 24:3248–3251.
15. Saito, A., Miyamura, Y., Ishikawa, Y., Murase, J., Akai-Kasaya, M., and Kuwahara, Y. (2009). Reproduction, mass production, and control of the *Morpho* butterfly's blue. *Proc. SPIE* 7205:1–9.
16. Taflove, A., and Hangness, S.C. (2005). *Computational electrodynamics: finite-difference time-domain method*. 3rd ed. Artech House, Norwood, MA.
17. Berenger, J.P. (1994). A perfectly matched layer for the absorption of electromagnetic waves. *J. Comp. Phys.* 114:185–200.
18. Lopez, C. (2008). A little disorder is just right. *Nat. Phys.* 4:755.
19. Harun-Ur-Rashid, M., Imran, A.B., Seki, T., Ishii, M., Nakamura, H., and Takeoka, Y. (2010). Angle-independent structural color in colloidal amorphous arrays. *ChemPhysChem* 11:579–583.
20. Parker, A.R., and McKenzie, D.R. (2003). The cause of 50 million-year-old colour. *Proc. Biol. Sci.* 270:S151–S153.

Index

A

abalones, 26
abathochroal eye, 28
ABC, *see* Absorbing boundary condition (ABC)
abiogenic minerals, 26
abiogenic opal, 29
absorbing boundary condition (ABC), 193
absorption, ix
acantharians, 20
Acanthopleura granulata, 29
Acheta domestica (crickets), 120, 133
adaptive spatiotemperal accumulation filter, 175
adaptive weighted averaging (AWA) filter, 182
advantages, structural color, 113
AFM, *see* Atomic force microscopy (AFM)
AG, *see* Antiglare (AG) films
air gap, 7
air inclusion and bubbles, 75
ALD, *see* Atomic layer deposition (ALD)
algae, *see also* Diatoms
- blue iridescence, 12
- sponge fibers, 37

alginate, 81
algorithm, night vision
- central processing unit implementation, 181
- color consideration, 180
- comparisons, 182
- contrast enhancement, 174
- noise reduction, 176–180
- overview, x, 165–166, 172–174, 185
- results, 182

alumina butterfly scale replica, 100–101
aluminum oxide, 39
ambient pressure, infrared receptor model, 122–123
ambient temperature change, 124, 127–128
analyses, FDTD application, 211–222
anatase, 23
anatomy, flowers, 2–5
Anchusa oficinalis (Common Bugloss), 4
angular dispersion, 108–109
anisotropic filtering, 177
anodic porous alumina, moth-eye antireflective surface, *see also* Antireflectors
- characteristics, 160–162
- films, diminishing reflection, 156
- overview, x, 155, 162–163
- production process, 157–160

Anoplagnathus parvulus, 83–84
anthocyanins, 2
anticounterfeiting structures, 61
antiglare (AG) films, 146
antireflectors and antireflection (AR), *see also* Moth-eye structure and antireflective surface
- commercial films, 156–162
- engineering, optical nanostructures, 57, 59

Antirrhinum maius, 3
anvil-shaped calcite plates, 25
Aphrodita sp., 60–61
Apis mellifera, 168–169, 172
applications, structural color
- beetle reflectance, 83–86
- fishes, 87–88
- helical Bouligand metallic reflectance, 85–86
- insect shell structures, 73–74
- light reflection principles, 74–82
- metallic appearance, 82–88
- multilayer materials and coatings, 88, 90–93
- multilayer metallic reflectors, 83–85
- narrowband reflection, 76–82
- overview, x, 72–73

AR, *see* Antireflectors and antireflection (AR)
arachidic acid, 45
aragonite
  biomineralization, 25–26
  calcite-based imaging, 29
  calcium carbonate, 21–22, 44
  comparisons, 27, 28
areolae, 30
artificial biominerals, 23
Aspidomorpha tecta, 78, 84
ASTA-filter, *see* Adaptive spatiotemperal accumulation filter
atomic force microscopy (AFM), 32, 107
atomic layer deposition (ALD), 41, 100–101
Aubrieta deltoidea, 4
AWA, *see* Adaptive weighted averaging (AWA) filter

B

Bacillariophyta class, 30
backscattering, *see* Retroreflection
backward- and forward-propagating monochromatic solutions, 197
Bakelite, viii
ball-rolling dung beetles, 166
bamboo, 29
Bayer pattern, 180
bees, aerodynamic, 166, *see also Megalopta genalis*
beetles, *see also* Pyrophilous beetles and bugs
  color shifts, 78, 143
  iridescent device engineering, 59–60
  matching color patterns, 83
  metallic appearance, 82
  narrowband reflection, 77
  paint pigment, 81
  reflectance, metallic appearance, 83–86
Begonia pavonina, 12
Bennett's method, 174
bidirectional reflectance distribution function (BRDF), 210
bifocal imaging, 28–29
bilateral filtering, 175, 177
biogenic minerals, 26
biogenic porous silica, 67
bioinspiration, viii
bioluminescence gene, 38
biomimetics defined, 57
biomimicry, viii
biomineralization
  artificial biominerals, 23
  biomineral replicas, 42–43
  calcite-based imaging, 28–29
  calcium carbonate, 21–22, 44
  diatoms, 30–36
  examples from nature, 26–38
  functionalizing biological structures, 38–43
  metal sulfides and metals, 45–46
  mother-of-pearl example, 26–28
  optical materials, 43–46
  optical properties, 20–23
  overview, x, 19–20, 46
  principles, 23–26
  silica, 22–23, 39, 44–45
  sponge fibers, 36–38
  surface functionalization, 41
  tabashir/plant silica, 29
  titanium dioxide, 45
  in vivo fluorochromation, 39, 41
  zinc oxide, 45
biomineral replicas, 42–43
bionics, viii
birefringence
  artificial optical biominerals, 23
  calcite-based imaging, 29
  calcium carbonate, 21
  colorless metallic sheen, 93
  light reflection principles, 75–76
  multilayer materials and coatings, 90
Bivalvia class, 26
black-body radiation, 132
blowflies, 78
blue fluorescence, 36
blue iridescence, *see also Morpho* butterfly structural color
  Danaea nodosa, 11–12
  Selaginella willdenowii, 11–12
  understory plants, 11–12
Blue Marble Tree, *see Elaeocarpus angustifolius*
Blue Spruce, *see Pseudotsuga menziesii*
Bombyx mori, 45

bottom-up strategies, 20, 162–163
Bouligand reflector, 91
Bouligand structures, 75–76
"bowls," 28
Bragg properties (diffraction, reflectors, refraction, resonance)
  colloidal crystal gels, 150–152
  iridescent device engineering, 60
  iridoviruses, 68
  nacre, 26
  opals, 145
  3D photonic crystal materials, 146, 147, 149
  tunable structure color, 144
BRDF, *see* Bidirectional reflectance distribution function (BRDF)
Brewster angle, 90
Brewster's law, 219
Brillouin scattering, ix
brittlestars, 29
brookite, 23
Brownian motion noise, 129
Bruggemann approximation, 75
bugs, *see* Pyrophilous beetles and bugs
bumblebees, 10, *see also Megalopta genalis*
buprestid beetles, *see* Pyrophilous beetles and bugs
Buttercup, *see Ranunculus repens*
butterflies, *see also Morpho* butterfly
  antireflector engineering, 57
  iridescent device engineering, 59
  narrowband reflection, 77, 79

C

cadmium sulfide, 20, 23
Calcarea class, 36
calcite
  biomineralization, 25–26
  calcium carbonate, 21–22, 44
  eyes, 28
  imaging, biomineralization example, 28–29
calcium carbonate
  biomineralization, 25
  optical materials approach, 44
  optical properties approach, 21–22
Calliphoridae family, 78
Cambrian epoch, 1
capacitance-to-digital converter (CDC), 135
capacitive detectors, 121
carboxylated polyaniline, 44
carotenoids, 2
Cartesian coordinates, 215
cascade, directional filters, 176
cavity fluid influence, 133–135
cavity pressure increase, 122–128
CCD, *see* Charge-coupled devices (CCDs)
ccp, *see* Cubic close packing (ccp)
CDC, *see* Capacitance-to-digital converter (CDC)
cell culture, 63
Centrales, 30–31
Central Secretariat of the International Organization for Standardization (ISO), viii
centric diatoms, 33
Cephalopoda class, 26, 142
cetyl trimethylammonium bromide, 44–45
chalk dudleya, 1
chalogenides, 45
Chamaeleonidae family, 78
chameleons, 78
chaos, xi
charge-coupled devices (CCDs), 180
Chilo sp., 68
chin illumination, 6, 7
chirped stack of layers, 84
chitin
  diatoms, 31
  insect shell structures, 73
  multilayer metallic reflectors, 83
  nacre, 27
  refractive index, 141
Chitonidae family, 29
chitosan matrix, 44
chlorocyclohexane droplets, 45
cholesteric liquid crystals (CLCs), 82, 91–93
cingulum, 30
Cistus cyprius, 4
citrate, 44
citric acid, 44
CLC, *see* Cholesteric liquid crystals (CLCs)

Clupea harengus, 88
Clupea sprattus (sprat), 88
CMOS, *see* Complimentary metal oxide semiconductor (CMOS) sensor
Coalinga, California fire, 120, 132
Cobalt Blue damselfish, 142
coccolithophores, 64, 67–68
coccoliths, 25
cockroaches, 170, 172
colloidal crystal gels, 149–152
colloidal photonic crystal, tunable structural color
  colloidal crystal gels, 149–152
  future outlook, 152
  mechanical stress application, 149–152
  1D photonic crystal materials, 144
  overview, x, 141–143
  swelling, 147–149
  3D colloidal crystals, 144–145
  3D photonic crystal materials, 146–149
colors, *see also* Tunable structural color
  death, 78, 84
  durability of, 72, 96–97, 106
  matching, dominant patterns, 83
  narrowband reflection, 77
  principles, 97–99
  production, 96
  reversible shifts, 78, 149
  textile reproduction, 99
compensation leak, 127–128
complimentary metal oxide semiconductor (CMOS) sensor, 180
components, number of, 236, 238
computer experts, vii
constellations of stars, 166
control, optical properties, 108–109
"cores," 28
cornea, preserved, 59
Coscinodiscus genus, 32–34
Coscinodiscus granii, 34, 41, 64
Coscinodiscus wailesii, 34, 36
cosmetics, 106, 113, 239
CPU/GPU algorithm, 181
crane flies, 170
crayfish, *see Procambarus clarkii*
crickets, *see Acheta domestica*
cristobalite, 23
cross-fertilization, 2
crystalline silica polymorphs, 23
crystals
  double-refraction-free orientation, 28
  fibers, 60
  growth, 25
  structure, biomineralization, 25
cubic close packing (ccp), 145
cucurbit[7]uril, 46
Curculionoidea superfamily, 144
cuticle, beetles, 60, 73
cyanobacteria, 37
Cylindrotheca fusiformis, 45

D

Dalmanitina socialis, 28
damselfish, 142
Danaea nodosa, 2, 11–12
dark noise, 168
death, color changes, 78, 84
decorations, 106, 113, 239
Delarbrea michieana, 13
demonstration, nonstandard method, 205, 207
demosaicing, 180
Demospongiae class, 36
denoising image data, 180, *see also* Noise
depth, nanopattern, 108–109
detectivity, Golay cells, 131
diatoms
  amorphous biosilica, 29
  biomineralization example, 30–36
  engineering, optical nanostructures, 64, 67–68
  narrowband reflection, 77
dichroitic mirrors, 26
Dicranostigma leptopodum (Yellow Poppy), 4
diffraction
  diatoms, 36
  fundamentals, ix
  silica, 44
diffusion equations, 177
diminishing reflection, 156
dim light, vision unreliability, 167–169
Diplazium tomentosum, 12
directional filter cascade, 176

directional scattering, 5–7
discoloration resistance, 72, 96–97, 106
discrete multilayers reproduction, 101–105
dispersity, nanosphere, 44
DNA nucleotides and sequences, 41
double-refraction-free orientation, 28
dry etching, 107, 239
Dudleya brittonii, 11
durability of colors, 72, 96–97, 106
dye-free, structurally color fibers, 99

E

ecdysis, 73–74
Echinodermata phylum, 29
Edelweiss, *see Leontopodium nivale*
eigenvalue analysis, 178–179
Elaeocarpus angustifolius (Blue Marble Tree), 2, 13–14
electroless deposition, 41
electroluminescence spectrum, 39
electron beam (EB) lithography
  butterfly scale structure, 239
  mass production, 106–107
  moth-eye antireflection film, 157
  reproduction, discrete multilayers, 101–102
electrooptical properties, 43
Elliott photospectrometer, 92
engineering, optical nanostructures
  antireflectors, 57, 59
  cell culture, 63
  coccolithophores, 64, 67–68
  diatoms, 64, 67–68
  future research, 69–70
  iridescent devices, 59–61, 63
  iridoviruses, 68
  mechanisms, natural engineering, 69–70
  overview, x, 56–57
epitheca, 30
Equisetum sp., 29
ethanol (EtOH), 148–149
Euplectella aspergillum, 37
Euplectella sp., 37
European honeybee, *see Apis mellifera*
examples from nature, biomineralization
  calcite-based imaging, 28–29
  diatoms, 30–36
  mother-of-pearl example, 26–28
  sponge fibers, 36–38
  tabashir/plant silica, 29
experimental results and devices, IR receptors, 135–137
extracellular biomineralization, 24

F

fabrication, nanosized color-producing part, 100–101
Falco tinnunculus (kestrals), 77, 80
false color noise, 182
FIB-CVD, *see* Focused-ion-beam chemical vapor deposition (FIB-CVD)
field emission (FE)-SEM, 159
filling of cavity, 133–135
films, 156–162
fingerprints, 79
finite-difference time-domain (FDTD) method
  analyses, 211–222
  computer platform, 228–229
  demonstration, nonstandard method, 205, 207
  formulations, 193–200
  general descriptions, 193
  interference fringe prevention, 230
  microscopic structure, 207–211
  Morpho butterfly scale application, 207–222
  nonstandard method, 200–205, 207
  optical measurements, 207–211
  optical properties control, 109–110
  overview, x–xi, 192, 222–223, 227
  randomness, 111
  software, 228–229, 234
  transverse electric mode, 196–197, 198–200
  transverse magnetic mode, 196–198
  Yee algorithm, 197–200
finite-difference time-domain method, 195
fireflies, 172
fishes
  changing structural color, 141–142

iridescent device engineering, 59
metallic appearance, 82, 87–88
"flash coloration," 83
flavonoids, 2
flexural stiffness, membrane, 127
flowers, photonic structures
anatomy, 2–5
directional scattering, 5–7
glossiness, 5–7
Hibiscus trionum, 8–10
optical response, 2–5
overview, 2
Ranunculus repens, 5–7
fluorescein isothiocyanate, 44
fluorescent marker dyes, 39, 41
fluorochromation, 39, 41
focused-ion-beam chemical vapor deposition (FIB-CVD), 100, 101
forest fires, 117–120, 132
formulations, FDTD method, 193–200
forward- and backward-propagating monochromatic solutions, 197
Fraunhofer diffraction, 108
fruits, 13
frustules
abnormalities, 69
blue fluorescence, 36
fundamentals, 30
narrowband reflection, 77
photonic properties, 32
surface functionalization, 41
frustulines, 31
full width half maximum (FWHM) value, 108
functionalizing biological structures
biomineral replicas, 42–43
overview, 38–39
silica replacement, 39
surface functionalization, 41
in vivo fluorochromation, 39, 41
future outlook and research
engineering, optical nanostructures, 69–70
fabrication, butterfly scale replica, 101
Morpho butterfly structural color, 111–112
optical role of random arrangement, 239
tunable structural color, 152
fuzziness, xi
FWHM, *see* Full width half maximum (FWHM) value

G

gas, filling of cavity, 134
gaseous silicon fluoride, 43
gas sensors, 112
Gastropoda class, 26
Gaussian distribution and function
noise reduction, 179
photoreceptor strategies, 169
reproduction, discrete multilayers, 101
Gazania tenuifolia, 4
GBO, *see* Giant birefringent optics (GBO) gene, bioluminescence, 38
geometrical confinement, 24
Geranium endressii, 4
germanium oxide
diatoms, 67
silica replacement, 39
giant birefringent optics (GBO), 90
glass sheets, 158
Glaucium avum (Yellow Horned Poppy), 4
glossiness, 5–7
Golay cells
detectivity, 131
experimental results and devices, 135–137
mechanical-thermal noise, 129–130
overview, 128
pneumatic detector setup, 121, 122
readout noise, 130
sources of noise, 128–132
temperature fluctuation noise, 129
total noise, 131
gold, 41, 64, 68
golden scarabid beetles, *see Plusiotis resplendens*
graphics processing unit (GPU) algorithm, 181
grasses, 29
green algae, 37, *see also* Diatoms
guanine, 141

H

hair cells, inner ear, 133
Halictidae family, 166
Haliotidae family, 26
Haliotis genus, 26
hardening processes, 83
hawkmoths, 57, 172
HCK-123, 41
helical Bouligand metallic reflectance, 83, 85–86
helicoidal Bouligand structure, 75
Heliocidaris erythograma, 43
herring, 87–88
Heterorrhina sp., 77
Hexactinellida class, 36
hexactinellid genus, 37
Hibiscus trionum (Venice Mallow), 2, 8–10
higher-dimensionally ordered structures, ix
higher-order strategies, 171–172
high reflective mirrors, 88, 90
histogram equalization, 174
holochroal eye, 28
holography, 59, 106, 113
honeybee, *see Apis mellifera*
Hooke, Robert, 56
horsetail, *see Equisetum* sp.
HSV, *see* Hue, saturation, value (HSV) color space
hue, saturation, value (HSV) color space, 180
hummingbird feather barbs, 60
Hyalonema sieboldii, 38
hydration levels, 23
hydrocarbons, filling of cavity, 134
hypotheca, 30

I

implicit equation, nonlinear case, 127
incident light
  color changes, 142
  films, diminishing reflection, 156–157
  light reflection principles, 75
incoherent light, 234–238
industrial production
  characteristics, 160–162
  colloidal crystal gels, 149–152
  films, diminishing reflection, 156
  future outlook, 152
  mechanical stress application, 149–152
  moth-eye antireflective surface, anodic porous alumina, 155–163
  1D photonic crystal materials, 144
  production process, 157–160
  swelling, 147–149
  3D colloidal crystals, 144–145
  3D photonic crystal materials, 146–149
  tunable structural color, 141–152
industrial revolution, viii
inner ear hair cells, 133
insects, *see also specific type*
  changing structural color, 141
  shell structures, 73–74
interference
  effect, contradiction, 98
  fringes, prevention, 230
  fundamentals, ix
interferometric refractive index, 37
intracellular biomineralization, 24
intralensar "bowls" and "cores," 28
inverse opal, iridescent device engineering, 60
inverse opal photonic crystals, 151
in vivo fluorochromation, 39, 41
iridescence
  devices, 59–61, 63
  leaves, 11–12
  nacre, 28
iridoviruses, 68
isobaric thermal expansion coefficient, 123
isopropanol, 147–149

K

kestrels, *see Falco tinnuncullus*
kTC noise, 130

L

lamina ganglionaris, 172
Lasioglossum leucozonium, 170–171, 172

lateral randomness
  industrial applications, 110
  optical role, 234
LCP, *see* Left circular polarized (LCP) light
leaves
  blue iridescence, understory plants, 11–12
  Danaea nodosa, 11–12
  main function, 10
  overview, 10
  Selaginella willdenowii, 11–12
  UV protection mechanism, 10–11
left circular polarized (LCP) light, 86
Leontopodium nivale (Edelweiss), 1–2, 10–11
light emission, 36
"light line," 35
light reflection principles
  applications, 79–82
  narrowband reflection, 76–78
  overview, 76–78
light scattering, ix
light waves, 21
Lindsaea lucida, 12
linear minimum mean-square-error filtering, 176
Linnean classification system, vii
lithium tantalate (LT), 135
long-chain polyamides, 31–32
LT, *see* Lithium tantalate (LT)
Lucilia caesar, 78
luminescence
  bioluminescence gene, 38
  electroluminescence spectrum, 39
  photoluminescence spectrum, 39, 45
  porous silicon, 64
  solr concentrators, 67
Lysosensor DND-160, 39

M

mackerel, 87
magnesium oxide (MgO), 42–43, 67
magnetic iron oxide chains, 20
manuka scarab beetle, *see* *Pyronota festiva*
Margaritaria nobilis, 2, 13
marine animals, 59, *see also specific type*
mass production, 106–107
matching color patterns, 83
material scientists, vii
mathematicians, vii
Maxwell's equations, 192–195, 204, 229
measurement, role of random arrangement, 228
mechanical stress application, 149–152
mechanical-thermal noise, 129–130
mechanisms, natural engineering, 69–70
Megalopta genalis
  dim light, vision unreliability, 167–169
  higher-order strategies, 171–172
  neural strategies, vision reliability, 169–172
  overview, 166–167, 185
  photoreceptor strategies, 169–171
  vision and visual processing, 166–169
Melanophila acuminata, 119–120
Melanophila genus, x, 117–118, *see also* Pyrophilous beetles and bugs
Melosira sp., 36
membrane, flexural stiffness, 127
membrane deflection, 122–128
mesostructured biominerals, 41
mesostructures, 44
metal fluorides, 43
metallic appearance
  beetle reflectance, 83–86
  fishes, 87–88
  helical Bouligand metallic reflectance, 85–86
  multilayer materials and coatings, 88, 90–93
  multilayer metallic reflectors, 83–85
  overview, 82–83
metal oxides, 102
metal sulfides and metals, 45–46
MgO, *see* Magnesium oxide (MgO)
microscopic structure, 207–211
Mie scattering, ix
mirrors, 88, 90
modeling and simulation, structural colors
  demonstration, nonstandard method, 205, 207

finite-difference time-domain method, 192–223
formulations, 193–200
future applications, 239
general descriptions, 193
incoherent light, 234–238
interference fringes, prevention, 230
lateral randomness, 234
measurement, 228
Morpho butterfly scale application, 207–222
nonstandard method, 200–205, 207
number of components, 236, 238
optical coherence consideration, 231, 233–234
optical properties, 227–228
optical role of random arrangement, 226–241
smooth angular dependence without fringes, 229–234
software/computer platform, 228–229
transverse electric mode, 196–197, 198–200
transverse magnetic mode, 196–198
vertical randomness, 234, 236
Yee algorithm, 197–200
Mollusca phylum, 26, 29
monochromatic solutions, 197
Monorphapsis sp., 38
Morpho butterfly structural color, modeling and simulation
angular dispersion, 108–109
color principles, 97–99
demonstration, nonstandard method, 205, 207
discrete multilayers reproduction, 101–105
fabrication, nanosized color-producing part, 100–101
finite-difference time-domain method, 192–223
formulations, 193–200
future applications and outlook, 111–112, 239
general descriptions, 193
incoherent light, 234–238
interference fringes, prevention, 230
lateral randomness, 234
mass production, 106–107
measurement, 228
microscopic structure, 207–211
nonstandard method, 200–205, 207
number of components, 236, 238
optical coherence consideration, 231, 233–234
optical measurements, 207–211
optical properties, 108–109, 227–228
optical role of random arrangement, 226–241
overview, x–xi, 96–97, 192, 222–223, 227
progress, 109–110
smooth angular dependence without fringes, 229–234
software/computer platform, 228–229
spectra, 109
textile reproduction, 99
transverse electric mode, 196–197, 198–200
transverse magnetic mode, 196–198
vertical randomness, 234, 236
Yee algorithm, 197–200
Morpho didius, 227
Morpho rhetenor
analyses, 211–222
microscopic structure, 207–211
optical measurements, 207–211
overview, 207
mother-of-pearl, 26–28
moth-eye structure and antireflective surface, *see also* Antireflectors
antireflector engineering, 57, 59
characteristics, 160–162
films, diminishing reflection, 156
overview, x, 155, 162–163
production process, 157–160
solar cells, 112
moths, aerodynamic, 166
multichannel spectral analyzer system, 100
multilayered, periodic structures, ix
multilayer materials and coatings, 88, 90–93
multilayer metallic reflectors, 83–85

N

nacre (mother-of-pearl), 26–28
nanoimprinting lithography (NIL), 106–107, 111, 157
nanopattern depth, 108–109
narrowband reflection, 76–82
natural quartz, 23
nature examples, biomineralization
  calcite-based imaging, 28–29
  diatoms, 30–36
  mother-of-pearl example, 26–28
  sponge fibers, 36–38
  tabashir/plant silica, 29
Nautilidae family, 26
Nautilus genus, 26
Nautilus pompilius, 27
Near-to-far-field (NTFF) transformation, 193, 205, 207
neon tetra, *see Paracheirodon innesi*
NEP, *see* Noise-equivalent power (NEP)
neural strategies, vision reliability, 169–172
Newton, Isaac, 56
night vision algorithm
  central processing unit implementation, 181
  color consideration, 180
  comparisons, 182
  contrast enhancement, 174
  noise reduction, 176–180
  overview, x, 165–166, 172–174, 185
  results, 182
night vision, *Megalopta genalis*
  dim light, vision unreliability, 167–169
  higher-order strategies, 171–172
  neural strategies, vision reliability, 169–172
  overview, 166–167, 185
  photoreceptor strategies, 169–171
  vision and visual processing, 166–169
NIL, *see* Nanoimprinting lithography (NIL)
Nitzschia liebethrutti, 69
noise, *see also* Golay cells
  color consideration, 180
  false color noise, 182
  radiation measurement below, 132–133
  reduction, 176–180
  sources, 131
  visual, 167
noise-equivalent power (NEP), 131, 136
nonadiabatic cavity, 124–125
non-neat minerals, 26
nonstandard finite-difference time-domain (NS-FDTD) method, 193, 200–207, 214, 216
NTFF, *see* Near-to-far-field (NTFF) transformation
nucleophiles, 64–65
number of components, 236, 238
Nylon, viii

O

ocelli, 29
octopus, 142
oil fires, 120, 132
1D photonic crystal materials, 144
one-dimensional ordered, layered structures, ix
opal, 22–23, 60, 145, *see also* Silica
Ophiocoma pumila, 29
Ophiocoma wendtii, 29
Ophiuroidea class, 29
optical coherence consideration, 231, 233–234
optical materials, biomimetic approaches
  calcium carbonate, 44
  metal sulfides and metals, 45–46
  overview, 43–44
  silica, 44–45
  titanium dioxide, 45
  zinc oxide, 45
optical measurements, 207–211
optical nanostructures
  engineering, 56–70
  structural color applications, 72–93
  structural color industrial applications, 96–112
optical properties
  artificial optical biominerals, 23
  calcium carbonate, 21–22
  control, 108–109
  optical role of random arrangement, 227–228

overview, 20–21
randomness in structures, 227–228
reproduction, discrete multilayers, 103–104
silica, 22–23
optical response, 2–5
optical role of random arrangement
future applications, 239
incoherent light, 234–238
interference fringes, prevention, 230
lateral randomness, 234
measurement, 228
number of components, 236, 238
optical coherence consideration, 231, 233–234
optical properties, 227–228
origin, smooth angular dependence without fringes, 229–234
overview, x–xi, 226–227, 239, 241
reproduction, discrete multilayers, 104–105
smooth angular dependence without fringes, 229–230
software/computer platform, 228–229
vertical randomness, 234, 236
optical wave coupling schemes, 36
optical waveguides, 37
optic ganglion, 172
origin, smooth angular dependence without fringes, 229–234
orthotitanate precursors, 45
Ostwald's step rule and ripening, 25, 145
oxide removal, 43
oxoanions, 39

P

paints, 81, 106, 239
panels, 113
Paracheirodon innesi (neon tetra), 142
parrots, *see Psittaciformes* sp.
PCB, *see* Printed circuit boards (PCBs)
PDMPO dye, 39
PDMS, *see* Polydimethylsilicone (PDMS) elastomer
pearl oysters, 26
Pennales, 31
pentafluorophenyl acrylate, 44–45
perfectly matched layer (PML), 193, 229
perfect order, xi
periodic nanostructure, 142
perovskites, 43
PET, *see* Poly(ethylene terephthalate) PET
petals and petal anatomy, 3, 6
Philocteanus rubroaureus, 72
photoluminescence (PL)
diatoms, 64
silica replacement, 39
zinc oxide, 45
photoluminescent porous nanocrystalline silicon replica, 43
photomechanic infrared receptors
cavity fluid influence, 133–135
cavity pressure increase, 122–128
detectivity, 131
experimental results and devices, 135–137
Golay cell noise, 128–132
mechanical-thermal noise, 129–130
Melanophila sp., 117–120
membrane deflection, 122–128
noise sources, 131
radiation measurement below noise level, 132–133
readout noise, 130
sensilla, structure and function, 118–119
sensitivity, infrared receptor, 119–120
sensor based on infrared receptor, 121–128
sensor model, 121–122
sources of noise, 128–131
temperature fluctuation noise, 129
total noise, 131
photonic paper, 151
photonic structures, plants
anatomy, 2–5
blue iridescence, understory plants, 11–12
Danaea nodosa, 11–12
directional scattering and glossiness, 5–7
flowers, 2–10
fruits, 13

Hibiscus trionum, 8–10
leaves, 10–12
optical response, 2–5
overview, ix–x, 1–2, 14
Ranunculus repens, 5–7
Selaginella willdenowii, 11–12
UV protection mechanism, 10–11
photoreceptor strategies, 169–171
photorefractive properties, 43
Phyllagathis rotundifolia, 12
Pinctada genus, 26
P-ink, 151
pitch, 156–157
PL, *see* Photoluminescence (PL)
plants, photonic structures
anatomy, 2–5
blue iridescence, understory plants, 11–12
Danaea nodosa, 11–12
directional scattering and glossiness, 5–7
flowers, 2–10
fruits, 13
Hibiscus trionum, 8–10
leaves, 10–12
optical response, 2–5
overview, ix–x, 1–2, 14
Ranunculus repens, 5–7
Selaginella willdenowii, 11–12
UV protection mechanism, 10–11
plant silica, 29, *see also* Silica
plastic packaging products, 81
pleuralins, 31
Plusiotis resplendens (golden scarabid beetles), 85–86, 92
PML, *see* Perfectly matched layer (PML)
PMMA, *see* Poly(methyl methacrylate)
Poaceae family, 29
Poisson's ratio, 123, 130
polarizability, light waves, 21
pollination, 2, 10
polyacrylamide matrix, 151
poly(allylamine), 44
polyamines, 44
polyaspartate, 44
polydimethylsilicone (PDMS) elastomer, 146, 147–148, 150
poly(ethylene terephthalate) PET, 158–159
polymers, viii
poly(methyl methacrylate) (PMMA), 90, 158
Polyplacophora class, 29
polystyrene (PS), 90, 145
polyvinyl acetate (PVAC), 90–91
Pomacentridae family, 142
pores, diatoms, 30
Porifera phylum, 36
position indices, 195, 197
pressure, infrared receptor model, 122–123
prevention, interference fringes, 230
printed circuit boards (PCBs), 135
Procambarus clarkii (crayfish), 133
production process, 157–160
protein chains, 27
PS, *see* Polystyrene (PS)
Pseudotsuga menziesii (Blue Spruce), 1, 11
Psittaciformes sp., 77
Pteriidae family, 26
PVAC, *see* Polyvinyl acetate (PVAC)
pyridine, 64
Pyronota festiva (manuka scarab beetle), 91
pyrophilous beetles and bugs, infrared receptors
cavity fluid influence, 133–135
cavity pressure increase, 122–128
detectivity, 131
experimental results and devices, 135–137
Golay cell noise, 128–132
mechanical-thermal noise, 129–130
Melanophila sp., 117–120
membrane deflection, 122–128
noise sources, 131
radiation measurement below noise level, 132–133
readout noise, 130
sensilla, structure and function, 118–119
sensitivity, infrared receptors, 119–120
sensor model, 121–122
sensors, infrared receptors, 121–128
sources of noise, 128–131
temperature fluctuation noise, 129
total noise, 131

Q

quality standards, 108
quantum bumps, 170
quarter-wave stacks, 74, 91
Queen of the Night Tulip cultivar, 2, 8–10

R

radiation measurement, 132–133
radiation noise, 129, 131
radiolarians
  amorphous biosilica, 29
  optical properties, 20
  silica, 45
Raman scattering, 41
randomness
  incoherent light, 234–238
  integral part, xi
  lateral randomness, 234
  optical properties, 227–228
  smooth angular dependence without fringes, 229–230
  vertical randomness, 234, 236
Ranunculus repens (Buttercup)
  directional scattering, 5–7
  glossiness, 1, 5–7
  petal epidermal cell shape, 4
rapid prototyping, 60
Rayleigh scattering, ix, 1
RCP, *see* Right circular polarized (RCP) light
readout noise, 130
recycling, 81
reflectance, 6
reflection, ix
reflectivity
  light reflection principles, 74–75
  reproduction, discrete multilayers, 103–104
refraction, ix
refractive index
  biomineralization, 19
  calcite-based imaging, 28
  calcium carbonate, 21–22
  chitin, 141
  color changes, 142
  colorless metallic sheen, 93
  diatoms, 32, 34, 67
  films, diminishing reflection, 157
  fundamentals, ix
  guanine, 141
  light reflection principles, 74
  metallic appearance, fishes, 88
  multilayer metallic reflectors, 83
  nacre, 27, 28
  opal silica, 23
  optical properties, 20
  reproduction, discrete multilayers, 102
  sponge fibers, 37
  tabashir, 29
  3D colloidal crystals, 145
  tinania, 23
replicas, biomineral, 42–43
retroreflection (backscattering), 208–212, 215–222
reversible color shifts, 78, 149, *see also* Tunable structural color
rewritable full-color media, 93
Rhodamine B and 6G, 68
rhombohedral shape, 44
right circular polarized (RCP) light, 86
rose beetles, 81
Rosella racovitzae, 37
rotation matrix, noise reduction, 179
rutile, 23

S

salmon, 87
saturation, *see* Hue, saturation, value (HSV) color space
scaling matrix, noise reduction, 179
scanning electron microscopy (SEM)
  anatomy and optical response, 3–4
  directional scattering and glossiness, 5–6
  iridescence, 8–9
  mass production, 107
  Morpho color principles, 98
  opal film, 145
  reproduction, discrete multilayers, 102–105
  3D photonic crystal materials, 146
  ultraviolet protection mechanism, 10
Scarabaeidae family, 86

scarabid beetles, *see Plusiotis resplendens*
scattered field (SF), 193
schizochroal eye, 28
Schmitt, Otto, viii
SDV, *see* Silica deposition vesicles (SDVs)
sea mouse, 60–61
sea urchins, 25, 43
second moment matrix, 177–178
Selaginella sp., 1
Selaginella willdenowii, 2, 11–12
self-assembled copolymer blends, 82
self-fertilization, 2
self-organization, 20, 162–163
SEM, *see* Scanning electron microscopy (SEM)
sensilla, structure and function, 118–119
sensitivity, infrared receptors, 119–120
sensor model, 121–122
sensors, infrared receptors, 121–128
SF, *see* Scattered field (SF)
S-FDTD, *see* Standard finite-difference time-domain (S-FDTD) method
shape-preserving displacement reactions, 42
signal-to-noise ratio (SNR), 128, 171–172
silacidines, 31
silaffines, 31–32, 44
silica
  biogenic porous, 67
  biomineralization, 25
  optical materials approach, 44–45
  optical properties approach, 22–23
  replacement, 39
  spinodal decomposition, 24
silica deposition vesicles (SDVs), 32, 69
silica dioxide (SiO_2), 44–45, 109
silicalemma, 32
silicatein
  grafting, 44
  sponge fibers, 36–37
  titanium dioxide, 45
silicon dioxide (SiO_2), 22
silintaphin-1, 36–37
silk worms, *see Bombyx mori*
silver-binding peptides, 46
single-step structure-sensitive adaptive smoothing kernels, 177
SiO_2, *see* Silica dioxide (SiO_2)
sixfold diffraction patterns, 36
size, nanosphere, 44
Skeletonema costatum, 69
smooth angular dependence without fringes, 229–234
smoothing techniques, 177
Snell's law, 145
software/computer platform, 228–229, 234
solar cells, 113
solar panel application, 59
solvent exchange, 149
sources of noise, 128–131, *see also* Noise
spatiotemperal weighted averaging, 176
specific detectivity D^*, 131
spectra, 109
speed, time response, 171
spiders, 170
spinodal decomposition, 24
sponge fibers, 36–38
sponges, 29
sprat, *see Clupea sprattus* (sprat)
squid, 142
stainless steel, 79
standard finite-difference time-domain (S-FDTD) method, 200–205
standardization, viii
standard of quality, 108
starch layer, 6–7
star constellations, 166
starlings, *see Sturnus vulgaris*
star-shaped cells, 45
"step coverage" effect, 102
stochastic resonance, 133
strontium sulfate, 20
structural color, applications
  advantages, 113
  beetle reflectance, 83–86
  fishes, 87–88
  helical Bouligand metallic reflectance, 85–86
  insect shell structures, 73–74
  light reflection principles, 74–82
  metallic appearance, 82–88
  multilayer materials and coatings, 88, 90–93
  multilayer metallic reflectors, 83–85
  narrowband reflection, 76–82
  overview, x, 72–73, 113
structural color, industrial applications

angular dispersion, 108–109
color principles, 97–99
color textile reproduction, 99
discrete multilayers reproduction, 101–105
fabrication, nanosized color-producing part, 100–101
future outlook, 111–112
mass production, 106–107
optical property control, 108–109
overview, x, 96–97
progress, 109–110
spectra, 109
structural color, modeling and simulation
demonstration, nonstandard method, 205, 207
finite-difference time-domain method, 192–223
formulations, 193–200
future applications, 239
general descriptions, 193
incoherent light, 234–238
interference fringes, prevention, 230
lateral randomness, 234
measurement, 228
Morpho butterfly scale application, 207–222
nonstandard method, 200–205, 207
number of components, 236, 238
optical coherence consideration, 231, 233–234
optical properties, 227–228
optical role, random arrangement, 226–241
smooth angular dependence without fringes, 229–234
software/computer platform, 228–229
transverse electric mode, 196–197, 198–200
transverse magnetic mode, 196–198
vertical randomness, 234, 236
Yee algorithm, 197–200
structural model, 98–99
structure-adaptive anisotropic filtering, 177
structures
functionalizing biological, 38–43
photonic, in plants, 1–14
structure tensor, 177–178
Sturnus vulgaris (starlings), 77
Suberites domuncula, 38
summation strategies, 171–172
surface-enhanced Raman scattering, 41
surface functionalization, 41
sweat bees, *see Megalopta genalis*
swelling, tunable structural color, 147–149
synthetic amorphous quartz, 23

T

tabashir/plant silica, 29
tanning processes, 83
Taylor series, 204
TE, *see* Transverse electric mode (TE)
technology, processes demanded from, 111
telecommunications application, 61
TEM, *see* Transmission electron microscopy (TEM)
temperature
fluctuation noise, 129
regulation, 112
templating techniques, 43
Tethya aurantia, 37, 45
Tethya aurantium, 37
tetraethoxysilane, 45
tetrahydroforan (THF), 90
textiles, 79, 99, 113
textures, 106, 239
TF, *see* Total field (TF)
TF/SF, *see* Total field/scattered field (TF/SF)
Thalassiosira pseudonana, 31
Thalassiosira rotula, 64
theca, 30
thermal issues, *see* Temperature
thermoelastic noise, 129
Thermonectus marmoratus, 28
THF, *see* Tetrahydroforan (THF)
thin coatings, 79
thin-film technologies, 97
3D colloidal crystals, 144–145
3D photonic crystal materials, 146–149
three-step transformation series, 25
tiles, 113

time response, 170–171
titania
  biomineralization, 26
  optical properties, 20
  silica replacement, 39
titanium dioxide (TiO_2)
  artificial optical biominerals, 23
  biomineral replicas, 43
  diatoms, 67
  optical materials approach, 45
  optical properties control, 109
  surface functionalization, 41
TM, *see* Transverse magnetic mode (TM)
tone mapping, 174, 176, 182
tortoise beetles, 143
total field/scattered field (TF/SF), 193, 229
total field (TF), 193
total noise, 131
trabeculae, 28
transducer noise, 168
trans-Golgi-derived vesicles, 69
transmission electron microscopy (TEM)
  blue iridescence, understory plants, 11–12
  directional scattering and glossiness, 5–6
  finite-difference time-domain method, 212
transverse electric mode (TE), 196–200, 208, 230
transverse magnetic mode (TM), 196–198, 208, 230
tribolites, 28
tridymite, 23
Tulipa kaufmanniana, 8
Tulipa kolpakowskiana, 2, 8–10
tunable structural color
  colloidal crystal gels, 149–152
  future outlook, 152
  mechanical stress application, 149–152
  1D photonic crystal materials, 144
  overview, x, 141–143
  swelling, 147–149
  3D colloidal crystals, 144–145
  3D photonic crystal materials, 146–149
tunneling displacement transducer, 122
Tyndall scattering, ix, 78

U

ultraviolet (UV) rays
  diatoms, 31, 64
  Edelweiss reflection, 1, 10–11
  eye sensitivity to, 6
  narrowband reflection, 77
  polymer identification, 80–81
  protection from, 10–11, 77
  wavelengths, diatoms, 36
ultraviolet (UV-VIS) characterization, 7
understory plants, 11–12
uniaxial negative birefringence, 21
unreliability of vision, dim light, 167–169
unusually sculpted 3D architectures, 60
urease, 46

V

value, *see* Hue, saturation, value (HSV) color space
valva, 30
vaterite
  biomineralization, 25
  calcium carbonate, 21, 44
Venice Mallow, *see Hibiscus trionum*
Venus flower basket, *see Euplectella aspergillum*
vertical randomness
  industrial applications, 110
  optical role, 234, 236
vision and visual processing, 166–169, *see also* Night vision algorithm
visual speed, 171

W

water, filling of cavity, 133–134
water-repelling surfaces, 112
waveguiding and waveguide modes
  alumina butterfly scale replica, 100
  diatoms, 32–36, 64
  sponge fibers, 37
wavelengths

iridescent device engineering, 60
light reflection principles, 74–75
weevils, 144
white noise voltages, 131
Wiseana sp., 68
Wright-Patterson Air Force Base, 68

X

xylene, 64

Y

Yee algorithm, 194, 197–200
Yellow Horned Poppy, *see Glaucium avum*
Yellow Poppy, *see Dicranostigma leptopodum*
Young's modulus, 123, 130

Z

zincite, 23
zinc oxide (ZnO)
artificial optical biominerals, 23
optical materials approach, 45
surface functionalization, 41
Zinngraut sp., 29

T - #0219 - 111024 - C0 - 234/156/14 - PB - 9780367576653 - Gloss Lamination